MONITORING AND PREDICTING AGRICULTURAL DROUGHT

Monitoring and Predicting Agricultural Drought

A Global Study

Edited by

Vijendra K. Boken, Arthur P. Cracknell,
and Ronald L. Heathcote

Sponsored by the World Meteorological Organization

2005

OXFORD
UNIVERSITY PRESS

Oxford University Press, Inc., publishes works that further
Oxford University's objective of excellence
in research, scholarship, and education.

Oxford   New York
Auckland   Cape Town   Dar es Salaam   Hong Kong   Karachi
Kuala Lumpur   Madrid   Melbourne   Mexico City   Mumbai   Nairobi
New Delhi   Shanghai   Taipei   Tokyo   Toronto

With offices in
Argentina   Austria   Brazil   Chile   Czech Republic   France   Greece
Guatemala   Hungary   Italy   Japan   Poland   Portugal   Singapore
South Korea   Switzerland   Thailand   Turkey   Ukraine   Vietnam

Copyright © 2005 by Oxford University Press, Inc.

Published by Oxford University Press, Inc.
198 Madison Avenue, New York, New York 10016

www.oup.com

Oxford is a registered trademark of Oxford University Press

All rights reserved. No part of this publication may be reproduced,
stored in a retrieval system, or transmitted, in any form or by any means,
electronic, mechanical, photocopying, recording, or otherwise,
without the prior permission of Oxford University Press.

Library of Congress Cataloging-in-Publication Data

Monitoring and predicting agricultural drought : a global study / edited by Vijendra K. Boken, Arthur P.
Cracknell, and Ronald L. Heathcote : sponsored by the World Meteorological Organization.
p. cm.
Includes bibliographical references and index.
ISBN-13 978-0-19-516234-9
ISBN 0-19-516234-X
1. Droughts. 2. Drought forecasting. 3. Climatic changes. 4. Meteorology, Agricultural—Remote
sensing. I. Boken, Vijendra K. II. Cracknell, Arthur P. III. Heathcote, Ronald L. IV. World Meteorological Organization.
S600.7.D76A36 2004
363.34'92963—dc22   2004000102

9 8 7 6 5 4 3 2 1
Printed in the United States of America
on acid-free paper

Foreword

MICHEL JARRAUD

Secretary-General, World Meteorological Organization, Geneva

Of the many climatic events that influence the Earth's environment, drought is perhaps the one that is most linked with desertification. Drought is the consequence of a natural reduction in the amount of precipitation received over an extended period, usually a season or more in length.

Drought disrupts cropping programmes, reduces breeding stock, and threatens permanent erosion of the capital and resource base of farming enterprises. Continuous droughts stretching over several years in different parts of the world in the past significantly affected productivity and national economies. In addition, the risk of serious environmental damage, particularly through vegetation loss and soil erosion, as has happened in the Sahel during the late 1960s and early 1970s, has long-term implications for the sustainability of agriculture. Bush fires and dust storms often increase during the dry period.

As the United Nations' specialized agency with responsibility for meteorology, operational hydrology, and related geophysical sciences, the World Meteorological Organization (WMO), since its inception, has been addressing the issue of agricultural droughts. The fight against drought receives a high priority in the WMO Long-Term Plan, particularly under the Agricultural Meteorology Programme, the Hydrology and Water Resources Programme, and the Technical Co-operation Programme. WMO involves actively the National Meteorological and Hydrological Services (NMHSs), regional and sub-regional meteorological centres, and other bodies in the improvement of hydrological and meteorological networks for systematic observation, exchange, and analysis of data for better monitoring of droughts and use of medium- and long-range weather forecasts, and assists in the transfer of knowledge and technology.

In order to provide leadership in addressing related issues, WMO has been in the forefront of research on interactions of climate, drought, and desertification from its beginnings in the mid-1970s, when it was suggested that human activities in drylands could alter surface features that would lead to an intensification of desertification processes and trends. The urgent need to predict interannual climate variations is impelled by the socio-economic developments, especially in Africa, over the past few decades. Research into the causes and effects of climate variations and long-term climate predictions with a view to providing early warnings is an essential component of this effort. WMO improves the climate prediction capability through the Climate Variability (CLIVAR) project of the World Climate Research Programme (WCRP). At a global level, the WMO's World Weather Watch and Hydrology and Water Resources Programmes provide a sound operational framework on which to build improved warning capacity for droughts.

WMO is pleased to sponsor this book on *Monitoring and Predicting Agricultural Drought: A Global Study*, which has been ably edited by Drs. V.K. Boken, A.P. Cracknell, and R.L. Heathcote. I believe that this publication will contribute to current efforts in developing effective drought preparedness and drought management strategies.

Foreword

JACQUES DIOUF

Director-General, Food and Agriculture Organization, Rome

Ensuring food security for all is at the heart of the mission of the Food and Agriculture Organization of the United Nations (FAO). While scientific research, particularly during the second half of the 20 century, in genetics, plant pathology, water management techniques, and weather forecasting has contributed to increasing yields, in many semi-arid developing countries, it is mainly the expansion of cultivated areas that has enabled food production to keep pace with the growing population. The increased use of marginal land and the inherent variability of rainfall often prevent crop yields from reaching their potential in many parts of the world. In semi-arid areas, agricultural water shortage occurs almost every year and causes significant variability and reductions in crop yields and food production.

Drought is one of the most complex natural hazards because its impacts may affect large areas over several years in a row. If farmers can react as soon as a drought appears, the downward effect on food production can be minimized. Therefore, an efficient mechanism for monitoring agricultural droughts on a global scale, using both traditional and modern tools, is of utmost significance. FAO's Global Information and Early Warning Systems for food and agriculture thus pays particular attention to drought. The Organization has played a pioneering role in the use of agrometeorological crop monitoring and the use of satellite imagery to assess drought impacts. Significant efforts have been made, particularly in Africa, to establish National Early Warning Systems and to train national staff in the use of modern tools.

Many farmers, particularly in societies that have been experiencing frequent droughts for years, have adopted strategies to cope with them. The techniques evolved in one part of the globe are not necessarily available for evaluation, testing, and adoption for communities living in other areas.

Clearly, more efforts are needed to extend the benefits of satellite and crop modelling techniques among farmers and planners in developing countries.

This book provides an overview of methods and techniques for effective monitoring and mitigation of the effect of drought on agriculture. It is intended to assist readers, including researchers and planners, to enhance their knowledge of drought monitoring or early warning systems. This in turn would help sustain food security for current and future generations.

Acknowledgments

Editing this book took more than two years. During this period, many people helped, directly or indirectly. All of the chapters were reviewed and edited by the editors. Nevertheless, selected chapters were sent to external reviewers for general comments. The external reviewers were Kenneth J. Boote of the University of Florida, Anthony W. England of the University of Michigan, Chester F. Ropelewski of Columbia University, Tribeni C. Sharma of St. Francis Xavier University, Masato Shinoda of Tokyo Metropolitan University, Olga Wilhelmi of the National Center for Atmospheric Research, Zhi-Yong Yin of Georgia State University, and Jakob van Zyl of the Jet Propulsion Laboratory. The United Nations World Meteorological Organization sponsored the book and provided some financial support.

The editors also thank these scholars for reviewing the book outline: William A. Dando of Indiana State University, Thomas R. Loveland of EROS Data Center (U.S. Geological Survey), and David Stea and John P. Tiefenbacher of Texas State University. In addition, the following people helped identify a suitable author or a reviewer: Connie Falk of New Mexico State University, Michael Glantz of the National Center for Atmospheric Research, Gerrit Hoogenboom of the University of Georgia, M.V.K. Sivakumar of the World Meteorological Organization, and Fawwaj T. Ulaby of the University of Michigan. Gregory L. Easson and Richard P. Major of the University of Mississippi extended support to help bring the final phase of the book editing to a close. Sangeeta, Astitva, and Agrini deserve special thanks for extending immense support and understanding throughout the editing period.

Contents

Foreword by Michel Jarraud v
Foreword by Jacques Diouf vii
Contributors xv

PART I: BASIC CONCEPTS AND DROUGHT ANALYSIS

Chapter One: Agricultural Drought and Its Monitoring and Prediction: Some Concepts 3
Vijendra K. Boken

Chapter Two: Drought-Related Characteristics of Important Cereal Crops 11
Keith T. Ingram

Chapter Three: Monitoring Agricultural Drought Using El Niño and Southern Oscillation Data 28
Lino Naranjo Díaz

Chapter Four: Techniques to Predict Agricultural Droughts 40
Zekai Şen and Vijendra K. Boken

PART II: REMOTE SENSING

Chapter Five: Monitoring Drought Using Coarse-Resolution Polar-Orbiting Satellite Data 57
Assaf Anyamba, Compton J. Tucker, Alfredo R. Huete, and Vijendra K. Boken

Chapter Six: NOAA/AVHRR Satellite Data-Based Indices for Monitoring Agricultural Droughts 79
Felix N. Kogan

Chapter Seven: Passive Microwave Remote Sensing of Soil Moisture and Regional Drought Monitoring 89
Thomas J. Jackson

Chapter Eight: Active Microwave Systems for Monitoring Drought Stress 105
Anne M. Smith, Klaus Scipal, and Wolfgang Wagner

PART III: THE AMERICAS

Chapter Nine: Monitoring Drought in the United States: Status and Trends 121
Donald A. Wilhite, Mark D. Svoboda, and Michael J. Hayes

Chapter Ten: Agricultural Drought in North-Central Mexico 132
José Alfredo Rodríguez-Pineda, Lorrain Giddings, Héctor Gadsden, and Vijendra K. Boken

Chapter Eleven: Monitoring Agricultural Drought in the West Indies 144
Abraham Anthony Chen, Trevor Falloon, and Michael Taylor

Chapter Twelve: Agricultural Drought Phenomenon in Latin America with Focus on Brazil 156
Orivaldo Brunini, Pedro Leite Da Silva Dias, Alice M. Grimm, Eduardo Delgado Assad, and Vijendra K. Boken

PART IV: EUROPE, RUSSIA, AND THE NEAR EAST

Chapter Thirteen: Monitoring Agricultural Drought in Poland 171
Zbigniew Bochenek, Katarzyna Dabrowska-Zielinska, Andrzej Ciolkosz, Stanislaw Drupka, and Vijendra K. Boken

Chapter Fourteen: Monitoring Agricultural Drought in Mainland Portugal 181
Fátima Espírito Santo, Rita Guerreiro, Vanda Cabrinha Pires, Luís E.V. Pessanha, and Isabel M. Gomes

Chapter Fifteen: Monitoring Agricultural Drought in Russia 196
Alexander D. Kleschenko, Erodi K. Zoidze, and Vijendra K. Boken

Chapter Sixteen: Monitoring Agricultural Drought in the Near East 208
Eddy De Pauw

PART V: AFRICA

Chapter Seventeen: Agricultural Drought in Ethiopia 227
Engida Mersha and Vijendra K. Boken

Chapter Eighteen: Monitoring Agricultural Drought: The Case of Kenya 238
Laban A. Ogallo, Silvery B. Otengi, Peter Ambenje, William Nyakwada, and Faith Githui

Chapter Nineteen: Drought Monitoring Techniques for Famine Early Warning Systems in Africa 252
James Rowland, James Verdin, Alkhalil Adoum, and Gabriel Senay

Chapter Twenty: Monitoring Agricultural Drought in Southern Africa 266
Leonard S. Unganai and Tsitsi Bandason

Chapter Twenty-One: Harnessing Radio and Internet Systems to Monitor and Mitigate Agricultural Droughts in Rural African Communities 276
Marion Pratt, Macol Stewart Cerda, Mohammed Boulahya, and Kelly Sponberg

Chapter Twenty-Two: Livestock Early Warning System for Africa's Rangelands 283
Jerry W. Stuth, Jay Angerer, Robert Kaitho, Abdi Jama, and Raphael Marambii

PART VI: ASIA AND AUSTRALIA

Chapter Twenty-Three: Monitoring and Managing Agricultural Drought in India 297
A.S. Rao and Vijendra K. Boken

Chapter Twenty-Four: Agricultural Drought in Bangladesh 313
Ahsan U. Ahmed, Anwar Iqbal, and Abdul M. Choudhury

Chapter Twenty-Five: A Drought Warning System for Thailand 323
Apisit Eiumnoh, Rajendra P. Shrestha, and Vijendra K. Boken

Chapter Twenty-Six: Agricultural Drought in Indonesia 330
Rizaldi Boer and Arjunapermal R. Subbiah

Chapter Twenty-Seven: Agricultural Drought in Vietnam 345
Nguyen van Viet and Vijendra K. Boken

Chapter Twenty-Eight: Monitoring Agricultural Drought in China 354
Guoliang Tian and Vijendra K. Boken

Chapter Twenty-Nine: Monitoring Agricultural Drought in Australia 369
Kenneth A. Day, Kenwyn G. Rickert, and Gregory M. McKeon

Chapter Thirty: Monitoring Agricultural Drought in South Korea 386
Hi-Ryong Byun, Suk-Young Hong, and Vijendra K. Boken

PART VII: INTERNATIONAL EFFORTS AND CLIMATE CHANGE

Chapter Thirty-One: World Meteorological Organization and Agricultural Droughts 401
Mannava V.K. Sivakumar

Chapter Thirty-Two: Food and Agriculture Organization and Agricultural Droughts 411
Elijah Mukhala

Chapter Thirty-Three: International Activities Related to Dryland Degradation Assessment and Drought Early Warning 421
Ashbindu Singh

Chapter Thirty-Four: Climate Change, Global Warming, and Agricultural Droughts 429
Gennady V. Menzhulin, Sergey P. Savvateyev, Arthur P. Cracknell, and Vijendra K. Boken

Index 451

Contributors

ALKHALIL ADOUM, USGS FEWS-NET—Sahel, AGRHYMET Center, B.P. 11011, Niamey, Niger

AHSAN U. AHMED, Water and Environment Division, Bangladesh Unnayan Parishad (BUP), House 50, Block D, Niketon, Guishan-1, Dhaka-1212, Bangladesh

PETER AMBENJE, Drought Monitoring Center, P.O. Box 10304, 00100-Nairobi, Kenya

JAY ANGERER, Department of Rangeland Ecology and Management, Texas A&M University, College Station, Texas 77843-2126, United States

ASSAF ANYAMBA, NASA/Goddard Space Flight Center, Laboratory for Terrestrial Physics, Biospheric Science Branch, Greenbelt, Maryland 20771, United States

EDUARDO DELGADO ASSAD, Informática na Agricultura—Empresa Brasileira de Pesquisa Agropecuária, Campinas, São Paulo State, Brazil

TSITSI BANDASON, Department of Meteorological Services, P.O. Box BE150, Belvedere, Harare, Zimbabwe

ZBIGNIEW BOCHENEK, Institute of Geodesy and Cartography, Remote Sensing Department, ul. Modzelewskiego 27, 02-679 Warsaw, Poland

RIZALDI BOER, Department of Geophysics and Meteorology, Faculty of Mathematics and Natural Sciences, Bogor Agricultural University, Jalan Raya Pajajaran, Bogor-16144, Indonesia

VIJENDRA K. BOKEN, Department of Geology and Geological Engineering, The University of Mississippi Geoinformatics Center, University, Mississippi 38677, United States

MOHAMMED BOULAHYA, African Center for Meteorological Applications for Development, BP 13184, Niamey, Niger

ORIVALDO BRUNINI, Centro de Ecofisiologia e Biofisica, Instituto Agronomico, Avenida Barão de Itapura-1481, Bairro Guanabara, 13.020-430-Campinas, São Paulo State, Brazil

HI-RYONG BYUN, Department of Environmental Atmospheric Sciences, Pukyong University, 599-1 Daeyondong, Namgu, Pusan 608-737, South Korea

MACOL STEWART CERDA, NOAA Office of Global Programs, Climate and Societal Interaction Division, 1100 Wayne Ave, Suite 1210, Silver Spring, Maryland 20910, United States

ABRAHAM ANTHONY CHEN, Department of Physics, University of the West Indies, Mona, Kingston 7, Jamaica

ABDUL M. CHOUDHURY, Former Chairman, Bangladesh Space Research and Remote Sensing Organization (SPARRSO), Dhaka, Bangladesh

ANDRZEJ CIOLKOSZ, Institute of Geodesy and Cartography, Remote Sensing Department, ul. Modzelewskiego 27, 02-679 Warsaw, Poland

ARTHUR P. CRACKNELL, Division of Electronic Engineering and Physics, University of Dundee, Dundee DD1 4HN, Scotland

KATARZYNA DABROWSKA-ZIELINSKA, Institute of Geodesy and Cartography, Remote Sensing Department, ul. Modzelewskiego 27, 02-679 Warsaw, Poland

KENNETH A. DAY, Natural Resource Sciences, Department of Natural Resources and Mines and Energy, Indooroopilly, Queensland, Australia 4068

EDDY DE PAUW, International Center for Agricultural Research in Dry Areas (ICARDA), P.O. Box 5466, Aleppo, Syrian Arab Republic

PEDRO LEITE DA SILVA DIAS, Departamento de Ciências Atmosféricas, Instituto de Astronomia, Geofísica and Ciências Atmosféricas, Universidade de São Paulo, São Paulo State, Brazil

LINO NARANJO DÍAZ, Grupo de Fisica Nolineal, Facultade de Fisica, Campus Universitario Sur, 15782 Santiago de Compostela, A Coruna, Spain

STANISLAW DRUPKA, Institute for Land Reclamation and Grassland Farming, 05-090 Raszyn, Falenty, Poland

APISIT EIUMNOH, School of Environment, Resources and Development, Asian Institute of Technology, P.O. Box 4, Klong Luang, Pathumthani 12120, Thailand

TREVOR FALLOON, Department of Physics, The University of the West Indies, Mona, Kingston 7, Jamaica

HÉCTOR GADSDEN, Centro de Investigaciones Sobre la Sequía, del Instituto de Ecología, A.C., Km. 33.3, Carretera Chihuahua-Ojinaga, A.P. 28, C.P 32900, Aldama, Chihuahua, México

LORRAIN GIDDINGS, Instituto de Ecología, A.C., Km.2.5 Carretera. Antigua a Coatepec No. 351, Congregación El Haya, C.P. 91070, Xalapa,Veracruz, México

FAITH GITHUI, Drought Monitoring Center, P.O. Box 10304, 00100-Nairobi, Kenya

ISABEL M. GOMES, Climate and Environment Department, Institute of

Meteorology, Rua C, Lisbon Airport, 1749-077 Lisbon, Portugal
ALICE M. GRIMM, Departamento de Física, Universidade Federal do Paraná, Curitiba, Brazil
RITA GUERREIRO, Hydrometeorology and Agrometeorology Division, Institute of Meteorology, Rua C, Lisbon Airport, 1749-077 Lisbon, Portugal
MICHAEL J. HAYES, National Drought Mitigation Center, University of Nebraska, Lincoln, Nebraska 68583, United States
SUK-YOUNG HONG, Agricultural Environment Department, National Institute of Agricultural Science and Technology (NIAST), 249 Seodun-dong, Gwensun-gu, Suwon 441-707, South Korea
ALFREDO R. HUETE, Department of Soil, Water, and Environmental Sciences, University of Arizona, Tucson, Arizona 85721-0038, United States
KEITH T. INGRAM, Department of Biological and Agricultural Engineering, University of Florida, Gainesville, Florida 32611-0570, United States
ANWAR IQBAL, Bangladesh Agricultural Research Council, Dhaka, Bangladesh
THOMAS J. JACKSON, ARS Hydrology and Remote Sensing Lab, U.S. Department of Agriculture, Beltsville, Maryland 20705, United States
ABDI JAMA, Department of Rangeland Ecology and Management, Texas A&M University, College Station, Texas 77843-2126, United States
ROBERT KAITHO, Department of Rangeland Ecology and Management, Texas A&M University, College Station, Texas 77843-2126, United States
ALEXANDER D. KLESCHENKO, National Institute of Agricultural Meteorology, 82 Lenin St., Obninsk, Kaluga Region, 249030, Russia
FELIX N. KOGAN, National Oceanic and Atmospheric Administration/National Environmental Satellite Data and Information Services, Camp Springs, Maryland 20746, United States
RAPHAEL MARAMBII, Department of Rangeland Ecology and Management, Texas A&M University, College Station, Texas 77843-2126, United States
GREGORY M. MCKEON, Natural Resource Sciences, Department of Natural Resources and Mines and Energy, Indooroopilly, Queensland, Australia 4068
ENGIDA MERSHA, Ethiopian Agricultural Research Organisation, P.O.Box 2003, Addis Ababa, Ethiopia
GENNADY V. MENZHULIN, Research Center for Interdisciplinary Environmental Cooperation of Russian Academy of Sciences, 14-Kutuzova Embankment, 191187, St. Petersburg, Russia
ELIJAH MUKHALA, FAO/SADC Regional Remote Sensing Project, 43 Robson Manyika Ave., P.O. Box 4046, Harare, Zimbabwe
WILLIAM NYAKWADA, Kenya Meteorological Department, P.O. Box 30259, 00100-Nairobi, Kenya
LABAN A. OGALLO, Drought Monitoring Center, P.O. Box 10304,

00100-Nairobi, Kenya

SILVERY B. OTENGI, Department of Meteorology, University of Nairobi, P.O. Box 30197, 00100-Nairobi, Kenya

LUÍS E.V. PESSANHA, Institute of Meteorology, Rua C, Lisbon Airport, 1749-077 Lisbon, Portugal

VANDA CABRINHA PIRES, Climate and Environment Department, Institute of Meteorology, Rua C, Lisbon Airport 1749-077, Portugal

MARION PRATT, Office of Foreign Disaster Assistance, U.S. Agency for International Development (USAID), 1300 Pennsylvania Ave., NW, Washington, DC 20523, United States

AMJURI S. RAO, Division of Natural Resources and Environment, Central Arid Zone Research Institute, Jodhpur-342 003, India

KENWYN G. RICKERT, School of Natural and Rural Systems Management, The University of Queensland, Gatton Campus, Gatton, Queensland, Australia

JOSÉ ALFREDO RODRÍGUEZ-PINEDA, Jefe—Centro de Investigaciones Sobre la Sequía (CEISS), Instituto de Ecología, A.C. Km. 33.3 Carretera Chihuahua-Ojinaga, A.P. 28 Aldama, Chihuahua, México

JAMES ROWLAND, Science Department, Raytheon, EROS Data Center, Sioux Falls, South Dakota 57198, United States

FÁTIMA ESPÍRITO SANTO, Climate and Environment Department, Institute of Meteorology, Rua C, Lisbon Airport, 1749-077 Lisbon, Portugal

SERGEY P. SAVVATEYEV, Research Center for Interdisciplinary Environmental Cooperation of Russian Academy of Sciences, 14-Kutuzova Embankment, 191187, St. Petersburg, Russia

KLAUS SCIPAL, Institute of Photogrammetry and Remote Sensing, Vienna University of Technology, Gusshausstraße 27-29, 1040 Vienna, Austria

ZEKAI ŞEN, İstanbul Technical University, Civil Engineering Faculty, Hydraulics Division, Maslak 80626, İstanbul, Turkey

GABRIEL SENAY, SAIC, EROS Data Center, Sioux Falls, South Dakota 57198, United States

RAJENDRA P. SHRESTHA, School of Environment, Resources and Development, Asian Institute of Technology, P.O. Box 4, Klong Luang, Pathumthani 12120, Thailand

ASHBINDU SINGH, United Nations Environment Program's Division of Early Warning and Assessment—North America, 1707 H St., N.W., Suite 300, Washington, DC 20006, United States

MANNAVA V. K. SIVAKUMAR, Agricultural Meteorology Division, World Meteorological Organization, 7 Bis, Avenue de la Paix, CP No. 2300, CH-1211, Geneva 2, Switzerland

ANNE M. SMITH, Agriculture and Agri-Food Canada, Research Center, P.O. Box 3000, Lethbridge, Alberta T1J 4B1, Canada

KELLY SPONBERG, Climate Information Access Program (University Corporation for Atmospheric Research), NOAA Office of Global Programs, Climate and Societal Interaction Division, 1100 Wayne

Ave, Suite 1210, Silver Spring, Maryland 20910, United States
JERRY W. STUTH, Department of Rangeland Ecology and Management, Texas A&M University, College Station, Texas 77843-2126, United States
ARJUNAPERMAL R. SUBBIAH, Asian Disaster Preparedness Center, Asian Institute of Technology, P.O. Box 4, Klong Luang, Pathumthani 12120, Thailand
MARK D. SVOBODA, National Drought Mitigation Center, University of Nebraska, Lincoln, Nebraska 68583, United States
MICHAEL TAYLOR, Department of Physics, The University of the West Indies, Mona, Kingston 7, Jamaica
GUOLIANG TIAN, Institute of Remote Sensing Applications, Chinese Academy of Sciences, P.O. Box 9718, Beijing 100101, People's Republic of China
COMPTON J. TUCKER, NASA/Goddard Space Flight Center, Laboratory for Terrestrial Physics, Biospheric Science Branch, Greenbelt, Maryland 20771, United States
LEONARD S. UNGANAI, Department of Meteorological Services, P.O. Box BE150, Belvedere, Harare, Zimbabwe
NGUYEN VAN VIET, Agrometeorological Research Centre, 5162 Nguyen Chi Thanh St., Lang Trung, Dong Da, Hanoi, Vietnam
JAMES VERDIN, U.S. Geological Survey EROS Data Center, Sioux Falls, South Dakota 57198, United States
WOLFGANG WAGNER, Institute of Photogrammetry and Remote Sensing, Vienna University of Technology, Gusshausstraße 27-29, 1040 Vienna, Austria
DONALD A. WILHITE, National Drought Mitigation Center, University of Nebraska, Lincoln, Nebraska 68583, United States
ERODI K. ZOIDZE, National Institute of Agricultural Meteorology, 82 Lenin St., Obninsk, Kaluga Region, 249030, Russia

MONITORING AND PREDICTING
AGRICULTURAL DROUGHT

PART I

BASIC CONCEPTS AND DROUGHT ANALYSIS

CHAPTER ONE

Agricultural Drought and Its Monitoring and Prediction: Some Concepts

VIJENDRA K. BOKEN

Droughts develop largely due to below-average precipitation over a land area, and they adversely affect various economic sectors in a region. Some of these adverse effects include reductions in agricultural production, hydropower generation, urban and rural water supplies, and industrial outputs. These effects lead to other consequences, secondary and tertiary, that further impact an economy. For instance, when agricultural production declines, food and other commodities tend to cost more and cause economic inflation in a society. Chain effects of persistent droughts can shatter an economy and even cause famine and sociopolitical upheaval in some countries.

How does one define a drought? Usually, either precipitation or a form of drought impact is used to define a drought. Because precipitation and drought impacts vary spatially, there is a geographical dimension to definitions of drought. In Saudi Arabia or Libya, droughts are recognized after two to three years without significant rainfall, whereas in Bali (Indonesia), any period of six days or more without rain is considered drought (Dracup et al., 1980; Sen, 1990). In Egypt, any year in which the Nile does not flood is considered a drought year.

More than 150 definitions of drought are available in the literature (Gibbs, 1975; Krishnan, 1979; Dracup et al., 1980; Wilhite and Glantz, 1987). For example, a drought can be characterized as climatological, meteorological, water management, socioeconomic, absolute, partial, dry spell, serious, severe, multiyear, design, critical, point, or regional (Palmer, 1965; Herbst et al., 1966; Joseph, 1970a, 1970b; Askew et al., 1971; Beard and Kubik, 1972; Karl, 1983; Santos, 1983; Alley, 1984; Chang, 1990). Often, the difference between an estimated water demand and an expected water supply in a region becomes the basis to define a drought for that region (Kumar and Panu, 1997). A few of the chapters in this book provide a brief description of drought definitions that have been adopted in some countries.

Despite the variation in drought definitions, a drought is broadly categorized as meteorological, hydrological, agricultural, or socioeconomic. A meteorological drought is said to occur when seasonal or annual precipitation falls below its long-term average. A hydrological drought develops when the meteorological drought is prolonged and causes shortage of surface and groundwater in the region. An agricultural drought sets in due to soil moisture stress that leads to significant decline in crop yields (production per unit area), and a socioeconomic drought is a manifestation of continued drought of severe intensity that shatters the economy and sociopolitical situation in a region or country.

Meteorological drought is just an indicator of deficiency in precipitation, whereas hydrological and agricultural droughts are physical manifestations of meteorological drought. This book focuses on agricultural drought, which is the most complex category of drought. One can detect a hydrological drought by observing water levels in ponds, reservoirs, lakes, and rivers. However, such an instantaneous observation is not possible in the case of agricultural drought. Unlike a hydrological drought, an agricultural drought occurs over a large area, and its impact is not accurately assessed until after crops are harvested—a few months after the symptoms of agricultural drought begin to appear. The symptoms may be deficiency in precipitation, failure of rains/monsoon systems, and poor crop conditions.

Precipitation is one of several components of the hydrological cycle that is driven by atmospheric and oceanic circulations, together with gravity (Yevjevich et al., 1978). At times, the cycle experiences an abnormal change in its pattern due to changes in the patterns of atmospheric and ocean circulations. Such an atmospheric activity is called El Niño and Southern Oscillation (ENSO). Chapter 3 describes how this activity is linked to the occurrence of agricultural droughts. An agricultural drought at a global scale can cause severe food shortage or even famine in some countries, causing loss of both the human and livestock population. Chapters 21 and 22 focus on methods to help move livestock to greener pastures at times of droughts in Africa.

Monitoring and Predicting Agricultural Drought

Effective and timely monitoring of agricultural droughts can help develop an early warning system which, in turn, can minimize losses due to droughts. International organizations such as the World Meteorological Organization (WMO), the Food and Agriculture Organization (FAO), and the United Nations Environment Program (UNEP) keep a watch on the development of agricultural droughts and famines in the world. Chapters 31, 32, and 33 elaborate on operations of WMO, FAO, and UNEP, respectively. In addition, the Famine Early Warning System (FEWS; chapter 19), which is sponsored by the U.S. Agency for International Development (USAID), monitors drought conditions in Africa. Because agricultural droughts occur due to low crop yields, monitoring them requires monitoring the factors that affect crop yields.

Factors Affecting Crop Yield

According to Diepen and van der Wall (1996), factors influencing yield can be categorized as (1) abiotic factors, such as soil water, soil fertility, soil texture, soil taxonomy class, and weather; (2) farm management factors, such as soil tillage, soil depth, planting density, sowing date, weeding intensity, manuring rate, crop protection against pests and diseases, harvesting techniques, postharvest loss, and degree of mechanization; (3) land development factors, such as field size, terracing, drainage, and irrigation; (4) socioeconomic factors, such as the distance to markets, population pressure, investments, costs of inputs, prices of output, education levels, skills, and infrastructure; and (5) catastrophic factors that include warfare, flooding, earthquakes, hailstorms, and frost. Measuring or estimating some of these factors is often not feasible, and the influence of some other factors may be considered insignificant or constant in an economically stable region. It is therefore weather conditions alone that affect crop yield most significantly. Various weather parameters such as temperature, precipitation, humidity, solar radiation, cloudiness, and wind velocity affect crop yield, but temperature and precipitation are most significant.

A change in temperature causes a shift in planting dates that, in turn, shifts the commencement and termination of the phenological phases. The entire crop-growth period (period from planting to harvest) may shift, shrink, or expand due to a change in temperature pattern. McKay (1983) observed for wheat production in Canada that a drop in the mean annual temperature of 1°C accompanied by a 9- to 15-day reduction in the growing season could be critical. However, interannual variation in temperature is much less than that in precipitation. It is for this reason that precipitation becomes more important than temperature for monitoring crop yields. An intricate relationship exists between precipitation and crop yield because it is the soil moisture, and not precipitation, that ultimately contributes to crop growth and crop yield.

Soil moisture data are more important than precipitation data for monitoring agricultural droughts, but soil moisture data are not as readily available as precipitation data are. Unlike precipitation data that are routinely available via a network of weather stations, soil moisture data are collected only on an experimental basis or estimated using agrometeorological models. Monitoring agricultural droughts requires soil moisture data for large areas on spatial and temporal scales. With advances in microwave remote sensing, it is becoming possible to estimate soil moisture for large areas, as described in detail in chapter 7.

Crop Yield and Soil Moisture

Yield depends on spatial and temporal distribution of soil moisture over a crop-growth period. Soil moisture requirements by a crop vary during this period, which consists of three main phenological phases that develop in sequence: (1) the vegetative phase, which includes the period from planting to the complete leaf (or canopy) development, (2) the grain-filling/heading/reproductive phase, which includes the period of grain formation in the plant, and (3) the harvesting phase, during which leaves senesce, grains harden, and the crop becomes ready for harvest. Soil moisture requirements increase rather linearly during the vegetative phase, remain at the peak during the reproductive phase, and decline during the harvesting phase. Moisture deficiency during the reproductive phase affects crop yield most significantly (Mahalakshmi et al., 1988).

A crop needs an adequate amount of soil moisture on a continuous basis throughout the growth period. Irrigation can meet such needs. However, in the absence of irrigation facilities, crop growth relies on precipitation. The amount as well as the temporal and spatial distribution of precipitation influence crop yields. Lower but well-distributed precipitation may result in a higher crop yield as opposed to higher but poorly distributed precipitation. The variation in precipitation is one of the significant factors causing agricultural droughts (Liverman, 1990).

Crop Yield and Planting Dates

Planting dates significantly affect crop yield and probability of agricultural droughts (Mahalakshmi et al., 1988; Kumar, 1998). Each crop has an ideal time window for its planting. A crop planted early or late may not reach its potential yield. Chapter 2 describes physiological characteristics of some major food crops (e.g., wheat, rice, maize, and millet). An understanding of these characteristics is essential to studying the impact of precipitation on crop yield. The major factors that contribute to the occurrence of agricultural droughts are spatial and temporal anomalies in temperature, precipitation, and planting dates.

Use of Satellite Data

Weather data are commonly used to estimate crop yield and agricultural droughts. However, weather data are collected only on point locations and do not adequately represent the spatial coverage by crops. In contrast, satellite data have greater capability to monitor crop condition in a spatially continuous fashion and on a regular time interval and therefore have proven to be a dependable source for monitoring crop yields and agricultural droughts. Various types of satellite data are currently available for monitoring crop conditions. These can be categorized into: (1) visible and infrared data, (2) passive microwave data, and (3) active microwave data.

Chapters 5, 6, 7, and 8 provide details on how different types of the satellite data can contribute to monitoring agricultural droughts.

Crop Growth Models

Based on a typical crop growth period, the Julian dates are identified to acquire satellite data for monitoring crop conditions. Weather conditions, particularly during times of drought, cause a shift in the planting dates and, consequently, in the commencement and termination of various phenological phases. As a result, selecting dates to acquire satellite data becomes a challenge. A biometeorological time scale model can be used to determine commencement and termination of various phenological phases of a crop.

Biometeorological Time Scale Model

Robertson (1968) developed the following model to estimate the commencement and termination of five phenological phases (i.e., emergence, jointing, heading, soft dough, and ripening) of wheat:

$$1 = \sum_{S_1}^{S_2} \{[a_1(L-a_0) + a_2(L-a_0)^2][b_1(T_1-b_0) + b_2(T_1-b_0)^2 \\ + c_1(T_2-b_0) + c_2(T_2-b_0)^2]\} \quad [1.1]$$

where $a_0, a_1, a_2, b_0, b_1, b_2, c_0, c_1$, and c_2 are coefficients (table 1.1), L is the daily photoperiod (duration from sunrise to sunset, in hours), which can be estimated for a given location following a procedure by Robertson and Russelo (1968), T_1 is the daily maximum temperature (°F), T_2 is the daily minimum temperature (°F); and S_1 and S_2 refer to the commencement and the termination stages, respectively, for a phenological phase.

Kumar (1999) developed a computer program to apply Robertson's biometeorological time scale model for the prairie region to determine dates for the heading phase of wheat. The program helped select the satellite-data-based normalized difference vegetation index (NDVI) data for the heading phase. The average NDVI during the heading phase was a significant variable for predicting wheat yield (Boken and Shaykewich, 2002).

In addition to the biometerological time scale model, various crop models have been developed to simulate crop growth using various agrometeorological data. Some of the commonly used models are Decision Support System for Agrotechnology Transfer (DSSAT; Tsuji et al., 1994; Hoogenboom et al., 1999), Erosion Productivity Impact Calculator or Environmental Policy Integrated Climate (EPIC; Williams et al., 1989), and Agricultural Production Systems Simulator (APSIM; www.apsru.gov.au/apsru/Products/apsim.htm).

Predicting agricultural drought requires predicting crop yield. Chapter 4 describes some common techniques that can be used to predict crop yield and hence agricultural droughts. In this context, variables, based on weather and satellite data, play a pivotal role in the prediction process.

Table 1.1 Coefficients of the biometeorological time scale model

Coefficient	Development phase[a]				
	PE	EJ	JH	HS	SR
a_0	$[V_1 = 1]$[b]	8.413	10.93	10.94	24.38
a_1		1.005	0.9256	1.389	−1.140
a_2		0	−0.06025	−0.08191	0
b_0	44.37	23.64	42.65	42.18	37.67
b_1	0.01086	−0.003512	0.0002958	0.0002458	0.00006733
b_2	−0.0002230	0.00005026	0	0	0
c_1	0.009732	0.0003666	0.0003943	0.00003109	0.0003442
c_2	−0.0002267	−0.000004282	0	0	0

Source: Robertson (1968).
[a]PE = planting to emergence, EJ = emergence to jointing, JH = jointing to heading, HS = heading to soft dough, and SR = soft dough to ripening.
[b]For determination of the PE, $V_1 = [a_1(L\text{-}a_0) + a_2(L\text{-}a_0)^2] = 1.0$ in equation 1.3, thus not requiring use of a_0, a_1, and a_2 coefficients separately.

Various chapters in this book include variables that are used to predict agricultural droughts in different regions of the world.

In the years to come, a change in average patterns of temperature and precipitation will occur due to climate change. As a result, the existing patterns of drought-prone areas will undergo a transformation. Chapter 34 discusses effects of climate change and global warming on the occurrence of agricultural drought.

Conclusions

Agricultural drought is the most complex natural hazard, and so is its monitoring and prediction. Agricultural drought affects large areas and causes significant drops in the food production. If an effective early warning system for monitoring and predicting agricultrural droughts can be developed using weather-based variables, satellite data and crop growth models, human and livestock mortality and decline in food production can be minimized. Although some countries in the world have established or have initiated work on developing a drought monitoring or early warning system, others have not. Information and issues presented in the chapters of this book will help drought planners of any nation to develop or improve their drought monitoring or early warning system to tackle drought situations more effectively. This in turn will contribute to ensuring sustainability of food production for future generations.

References

Alley, W.M. 1984. Palmer drought severity index: limitations and assumptions. J. Climate Appl. Meteor. 23:1100–1109.

Askew, A.J., W.W. Yeh, and W.A. Hall. 1971. A comparative study of critical drought simulation. Water Resources Res. 7:52–62.

Beard, L.R., and H.K. Kubik. 1972. Drought severity and water supply dependability. J. Irrig. Drain. Div. ASCE 98:433–442.

Boken, V.K., and C.F. Shaykewich. 2002. Improving an operational wheat yield model for the Canadian Prairies using phenological-stage-based normalized difference vegetation index. Intl. J. Remote Sens. 23:4157–4170.

Chang, T.J. 1990. Effects of droughts on streamflow characteristics. J. Irrig. Drain. Div. ASCE 116:332–341.

Diepen, C.A., and T. van der Wall. 1996. Crop growth monitoring and yield forecasting at regional and national scale. In: J.F. Dallemend and P. Vossen (eds.), Proc. Workshop for Central and Eastern Europe on Agrometeorological Models: Theory and Applications, The MARS Project Ispra, Italy, November 21–25, 1994. European Commission, Luxembourg, pp. 143–157.

Dracup, J.A., K.S. Lee, and E.G. Paulson, Jr. 1980. On definitions of droughts. Water Resources Res. 16:297–302.

Gibbs, W.J. 1975. Drought—its definition, delineation and effects. In: WMO Spec. Environ. Rept. 5:11–39.

Herbst, P.H., D.B. Brendenkemp, and H.M. Barke. 1966. A technique for evaluation of drought from the rainfall. J. Hydrol. 4:264–272.

Hoogenboom, G., P.W. Wilkens, and G.Y. Tsuji, eds. 1999. DSSAT, version 3, vol. 4. University of Hawaii, Honolulu.

Joseph, E.S. 1970a. Probability distribution of annual droughts. J. Irrig. Drain. Div. ASCE 96:461–473.

Joseph, E.S. 1970b. Frequency of design droughts. Water Resources Res. 6:1199–1201.

Karl, T.R. 1983. Some spatial characteristics of drought duration in the United States. J. Climate Appl. Meteorol. 22:1356–1366.

Krishnan, A. 1979. Definitions of drought and factors relevant to specifications of agricultural and hydrological droughts. In: Proc. International Symposium on hydrological aspects of droughts, vol. I. Indian Institute of Technology, New Delhi, pp. 67–102.

(Boken) Kumar, V. 1998. An early warning system for agricultural drought in an arid region using limited data. J. Arid Environ. 40:199–209.

(Boken) Kumar, V., and U. Panu. 1997. Predictive assessment of severity of agricultural droughts based on agro-climatic factors. J. Am. Water Resources Assoc. 33:1255–1264.

(Boken) Kumar, V. 1999. Predicting agricultural droughts for the Canadian prairies using climatic and satellite data. Ph.D. thesis, Department of Geography, University of Manitoba, Winnipeg, Canada.

Liverman, D.M. 1990. Drought impacts in Mexico: climate, agriculture, technology, and land tenure in Sonora, and Puebla. Ann. Am. Assoc. Geogr. 80:49–72.

Mahalakshmi, V., F.R. Bidinger, and G.D.P. Rao. 1988. Training and intensity of water deficit during flowering and grain filling in pearl millet. Agron. J. 80:130–135.

McKay, G.A. 1983. Climatic hazards and Canadian wheat trade. In: K. Hewitt (ed.), Interpretations of Calamity from the View Point of Human Ecology. Allen and Unwin, Boston, pp. 220–228.

Palmer, W.C. 1965. Meteorological drought. Paper no. 45. Weather Bureau, U.S. Dept. of Commerce, Washington, DC, pp. 1–58.

Robertson, G.W. 1968. A biometeorological time scale for a cereal crop involving day and night temperatures and photoperiod. Intl. J. Biometeor. 12:191–223.

Robertson, G.W., and D.A. Russelo. 1968. Agrometeorological estimator for estimating time when the sun is at any elevation, time elapsed between the same elevations in the morning and afternoon, and hourly and daily values of solar energy. *Quarterly Agricultural Meteorology. Tech. Bull.* 14. Agrometeorology Section, Agriculture Canada, Ottawa, 22 pp.

Santos, M.A. 1983. Regional droughts: a stochastic characterization. J. Hydrol. 66:183–211.

Sen, Z. 1990. Critical drought analysis by second order Markov chain. J. Hydrol. 120:183–202.

Tsuji, G.Y., G. Uehava, and S. Balas. 1994. A decision support system for agrotechnology transfer. University of Hawaii, Honolulu.

Wilhite, D.A., and M.H. Glantz. 1987. Understanding the drought phenomenon. In: D.A. Wilhite, W.E. Easterling, and D.A. Wood (eds.), Planning for Drought: Toward a Reduction of Societal Vulnerability. Westview Press, Boulder, CO, pp. 11–27.

Williams, J.R., C.A. Jones, J.R. Kiniry, and D.K. Spanel. 1989. The EPIC crop growth model. Trans. ASAE 32:497–511.

Yevjevich, V., W.A. Hall, and J.D. Salas. (eds.). 1978. Drought Research Needs. Water Resources Publications, Fort Collins, CO.

CHAPTER TWO

Drought-Related Characteristics of Important Cereal Crops
KEITH T. INGRAM

Humans cultivate more than 200 species of plants, but this chapter reviews responses of 5 important cereal crops to drought. These crops are maize (*Zea mays* L.), rice (*Oryza sativa* L.), wheat (*Triticum aestivum* and *Triticum turgidum* L. var. durum), sorghum (*Sorghum bicolor* [L.] Moench), and pearl millet (*Pennisetum glaucum* [L.] R. Br), which provide the majority of food in the world. In general, farmers cultivate millet in the most drought-prone environments and sorghum where a short growing season is the greatest constraint to production. Some sorghum cultivars set grain in as short as 50–60 days (Roncoli et al., 2001). Rice is grown under a wide range of environments, from tropical to temperate zones, from deep water-flooded zones to nonflooded uplands. Rice productivity is limited mostly by water (IRRI, 2002). Drought limits, to a varying extent, the productivity of all of these crops. Although water is likely the most important manageable limit to food production worldwide, we should recognize that water management cannot be isolated from nutrient, crop, and pest management.

Crop Water Use

Life on earth depends on green plants, which capture solar energy and store chemical energy by the process of photosynthesis. Although plants use a small amount of water in the reactions of photosynthesis and retain small amounts of water in plant tissues, as much as 99% of the water that plants take up is lost through transpiration (i.e., gaseous water transport through the stomata of leaves). Stomata, which are small pores on leaf surfaces, must open to allow carbon dioxide to enter leaf tissues for photosynthesis and plant growth, but open stomata also allow water to escape.

In addition to transpiration, there are several other avenues of water loss from a crop system. Water may exit the crop system by evaporation from the soil, transpiration of weeds, deep drainage beyond the root zone, lateral flow beneath the soil surface, or runoff. We can sum the daily additions and losses of water to form a water balance equation:

$$S = G + P + I - E - T - T_w - D - L - R \qquad [2.1]$$

where all the terms are measured in millimeters, S is the amount of the soil water currently available to the crop, G is the amount of soil water available on previous day, P is precipitation, I is irrigation, E is water evaporated from soil surface, T is water transpired by crop, T_w is water transpired by weeds, D is water that drains below the root zone, L is lateral flow beneath the soil surface, and R is runoff or lateral flow above the soil surface.

It should be noted that D, L, and R are normally negative values, but sometimes they may also be positive. Soil water may move upward by capillary action to enter the root zone. In some fields, such as valley bottoms, subsurface lateral flow or surface runoff may constitute a major source of water entering the crop system.

The ratio of crop production to the water it uses is called the water use efficiency (WUE). The WUE is more commonly defined as the ratio of total dry matter produced to total water used by the crop during the growing season. Two other common measures of water use include the net WUE, which is the ratio of marketable yield to total water consumed (Jensen et al., 1990), and the irrigation WUE, which is the ratio of marketable yield to total water applied (Hillel, 1990).

Drought and Water Deficit Effects on Plants

Although all droughts lead to plant water deficit, not all plant water deficits result from drought. Disease or pests that damage roots can also cause plant water deficit that cannot be overcome by adding water to the crop. Visual symptoms of root pests and diseases may be indistinguishable from symptoms of drought. Similarly, plants with shallow root systems may suffer water deficit if upper soil layers dry, even though water is available in soil layers below the root zone.

As the basic medium of life, water plays many roles in plant growth and development. Similarly, water deficit affects numerous plant processes. For crop plants, it is convenient to consider four broad groups of responses to water deficit: germination processes, turgor-mediated phenomena, substrate-mediated phenomena, and desiccation effects.

Germination

Crops first require water for germination. Seeds must imbibe enough water to initiate biochemical processes needed to break down the storage compounds to provide the energy and substrate for growth. Seeds are very resistant to drought as long as they have not begun the process of germination. Once seeds begin breaking down the storage compounds, the radicle begins to elongate seedlings, which become highly susceptible to water deficit.

For the most part, there is a close relationship between crop transpiration and drought sensitivity (figure 2.1). It is only during the short period after seedling emergence that crops are highly susceptible to drought, even though their transpiration rates are low. A light rain may allow seeds to imbibe enough water to begin germination, but if additional rains or irrigation do not follow within a few days, either seedlings may not have enough water to emerge from the soil, or if they have emerged, roots may be too shallow to take up the water needed for the shoots. One of the most common economic losses to rain-fed farming systems results from farmers having to replant crops that fail to establish with the first rains.

Turgor-Mediated Phenomena

Once seedlings have established, water deficits tend to follow a cyclic pattern, with stress levels increasing as the time increases between successive rainfall or irrigation events (figure 2.2). The first water deficit effect, which occurs under relatively mild stress levels, is the slowing or inhibition of cell expansion. A reduction of expansive growth is reflected in smaller leaves and shorter plants. As stress becomes more severe, leaves wilt or roll. One turgor-mediated phenomenon is reduction in stomatal conductance (or stomatal closure). Although partial or complete closure of stomata will conserve moisture within the plant, it also reduces the uptake of carbon dioxide needed for photosynthesis. Similarly, leaf rolling and wilting may reduce the load of radiant energy on the plant and decrease transpiration

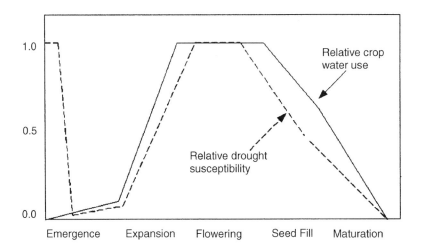

Figure 2.1 General pattern of changes in relative drought susceptibility and relative crop transpiration during crop development.

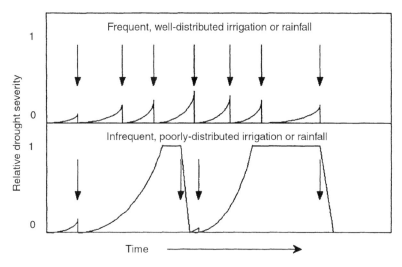

Figure 2.2 Idealized graphs showing the severity of drought and time between irrigation or rainfall events for crops receiving frequent, well-distributed (top), or infrequent, poorly distributed rainfall (bottom). Arrows indicate rainfall or irrigation events.

from leaf surface area, which also would conserve water at the expense of reduced capacity for assimilation.

In some plants, water deficits lead to accumulation of solutes within cells in a process called osmotic adjustment. Simple sugars are the principal solutes that contribute to osmotic adjustment, though any dissolved molecules will reduce the solute potential of a cell. An increased concentration of solutes may theoretically allow cells to maintain turgor at lower water potential levels, but there has been no documented case where osmotic adjustment has increased crop productivity under water-limited conditions.

In cereal crops, stems elongate rapidly to force emergence of the reproductive organ from the whorl of leaves. Water deficit during this period can delay or prevent the emergence of the reproductive organ, which in turn causes severe yield loss. Many studies have reported that water deficit increases the root/shoot ratio while reducing both roots and shoots because the deficit reduces shoot growth more than it reduces root growth.

Substrate-Mediated Phenomena

Although the photosynthetic apparatus of leaves is relatively drought resistant, carbon dioxide assimilation declines because of stomatal closure, leaf wilting, and leaf rolling, which occur as a result of moderate to severe water deficit conditions. Reductions in carbon dioxide assimilation reduce substrate available for cell division and growth. Lack of growth substrate may delay ontogenic development, which is seen in reduced rates of leaf appearance, tillering or branching, or in the delay of the reproductive growth phase.

Perhaps the most devastating substrate-mediated response to water deficit is lack of seed set. Near the time of pollination, depending on the availability of assimilates from photosynthesis or stored carbohydrates, plants establish the number of seeds that will develop further. If drought restricts carbon dioxide assimilation or the rate or transfer of stored assimilates, then fewer seeds will develop. Once a plant aborts a seed, the seed can grow no further and potential yield is irrevocably lost.

Desiccation Effects   Even at relatively moderate degrees of drought severity, plant extremities (such as leaf tips beginning to fire, turning brown, and desiccate) take place. As drought intensifies, leaves die and abscise. Eventually, shoots, and then roots, die. Only viable seeds can survive desiccation. The following sections explain the distribution of the selected major crops and their response to water deficits.

Maize

Distribution and Use

The center of origin for maize lies in Central America. It is grown widely throughout the world including both tropical and temperate countries, with the largest total production estimated at more than 604 million tons (table 2.1). Industrialized countries generally grow maize for livestock feed. Among the top five maize producers, significant amounts of maize are consumed directly by humans only in Mexico. In the developing countries of Africa and Asia, maize is mostly consumed directly by humans.

Drought and Water Use Efficiency

Maize has a C_4 photosynthetic mechanism, which gives it high potential productivity with record yields near 20 Mg/ha (Rhoads and Bennett, 1990). Estimates of seasonal WUE for maize range from 0.012 Mg ha^{-1} mm^{-1} in a relatively arid environment (Musick and Dusek, 1980) to 0.030 Mg ha^{-1} mm^{-1} in a humid environment (Hook, 1985).

With an intermediate level of drought tolerance as compared with other cereal crops, the majority of global maize production is under rain-fed conditions. In contrast, because maize has a high water requirement during tasseling and silking phases, drought during these phases may significantly reduce productivity. Thus, maize crops are often irrigated in areas that have frequent rainfall deficit during the early reproductive phase (Rhoads and Bennett, 1990).

Rather than applying water to maintain soil moisture at a high level, some farmers allow soils to dry appreciably before irrigation and then do not replenish the full soil reservoir, a practice called deficit irrigation. If properly maintained, a deficit irrigation schedule may improve irrigation WUE of maize. For example, Steele (1994) found that irrigation after soil

Table 2.1 Global maize production and area harvested

Country	Area (%)	Production (%)
United States	20.3	39.6
China	17.0	18.3
Brazil	9.0	6.9
Mexico	5.3	3.1
Argentina	2.0	2.5
France	1.4	2.7
Others	45.0	26.9
World total	138 Mha	604 Tg

Source: FAO (2001).

matric potential at 25 cm depth reached −0.1 MPa produced the greatest irrigation WUE in North Dakota, United States. Because the WUE depends on the water-holding capacity of the rooted soil layer, irrigation schedules must be optimized for each location.

Optimum WUE also depends on crop nutrient status. Sylvia et al. (1993) also found that they could improve maize grain yield at an average of 0.8 Mg/ha across a range of irrigation levels though inoculation with mycorhizae (*Glomus etunicatum* Becker and Gerdemann), a soil fungus that functions to increase the absorptive area of roots.

Mycotoxins

For maize, it is especially important to recognize that drought reduces crop quality as well as productivity. Under drought, fungi often attack maize ears and may produce mycotoxins, especially aflatoxins and fumonisins, which are among the most carcinogenic natural substances known. Mycotoxins have serious health implications for humans and livestock. They may cause liver cancer, thyroid cancer, reduced immune system function, slowed growth of livestock, and exacerbation of other illnesses (Galvano et al., 2002). Wherever and whenever maize crops are exposed to drought during grain filling, a special effort should be made to prevent consumption of these mycotoxins.

Crop Management

In western Kansas, United States, farmers grow rain-fed sorghum but generally do not grow maize because they believe it is insufficiently drought and heat tolerant for rain-fed production. In a five-year field experiment, Norwood and Currie (1997) found that no-till cultivation increased maize yields by 28% and net returns by 69% compared with conventional tillage. In contrast, no-till cultivation increased sorghum yields by only 11% and did not increase net return. Moreover, no-till maize yielded 28% more grain with 169% greater net return than no-till sorghum over the course of the experiment. The challenge is whether farmers are willing to accept low yields during dry years for the greater potential yields of maize during the wet years.

Rice

Distribution

Rice's center of origin is in Southeast Asia. Though rice ranks third in total production, it provides more than half of the daily dietary calories for the majority of the world's population. This is because most rice is consumed directly by humans in countries where it is produced. Generally, less than 5% of the global rice production is traded internationally. Although rice is cultivated from the equator to 55° latitude, one of the widest production zones of any crop, only 15 countries produce more than 90% of the annual total rice production. Less than 20% of total rice production lies outside of Asia (table 2.2).

The flood-irrigated lowland rice ecosystem accounts for 75% or more of the total rice production, though it occupies only 55% of the production area (IRRI, 2002). In contrast, rain-fed lowland and upland rice, which occupies about 38% of the total rice area, produces only 21% of the world's total rice production. Although there has been some success in efforts to improve the drought resistance of rice through breeding (Ingram et al., 1995), because of genetic limitations and because soils and soil management practices often restrict rice root growth, rice is arguably the most drought susceptible of the important food crops. Under tropical growth conditions rice plants have a transpiration WUE of about 0.010 Mg ha^{-1} mm^{-1} (Yambao and Ingram, 1988), and an evapotranspiration WUE about half of that, giving rice a WUE less than half of the most inefficient maize plants.

Drought in Rice Ecosystems

For the most part, irrigated rice only experiences drought when irrigation systems break down or when a regional drought restricts the amount of water available for irrigation. Though both rain-fed lowland and upland rice ecosystems suffer from frequent droughts, there are important differences between the two.

In the rain-fed lowland ecosystem, roots must be able to function under both flooded and drained soil conditions. Thus, roots of rain-fed lowland rice often have a mix of aerenchyma and non-aerenchymatous tissue (Ingram et al., 1994). Farmers generally puddle the soils of rain-fed lowland rice fields, like those of the irrigated lowlands. This puddling creates a hard soil layer about 10–15 cm below the soil surface. When these puddled soils drain, they often become very hard and develop deep cracks, thereby

Table 2.2 Global rice production and area harvested

Country	Area (%)	Production (%)
India	29.4	22.5
China	18.9	31.0
Indonesia	7.7	8.4
Bangladesh	7.2	5.9
Thailand	6.5	4.3
Vietnam	5.0	5.5
Myanmar	4.3	3.5
Others	21.0	18.9
World total	151 Mha	585 Tg

Source: FAO (2001).

damaging roots and restricting root growth (Ingram et al., 1994). Although the roots systems of direct-seeded lowland rice plants appear to be better able to supply water to the plants during periods of drying and rewetting, many farmers continue to transplant rice to improve weed management (Ingram et al., 1995). Upland rice roots generally do not form aerenchmya and perform much like the roots of other cereal crops. Upland rice soils often have low pH and high aluminum saturation, both of which restrict rice root growth and increase susceptibility to drought. Because of frequent water and nutrient deficits, upland rice generally yields 1 Mg/ha or less, with genetic improvement coming largely through growing rice varieties with developmental rates that match the average length of the rainy season (Ingram et al., 1995).

Rice is most susceptible to drought during panicle emergence, largely because this is the crop stage at which the leaf area has reached a maximum and plants transpire the most water (Yambao and Ingram, 1988; Singh et al., 2001). Rice varieties differ in sensitivity to drought during panicle emergence, which may be explained by the ability of some varieties to store nonstructural carbohydrate in vegetative tissues before drought and remobilize those substrates to maintain seed set and growth during drought (Ingram et al., 1995).

Water Production Functions

Yambao and Ingram (1988) developed a drought stress index for rice based on growth stage and the cumulative difference between actual and potential transpiration. They found that most of the difference in growth stage sensitivity to water deficit could be explained by the differences in crop water demands. Rice is most sensitive to water deficit during panicle emergence largely because that is when potential transpiration is greatest. Building on this concept, Singh et al. (2001) tested a range of water production functions for rice. In general, those that considered differential sensitivity to water deficit at different growth stages performed better than those that did not, but results differed greatly among cultivars. As for other crops, there is a strong relationship between evapotranspiration reduction and yield reduction by drought, but these drought stress indices and production functions must be calibrated for each variety and location rather than having a general applicable relationship.

Wheat

Distribution

At more than 575 million tons, wheat has the third largest global production of any food crop. The center of origin for wheat is in the Middle East between the Tigris and Euphrates rivers. As a cereal with C_3 photosynthesis metabolism, wheat productivity generally declines as temperatures increase above 30°C (Abrol and Ingram, 1996). Thus, wheat is predominantly produced in the temperate regions of the world (table 2.3). Thirty countries produce 90% of all wheat produced globally. A small portion of the wheat production lies in subtropical zones during winter.

Wheat has a dual role in the food system. It provides human nutrition both by direct consumption and through feeding to livestock. Although high temperatures restrict the productivity of wheat in the tropics, the importance of wheat for direct human consumption increases with national development as people develop a taste for breads and pastas. About half of the total world wheat production is traded internationally, with a large portion of that trade occurring from wheat produced in the industrialized nations and exported to developing countries.

Drought and WUE

Although the evapotranspiration WUE of wheat is only 0.010–0.017 Mg ha^{-1} mm^{-1} (Zhang et al., 1999), the wheat root system is generally very deep and the crop has a relatively high level of drought tolerance (Musick and Porter, 1990). As is the case for most cereals, wheat is most susceptible to yield reduction when stress occurs during the heading phase. In the North China Plain, wheat was most sensitive to water deficit from stem elongation to heading and from heading to milk dough phases (Zhang et al., 1999). Irrigation two to four times in North China increased evapotranspiration WUE from 0.010 to 0.012 Mg ha^{-1} mm^{-1}.

There is a wide range of drought tolerance among wheat varieties. Moustafa et al. (1996) found that a 10-day water deficit during heading reduced yield from 0 to 44% among four wheat cultivars. Although the physiological drought adaptations of wheat include osmotic adjustment, desiccation tolerance, and cellular elasticity, the principal contributor to drought resistance appears to be its deep root system (Musick and Porter, 1990). In Mexico, Calhoun et al. (1994) found that selecting wheat

Table 2.3 Global wheat production and area harvested

Country	Area (%)	Production (%)
India	11.7	11.9
China	11.5	16.4
Russian Federation	9.9	7.4
United States	9.3	9.3
Australia	5.2	3.5
Canada	5.2	3.6
Turkey	4.1	2.8
Pakistan	3.8	3.3
Argentina	3.2	3.1
Others	36.1	38.8
World total	212 Mha	576 Tg

Source: FAO (2001).

germplasm for high yield under full irrigation would lead to increased yield under drought conditions.

Irrigation

Though most wheat is produced under rain-fed conditions, the area of irrigated wheat is increasing. In the subtropical wheat production areas of South Asia, wheat is often grown during the cool season in rotation with irrigated lowland rice and may receive several irrigations during the season. Where declining water quality or water abundance restricts irrigation, some irrigated wheat production areas in the Great Plains of the United States are being converted to rain-fed production, although it is most economical to combine crop rotations (wheat-sorghum-fallow) with limited amounts of irrigation rather than to convert entirely to rain-fed production (Norwood, 1995). Irrigation generally increases wheat yield, but it may also lead to diseases. For example, in Denmark, Olesen et al. (2000) found that irrigation increased wheat yield, but it also increased the incidence of mildew disease.

Where wheat is irrigated, farmers generally apply irrigation amounts below levels of potential evapotranspiration. Schneider and Howell (1997) reported that deficit irrigation at a level of 33 or 66% replacement of the water lost by ET generally increased grain yield per unit irrigation more than did a late start or early termination of irrigation. Stegman and Soderlund (1992) used infrared thermometry to estimate a crop water satisfaction index (CWSI) for spring wheat. They found minimal yield reductions for CWSI less than 0.4–0.5 and maintaining available soil moisture above 50% of total root zone moisture availability. Garrot et al. (1994) showed that a CWSI could be used to schedule irrigation of durum wheat for increased WUE in an arid zone and recommended irrigation to maintain CWSI values less than 0.3–0.37. The relationship between yield and ET changes from year to year, and the CWSI for wheat must be calibrated separately for each variety and location (Stegman and Soderlund, 1992).

Water production functions have been developed widely as guides for the efficient use of irrigation water for wheat. In Saudi Arabia, Helweg (1991) modeled the water-yield relationship of wheat, and in Turkey, Ozsbuncuoglu (1998) tested water production functions to analyze the economic value of water. Though both Helweg (1991) and Ozsbuncouglu (1998) were able to develop accurate water production functions, mathematical relationships differed greatly between the two countries. To be useful, water production functions must be tested and calibrated separately for each location of interest.

Crop Management

Rain-fed wheat must be sown early enough to assure that the root system is well developed before there is a significant probability of water deficit. Saunders et al. (1997) found that sowing wheat in autumn under rain-fed conditions produced yields similar to sowing wheat in spring under irrigation in the Canterbury Plains of the United Kingdom. Thus, autumn sowing could promote the efficient use of scarce water resources. Similarly, delay in the sowing date from November to January consistently reduced yields and responses to inputs of supplemental irrigation and nitrogen for wheat cultivated in Syria (Oweis et al., 1998).

Plant nutrition also affects responses to water deficit. Application of fertilizer nitrogen generally increases the productivity of rain-fed wheat, although in some regions supplemental irrigation may be needed for wheat to fully benefit from the applied nitrogen (Frederick and Camberato, 1995). In contrast, it is important not to apply nutrients at levels so high that shoot growth increases more than root growth. In the Coastal Plain of the United States, increase in the amount of nitrogen applied to winter wheat increased the severity of drought stress during the grain-filling phase and reduced yields (Frederick and Camberato, 1995).

In hot and relatively arid environments such as Sudan and Mexico, mulch and irrigation increased wheat yields, but mulch and irrigation did not affect yields in hot and humid environments such as Bangladesh (Badaruddin et al., 1999). Farmyard manure, which both provides nutrients to the crop and increases the water-holding capacity of soils, increased yields in both arid and humid environments (Badaruddin et al., 1999).

Sorghum and Millet

Distribution

Global production of sorghum and millet is 85.8 million tons (tables 2.4 and 2.5), one-seventh of the total wheat production. Still, sorghum and

Table 2.4 Global sorghum production and area harvested

Country	Area (%)	Production (%)
India	24.0	12.9
Nigeria	16.6	13.4
United States	8.5	23.7
Mexico	4.6	10.8
Burkina Faso	2.9	1.7
Ethiopia	2.4	2.1
China	2.3	1.6
Argentina	1.5	5.1
Australia	1.4	2.3
Others	35.7	26.2
World total	41 Mha	57 Tg

Source: FAO (2001).

Table 2.5 Global millet production and area harvested

Country	Area (%)	Production (%)
India	34.1	33.4
Nigeria	15.8	21.5
Niger	14.2	7.7
China	3.7	8.6
Burkina Faso	3.1	2.6
Mali	2.8	2.8
Russian Federation	2.7	4.6
Others	23.5	18.7
World total	37 Mha	28 Tg

Source: FAO (2001).

millet are grown on about half of the total area covered by rice or maize and provide food for people living in some of the world's harshest climates, particularly those of semiarid South Asia and Sub-Saharan Africa. Although the United States is the world's largest producer of sorghum, nearly all of the sorghum produced in United States is fed to livestock.

Drought and Drought Resistance

Among the primary food cereals, sorghum and millet are grown under the driest and most variable climate conditions. Though both are more drought resistant than maize, millet is generally considered to be more drought resistant than sorghum. Because some sorghum varieties can mature as early as 65–70 days after sowing, farmers are increasing the area of sorghum cultivated in the Sahel-Sudan region, where length of the rainy season limits crop production (Roncoli et al., 2001).

Like maize, sorghum has a C_4 photosynthetic pathway. Beyond a minimum of 150 mm seasonal evapotranspiration, the evapotranspiration WUE of sorghum is about 0.015 Mg ha^{-1} mm^{-1} (Kreig and Loscano, 1990), which is similar to maize grown in arid or semiarid environments. The advantage of sorghum and millet lies in their capacity to yield something, albeit little, under conditions with insufficient moisture for maize to produce anything.

For sweet sorghum, in which biomass production is more important than grain production, Mastrorilli et al. (1999) found that the sensitivity to water deficit was higher during leaf production, which reduced biomass production by 30% and WUE by 16%. For grain sorghum, Camargo and Hubbard (1999) measured water use with a neutron probe in the field to estimate evapotranspiration. They found that the ratios of evapotranspiration and transpiration to potential evapotranspiration varied from 0.317 to 0.922 across growth stages and irrigation treatments. Calculating a drought index from the ratio of actual to potential yield, they found that sorghum grain yield was about five times as sensitive to drought during heading and grain-filling phases than it was to drought during the vegetative phase. Hybrid sorghum varieties show relatively broad adaptation to drought-prone environments. For example, Haussmann et al. (2000) found that hybrid sorghum out-yielded parents by 54% across eight semi-arid macro-environments of Kenya.

For millet produced in the Sahel, Winkel et al. (1997) found that drought, when it occurred before flowering or at the beginning of flowering phases, severely reduced both biomass and grain yields, whereas drought at the end of flowering phase did not affect either biomass or grain yield. Clearly, farmers must select appropriate varieties and planting periods for pearl millet to escape drought where there is a significant probability of severe drought. Mechanisms of drought resistance in sorghum and millet include leaf waxiness to prevent desiccation, plastic developmental responses to stress, and drought escape. Because they are already highly drought tolerant, efforts to improve sorghum and millet through breeding have focused more on resistance to parasites and diseases than on drought resistance (Krieg and Lascano, 1990).

Using a multiple regression approach, Kumar (1998) analyzed grain yield of pearl millet in India as a function of planting date and rainfall distribution. He found that yield was significantly reduced by the delay in sowing in the arid region of India.

Irrigation

If irrigation resources are available, farmers generally prefer to use them for more profitable crops, so sorghum is mostly irrigated only where it is grown as a component of a rotation system. Sweeney and Lamm (1993) found that irrigation increased sorghum grain yield by 1 Mg/ha for a three-year field experiment with different timings of irrigation. Because early irrigation increased seeds per head, whereas later irrigation increased weight per seed, irrigation timing did not affect grain yield.

Using crop-water production functions by growth stage for corn, sorghum, and wheat grown in the Texas High Plains area of the United States, Hoyt (1982) found that it would be possible to reduce irrigation by 20% without significantly reducing profitability, thereby extending the life expectancy of the Ogallala aquifer.

Crop Management

According to Payne (1997), soil fertility rather than water availability is the factor that most limits yield of pearl millet in the Sahel of West Africa. In a four-year experiment, Payne found that increasing plant population density to at least 20,000/ha and fertilizer applications to at least 40 kg N/ha and 18 kg P/ha increased total evapotranspiration by 50 mm. Moderate plant population densities of 10,000/ha and fertilizer rates of 20 kg N/ha and 9 kg P/ha tripled WUE and substantially increased yield. Similarly, over a four-year experiment in Niger, which included three dry years and one wet year, Sivakumar and Salaam (1999) found that P and N fertilizer application increased the WUE of pearl millet by an average of 84%. At the same time that P and N fertilizer applications may increase productivity of pearl millet, few farmers in the Sahel have access to the cash or credit needed to purchase such inputs. Instead, they apply small amounts of farmyard manure collected from their livestock (Roncoli et al., 2001).

Conclusions

The growth and productivity of all crops is directly related to the amount of water they transpire. Differences in productivity among the primary cereal crops under water-limited conditions may be explained by differences in physiology, morphology, crop duration, and management. Physiologically, the WUE ranges from 0.01 Mg ha^{-1} mm^{-1} or less for rice to 0.03 Mg ha^{-1} mm^{-1} for maize. Thus, maize can produce three times as much biomass as rice from the same amount of water. Wheat grows well under rain-fed conditions because its deep root system provides access to a larger soil volume. Millet and sorghum are grown in the most drought-prone environments, largely because they have either very rapid development to take advantage of a short growing season or because they produce something with relatively low input levels when other crops produce nothing at all. Plant breeding efforts have enabled some crop plants to escape drought by developing varieties whose duration better matches the length of the rainy season. Both plant breeding and crop management efforts have succeeded to some extent in improving root system depth, thereby allowing plants greater access to soil moisture.

Irrigation can alleviate drought stress in areas where water resources and infrastructure are available. Deficit irrigation generally increases both WUE and economic returns. Other management practices that can improve crop productivity under water-limited conditions include tillage to allow deeper root growth, fertilizer applications, inoculation with mychorrhizae, and appropriate times of planting. Crops cannot grow without water, but we certainly can improve their performance under drought-prone conditions by selecting appropriate crops and management practices.

References

Abrol, Y.P., and K.T. Ingram. 1996. Effects of higher day and night temperatures of growth and yields of some crop plants. In: F. Bazzaz and W. Sombroek (eds.), Global Climate Change and Agricultural Production. Wiley & Sons, West Sussex, England, pp. 123–140.

Badaruddin, M., M.P. Reynolds, and O.A.A. Ageeb. 1999. Wheat management in warm environments: effect of organic and inorganic fertilizers, irrigation frequency, and mulching. Agron. J. 91:975–983.

Calhoun, D.S., G. Gebeyehu, A. Miranda, S. Rajaran, and M. Van Ginkel. 1994. Choosing evaluation environments to increase wheat grain yield under drought conditions. Crop Sci. 34:673–678.

Camargo, M.B.P., and K.G. Hubbard. 1999. Drought sensitivity indices for a sorghum crop. J. Prod. Agric. 12:312–316.

FAO. 2001. Agricultural data. Food and Agriculture Organization, Rome, Italy. Available http://faostat.org/faostat/collections?subset=agriculture.

Frederick, J.R., and J.J. Camberato. 1995. Water and nitrogen effects on winter wheat in the southeastern Coastal Plain. I. Grain yield and kernel traits. Agron. J. 87:521–526.

Galvano, F., A. Russo, V. Cardile, G. Galvano, A. Vanella, and M. Renis. 2002. DNA damage in human fibroblasts exposed to fumonisin B-1. Food Chem. Toxicol. 40:25–31.

Garrot, D.J. Jr., M.J. Ottman, D.D. Fangmeier, and S.H. Husman. 1994. Quantifying wheat water stress with the crop water stress index to schedule irrigations. Agron. J. 86:195–199.

Haussmann, B.I.G., A.B. Obilana, P.O. Ayiecho, A. Blum, W. Schipprack, and H.H. Geiger. 2000. Yield and yield stability of four population types of grain sorghum in a semi-arid area of Kenya. Crop Sci. 40:319–329.

Helweg, O.J. 1991. Functions of crop yield from applied water. Agron. J. 83:769–773.

Hillel, D. 1990. Role of irrigation in agricultural systems. In: B.A. Stewart and D.R. Nielson (eds.), Irrigation of Agricultural Crops. American Society of Agronomy, Madison, WI., pp. 5–30.

Hook, J.E. 1985. Irrigated corn management for the coastal plain: Irrigation scheduling and response to soil water and evaporative demand. Univ. Georgia Agric. Exp. Sta. Res. Bull. 355.

Hoyt, P.G. 1982. Crop-water production functions and economic implications for the Texas High Plains region. Economic Research Services Staff Report. USDA, Washington, DC.

Ingram, K.T., F.D. Bueno, O.S. Namuco, E.B. Yambao, and C.A. Beyrouty. 1994. Rice root traits for drought tolerance and their genetic variation. In: G.J.D. Kirk (ed.), Rice Roots: Nutrient and Water Use. Intl. Rice Res. Inst., Los Baños, Laguna, Philippines, pp. 67–77.

Ingram, K.T., R. Rodriguez, S. Sarkarung, and E.B. Yambao. 1995. Germplasm evaluation and improvement for dry seeded rice in drought-prone environments. In: K.T. Ingram (ed.), Rainfed Lowland Rice: Agricultural Research

for High-risk Environments. International Rice Research Institute, Los Baños, Philippines, pp. 55–67.
IRRI. 2002. Rice Facts. International Rice Research Institute, Manila, Philippines.
Jensen, M.E., W.R. Rangeley, and P.J. Dielman. 1990. Irrigation trends in world agriculture. In: B.A. Stewart and D.R. Nielson (eds.), Irrigation of Agricultural Crops. American Society of Agronomy, Madison, WI, pp. 31–67.
Krieg, D.R., and R.J. Lascano. 1990. Sorghum. In: B.A. Stewart and D.R. Nielson (eds.), Irrigation of Agricultural Crops. American Society of Agronomy, Madison, WI, pp. 719–739.
(Boken) Kumar, V. 1998. An early warning system for agricultural drought in an arid region using limited data. J. Arid Environ. 40:199–209.
Mastrorilli, M., N. Katerji, and G. Rana. 1999. Productivity and water use efficiency of sweet sorghum as affected by soil water deficit occurring at different vegetative growth stages. Eur. J. Agron. 11:207–215.
Moustafa, M.A., L. Boersma, and W.E. Kronstad. 1996. Response of four spring wheat cultivars to drought stress. Crop Sci. 36:982–986.
Musick, J.T., and D.A. Dusek. 1980. Irrigated corn yield response to water. Trans. ASAE 23:92–98.
Musick, J.T., and K.B. Porter. 1990. Wheat. In: B.A. Stewart and D.R. Nielson (eds.), Irrigation of Agricultural Crops. American Society of Agronomy, Madison, WI, pp. 597–638.
Norwood, C.A. 1995. Comparison of limited irrigated vs. dryland cropping systems in the U.S. Great Plains. Agron. J. 87:737–743.
Norwood, C.A., and R.S. Currie. 1997. Dryland corn vs. grain sorghum in western Kansas. J. Prod. Agric. 10:152–157.
Olesen, J.E., J.V. Mortensen, L.N. Jorgensen, and M.N. Andersen. 2000. Irrigation strategy, nitrogen application and fungicide control in winter wheat on a sandy soil. I. Yield, yield components and nitrogen uptake. J. Agric. Sci. 134:1–11.
Oweis, T., M. Pala, and J. Ryan. 1998. Stabilizing rainfed wheat yields with supplemental irrigation and nitrogen in a Mediterranean climate. Agron. J. 90:672–681.
Ozsbuncuoglu, I.H. 1998. Production function for wheat: A case study of Southeastern Anatolian Project (SAP) region. Agric. Econ. 18:75–87.
Payne, W.A. 1997. Managing yield and water use of pearl millet in the Sahel. Agron. J. 89:481–490.
Rhoads, F.M., and J.M. Bennett. 1990. Corn. In: B.A. Stewart and D.R. Nielson (eds.), Irrigation of Agricultural Crops. American Society of Agronomy, Madison, WI, pp. 569–596.
Roncoli, M.C., K.T. Ingram, and P.H. Kirshen. 2001. The costs and risks of coping: drought and diversified livelihoods in Burkina Faso. Climate Res. 19:119–132.
Saunders, L.S., T.H. Webb, and J.R.F. Barringer. 1997. Integrating economic data with spatial biophysical data to analyze profitability and risks of wheat production on a regional basis. Agric. Syst. 55:583–599.
Schneider, A.D., and T.A. Howell. 1997. Methods, amounts, and timing of sprinkler irrigation for winter wheat. Trans. ASAE 40:137–142.
Singh, H., K.T. Ingram, and R.K. Jhorar. 2001. Comparison of some water production functions for rice. Trop. Agric. 78:95–103.
Sivakumar, M.V.K., and S.A. Salaam. 1999. Effect of year and fertilizer on water-use efficiency of pearl millet (*Pennisetum glaucum*) in Niger. J. Agric. Sci. 132:139–148.

Steele, D.D. 1994. Field comparison of irrigation scheduling methods for corn. Trans. ASAE 37:1197–1203.

Stegman, E.C., and M. Soderlund. 1992. Irrigation scheduling of spring wheat using infrared thermometry. Trans. ASAE 35:143–152.

Sweeney, D.W., and F.R. Lamm. 1993. Timing of limited irrigation and N-injection for grain sorghum. Irrig. Sci. 14:35–39.

Sylvia, D.M., L.C. Hammond, J.M. Bennett, J.H. Haas, and S.B. Linda. 1993. Field response of maize to a VAM fungus and water management. Agron. J. 85:193–198.

Winkel, T., J.F. Renno, and W.A. Payne. 1997. Effect of the timing of water deficit on growth, phenology and yield of pearl millet (*Pennisetum glaucum* (L.) R. Br.) growth in Sahelian conditions. J. Exp. Bot. 48:1001–1009.

Yambao, E.B., and K.T. Ingram. 1988. Drought stress index for rice. Philipp. J. Crop Sci. 13:105–111.

Zhang, H., X. Wang, M. You, and C. Liu. 1999. Water-yield relations and water-use efficiency of winter wheat in the North China Plain. Irrig. Sci. 19:37–45.

CHAPTER THREE

Monitoring Agricultural Drought Using El Niño and Southern Oscillation Data

LINO NARANJO DÍAZ

Almost all the studies performed during the past century have shown that drought is not the result of a single cause. Instead, it is the result of many factors varying in nature and scales. For this reason, researchers have been focusing their studies on the components of the climate system to explain a link between patterns (regional and global) of climatic variability and drought.

Some drought patterns tend to recur frequently, particularly in the tropics. One such pattern is the El Niño and Southern Oscillation (ENSO). This chapter explains the main characteristics of the ENSO and its data forms, and how this phenomenon is related to the occurrence of drought in the world regions.

El Niño

Originally, the name El Niño was coined in the late 1800s by fishermen along the coast of Peru to refer to a seasonal invasion of south-flowing warm currents of the ocean that displaced the north-flowing cold currents in which they normally fished. The invasion of warm water disrupts both the marine food chain and the economies of coastal communities that are based on fishing and related industries. Because the phenomenon peaks around the Christmas season, the fishermen who first observed it named it "El Niño" ("the Christ Child"). In recent decades, scientists have recognized that El Niño is linked with other shifts in global weather patterns (Bjerknes, 1969; Wyrtki, 1975; Alexander, 1992; Trenberth, 1995; Nicholson and Kim, 1997).

The recurring period of El Niño varies from two to seven years. The intensity and duration of the event vary too and are hard to predict. Typically, the duration of El Niño ranges from 14 to 22 months, but it can also be much longer or shorter. El Niño often begins early in the year and peaks in the following boreal winter. Although most El Niño events have many features in common, no two events are exactly the same. The presence of El Niño events during historical periods can be detected using climatic data interpreted from the tree ring analysis, sediment or ice cores, coral reef samples, and even historical accounts from early settlers. Many researchers are currently working to determine whether global warming would intensify or otherwise affect El Niño.

El Niño and the Southern Oscillation

Sea surface variations in the equatorial Pacific have a profound influence on the global atmospheric circulation and are an integral part of the ENSO phenomenon. The western equatorial Pacific is characterized by a region of heavy precipitation and an intense and thermally driven circulation with low-level easterly trade winds and upper-level westerly winds to the east of the date line. This circulation is called the Walker circulation after Sir Gilbert Walker, who discovered and named a number of global climate phenomena, including the Southern Oscillation, while trying to predict the Indian summer monsoon (Walker, 1924).

The easterly trade winds in the tropics are part of the low-level component of the Walker circulation. Typically, the trade winds bring warm and moist air toward the Indonesian region (figure 3.1), where the moist air moves over the normally very warm seas and rises to high levels of the atmosphere. The air at high altitudes traveling eastward meets the westerly currents before sinking over the eastern Pacific Ocean. The rising air in the western Pacific is associated with a region of low air pressure, towering cumulonimbus clouds, and rain, whereas the sinking air in the eastern Pacific is associated with high pressure and dry conditions. Generally, when pressure rises over the eastern Pacific Ocean, it tends to drop in the western Pacific and vice versa (Maunder, 1992). A large-scale pressure seesaw between the western and eastern Pacific Oceans is called the Southern Oscillation, which causes variations in rainfall and winds. When the Southern Oscillation is combined with variations in the sea temperatures, it is often called the El Niño and Southern Oscillation or ENSO.

The Southern Oscillation can be measured by the difference between eastern sea-level pressure (ESLP) at Tahiti (on a French island in the Central Pacific) and western sea-level pressure (WSLP) at Darwin (in Australia). This pressure difference is called the Southern Oscillation Index (SOI) or Tahiti-Darwin Index (TDI). Although the SOI is the most widely used indicator of the Southern Oscillation, there are other indices using east—west pairs involving the locations such as Jakarta in Indonesia or Santiago de Chile in South America.

Two important events, El Niño and La Niña, are closely linked to the SOI. While the term "El Niño" was extracted from the folklore of the Peruvian sailors (Glantz, 1996), the term "La Niña" was recently coined by the

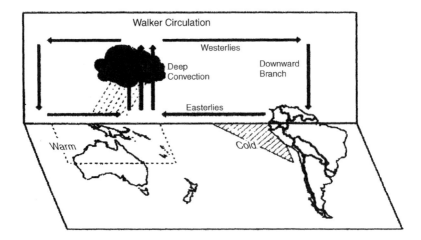

Figure 3.1 A schematic view of the normal atmospheric and oceanic features in the Pacific Ocean indicating the Walker circulation (from Glantz, 1996).

scientific community (Philander, 1989). When the SOI is negative, El Niño or ENSO warm phase occurs. During an El Niño event, the expanding warm waters tend to relax the easterly trade winds and weaken the Walker circulation. As a result, the Southern Oscillation is reversed and pressure rises in the west but falls in the east. When the SOI sustains high positive values, La Niña or ENSO cold phase occurs, which represents other extremes of the ENSO cycle. The intensification of the Walker circulation, in contrast, causes the eastern Pacific to cool down. These changes often bring widespread rain and flooding to western Pacific areas such as Australia and Indonesia.

ENSO and Droughts

The ENSO events have worldwide consequences. Many researchers have studied the relationships between ENSO events and weather anomalies around the globe to determine whether links between ENSO events and droughts exist. ENSO events perturb the atmosphere around the tropics to varying degrees, generating anomalous climatic patterns at the regional and local levels. As a result, possible connections between an ENSO event and, for example, drought in northeast Brazil, Australia, and Africa can be identified (figure 3.2).

Not all anomalies, even in ENSO years, are due to ENSO phenomenon. In fact, statistical evidence shows that ENSO accounts for only a small fraction of the interannual variance in rainfall. In many ENSO-affected places such as eastern and southern Africa, the ENSO accounts at most for about 50% of the variance in rainfall (Ogallo, 1994), but many of the more extreme anomalies, such as severe droughts, flooding and hurricanes, have strong teleconnections to ENSO events. (A "teleconnection" refers to a link between two events [e.g., ENSO, drought, flood] that are separated by large distances.)

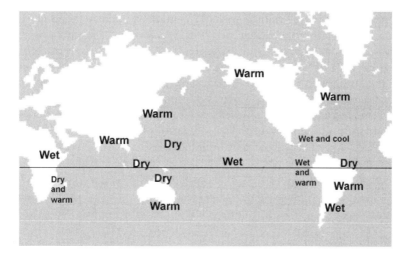

Figure 3.2 The main world regions impacted by the warm phase or ENSO (from Ropelewski and Halpert [1987]).

The 1982–83 El Niño was one of the strongest events in the recorded history. Virtually every continent was affected by this event. Some 1000–2000 deaths were blamed on the event and the disasters that accompanied it. The extreme drought in the Midwest Corn Belt of the United States during 1988 has been linked to the "cold event" of 1988 that followed the ENSO event of 1986–87 (Reibsame et al., 1990).

During an ENSO event, drought can occur virtually anywhere in the world. First, the shift of the deep convection areas to the east over the Pacific Ocean brings dry spells over much of the western Pacific countries, and then the global atmospheric anomalies (due to ENSO) induce droughts in places far from the Pacific Basin. The strongest connections between ENSO and intense drought can be found in Australia, India, Indonesia, the Philippines, parts of east and south Africa, the Western Pacific Basin islands (including Hawaii), Central America, and various parts of the United States.

Droughts occur due to below-normal precipitation, and a strong relationship exists between ENSO events and regional precipitation patterns around the globe (Ropelewski and Halpert, 1987). In northeastern South America, from Brazil up to Venezuela, El Niño brings heavy rains over Peruvian coast of South America. In addition, ENSO events affect regions in the lower latitudes, especially in the equatorial Pacific bordering tropical areas. The relationships in the mid-latitudes do not seem to be as pronounced or consistent. El Niño affects regional precipitation and results in wet or dry seasons. Indirectly, it produces significant impacts including agricultural droughts. For instance, during the 1998 El Niño, crop shortfalls from the worst drought ever recorded in Indonesia forced the government to import an unprecedented 6 million tons of rice to stave off social instability in the country. The Philippines, affected by the same drought, had to import more than 2 million tons of rice to meet the demand. Some agricultural experts predicted that rice shortages would increase in Asia and threaten regional stability. But by early 1999, the La Niña event brought heavy rainfall to much of Southeast Asia, and rice harvests in Indonesia and the Philippines recovered. Indonesia, the world's largest rice importer, bought less than half as much rice as it imported in 1998, while the Philippines enjoyed its highest rice stocks in 15 years. Major rice exporters such as Thailand and Vietnam reaped windfall profits from the 1997–98 El Niño as the export price of rice nearly doubled. But Thailand's rice exports declined in 1999, and rice prices plummeted. Global rice trade, normally averaging 18–20 million tons per year, expanded to nearly 27 million tons in 1998 (U.S. Department of Agriculture, 1999).

ENSO and Drought in Oceania

Australia and Oceania are severely affected by the lack of rains attributed to the warm phase of ENSO (Stone and Auliciems, 1992). Usually, El Niño events result in reduced rainfall across eastern and northern Australia, particularly during winter, spring, and early summer. However, the precise nature of the impact differs markedly from one event to another, even with similar changes and patterns in the Pacific Ocean. For example, the 1982–83 and 1997–98 events were both very strong, but their impacts in Australia were completely different. Eastern and southern Australia were gripped by severe droughts during 1982–83, while southern Australia experienced heat-wave conditions and bushfires, and virtually two-third of eastern Australia recorded extremely low rainfall. But in 1997, average or above-average rainfalls were common in May, and a dry spell over winter was broken by widespread, heavy rains in September.

In Papua New Guinea, the largest developing country in the southwest Pacific region, El Niño has been linked to the severe droughts of 1896, 1902,1914, 1972, and 1982 (Barr et al., 2001). During the 1997–98 El Niño, the lives of about 80,000–300,000 people were at risk due to the prolonged droughts, and rice cultivation was severely affected, forcing the government to import rice to insure the food availability. The1997–98 El Niño also caused drought in Fiji Islands, reducing sugarcane production by 50% (Kaloumaira et al., 2001).

ENSO and Drought in Southeast Asia

The ENSO produces a profound impact on climate and weather in southeast Asia (Indonesia, Brunei, Malaysia, Philippines, Vietnam, Myanmar, Cambodia, Laos, and Thailand) by causing large-scale changes in the atmospheric circulation in the region (Sirabaha and Caesar, 2000). In the Philippines, for instance, El Niño significantly impacts the main rainfall patterns. Due to the 1997–98 event, a dry spell with severe drought affected

68% of the country, compared to only 28% in 1972 and 16% in 1982. Water supply reduction affected thousands of hectares of rice and cornfields, bringing the agricultural output to the lowest level in 20 years.

In Vietnam, drought occurred in 1998, causing a total economic loss of about 5,000 billion Vietnamese dong. The lack of water increased the area affected by the saltwater intrusion in the Mekong Delta, one of the main agricultural areas of Vietnam. As a result, more than 4000 people almost starved in the mountain and central regions of Vietnam.

ENSO and Drought in Latin America

In Latin America, the onset of an El Niño event is associated with heavy rains along the Pacific coast because onshore winds carry more moisture as a result of warmer oceans, and with droughts in northeast Brazil (Ropelewski and Halpert, 1987; Grimm et al., 2000). In the 19th century, droughts during the 1877–79 period caused a famine in which 500,000 people may have died, and many others migrated. During 1982–83, yields of crops in this region of Brazil fell by more than 50%, and 28 million people and more than 1400 municipalities were affected (Rebello, 2000).

El Niño also causes droughts in other regions of Latin America including Colombia, Venezuela, and Mexico (Magaña and Quintanar, 1997; Poveda and Mesa, 1997). The drying up of reservoirs frequently disrupts hydroelectric energy supplies and drinking water supplies for cities such as Bogota, Colombia. Reduced rainfall from June to September hurt crops in an area stretching south from Mexico through Central America to Colombia and east to the Caribbean and northern Brazil.

ENSO and Droughts in Africa

In southern Africa, the frequency of drought is on the rise. Between 1988 and 1992, more than 15 drought events affected at least 1% of the population of this continent, compared to fewer than 5 such events between 1963 and 1967. This trend can be tied in part to the increased population growth and cultivation of marginal lands, and to some extent to the ENSO-related anomalies. Cane et al. (1994) noted a strong statistical relationship between Zimbabwean maize yields and sea surface temperatures in the equatorial Pacific. These studies generally show that, during ENSO episodes, large areas of southern Africa tend to experience drier than normal conditions. Between 1875 and 1978, there were 24 ENSO events, 17 of which corresponded to a decline in rainfall at least by 10% from the long-term median over this area (Rasmussen, 1987). Looking specifically at Zimbabwe's rainfall records for the 20th century, the trend indicates a clear increase in the number of below-average rainfall years and, since the 1960s, a more severe decline in rainfall in such years (Glantz et al., 1997). Figure 3.3 shows Zimbabwe's rainfall record from 1980 to 1992. It should be noted that water is the major limiting factor in Zimbabwe's

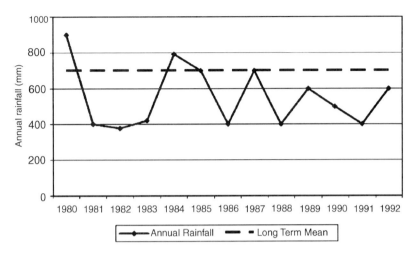

Figure 3.3 Zimbabwe's yearly average rainfall records (from Glantz et al., 1997).

agricultural production, given that its climate is fairly dry. In any given year, rainfall is generally not plentiful enough to allow adequate crop production throughout the country (Bratton, 1987).

Other regions in Africa have significant connections between ENSO and droughts. In Ethiopia, the major drought occurred following an El Niño that decreased the main June—September rainfall but boosted the small February—March rainfall. Drought is the dominant ENSO-related disaster in Ethiopia, which has led to the deaths of many people and animals during 1957–58, 1964–65, 1972–73, and 1983–84 (Tsegay et al., 2001).

ENSO and Droughts in the United States

In North America, particularly the United States, the impacts of ENSO are most dramatic in the winter. The variability induced by ENSO typically brings drought or rainy conditions to specific regions (Ropelewski and Halpert, 1986). In the Great Basin area of the western United States, during ENSO years, above-normal precipitation was recorded for the April–October season for 81% of the years. For the same percentage of the ENSO years in the southeastern United States and northern Mexico, above-normal precipitation was recorded during March—October. In the coastal west, the displacement of the jet stream can bring abnormally large amounts of rain and flooding to California, Oregon, and Washington. During the summer, heat waves and below-normal precipitation bring drought, crop failures, and even deaths. U.S. crop losses from the 1982–83 El Niño were projected to be in the neighborhood of $10–12 billion.

In the Southwest, El Niño years have a fairly consistent pattern of increased rainfall, with drought during La Niña. These impacts are also felt further inland, but the extent and magnitude is much more variable outside the Southwest.

Although El Niño events appear to have brought summer drought to the southeastern United States in the past 40–50 years, this pattern was not seen (and even reversed) during the late 19th and early 20th centuries. In fact, there seem to be substantial decadal changes in the impacts of ENSO on drought in the United States over the past 150 years (Cole et al., 1998).

ENSO-Related Data

An ENSO event usually continues for 12–18 months. If we could use ENSO-related data to monitor the development of an ENSO, it would be possible to predict regional precipitation and hence drought. Usually researchers make use of single "descriptor indices" to monitor an ENSO event. These indices are simple, easy to handle, and they capture relevant information about the events.

There are two main groups of ENSO indices: simple and composite. The simple indices are derived from single properties of either the atmosphere or the ocean. The most popular of such indices is the SOI, as defined earlier in this chapter. Surface temperature anomalies at different regions can be used to define these indices. Toward this end, the tropical Pacific has been divided into a number of regions such as Niño 1, 2, 3, 4, and 3.4 (which encompasses part of both regions 3 and 4). Niño 1 is the area defined by 80°–90° W and 5°–10° S, Niño 2 by 80°–90° W and 0°–5° S, Niño 3 by 90°–150° W and 5° N–5° S, Niño 4 by 150° W–160° E and 5° N–5° S, and Niño 3.4 by 120° W–170° W and 5° N–5° S (figure 3.4).

In addition, some research or monitoring centers all over the world use other more specific indices such as the outward long-wave radiation (OLR) to monitor the shift of the deep convection over the western Pacific, the zonal wind component at 850 Hpa level as a measure of the intensity of the trade winds, and the temperatures of the subsurface ocean layers. Simple indices may fail to capture a coupled ocean—atmosphere event, and their selection could lead to wrong results.

The composite (or coupled) indices combine characteristics of the ocean and atmosphere. The most successful index in this group is the multivariate ENSO index (MEI). The MEI is a composite index using sea-surface temperatures, surface air temperatures, sea-level pressure, zonal east–west surface wind, meridional north–south surface wind, and the total amount of cloudiness over the tropical Pacific. Positive MEI values are related to warm-phase or El Niño events and negative values to the cool-phase or La Niña events. The MEI can be understood as a weighted average of the main ENSO features, and it is calculated as the first unrotated principal component (PC) of all six observed fields combined. This is accomplished by normalizing the total variance of each field first and then performing the extraction of the first PC on the variance matrix of the combined fields

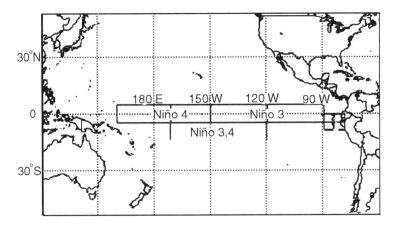

Figure 3.4 El Niño regions in the world.

(Wolter and Timlin, 1993). To keep the MEI comparable, all seasonal values are standardized with respect to each season and to the 1950–93 reference period. Further information about MEI and data sets can be found on the NOAA Web site (*http://www.cdc.noaa.gov/ENSO/enso.mei_index.html*; *http://www.cdc.noaa.gov/~kew/MEI/table.html*).

Recently, the Cuban Meteorological Service began to use a composite ENSO index, IE:

$$IE = S(MSSTA)(MSOI) \qquad [3.1]$$

where MSSTA is the three-month moving average of the sea surface temperature anomaly for El Niño 3 region, MSOI is the three-month moving average of the SOI index, and S is 1 or -1 depending on if the MSOI is positive or negative, respectively. Further information on the IE is available on a Web site (*http://www.met.inf.cu*). This composite index has proved better for the ENSO impact studies for Cuba than the traditional single indices (Cardenas and Naranjo, 1999).

By monitoring data for past events and the data for the months leading up to an event, scientists can use numerical models to help predict and/or simulate ENSO events. The dynamic coupled nature of the new models has allowed for prediction of ENSO events a year or more in advance. ENSO advisories are used to a lesser extent in planning in North America and other extratropical countries because the links between ENSO and weather patterns are less clear in these countries. As prediction models improve, the role of ENSO advisories for planning in the mid-latitude countries will increase.

Conclusions

El Niño and Southern Oscillation is one of the main features responsible for interannual climate variability. Although it is an ocean—atmospheric process over the tropical Pacific Ocean, its influence is worldwide because it modifies ocean circulation patterns influencing the weather and climate anomalies at a global scale.

ENSO-induced droughts are usually severe in nature and last long. Observational experience indicates that during an ENSO event drought conditions occur over many places in the world, causing severe economical and societal impacts.

The ENSO data have been used for long-term prediction of droughts. Usually researchers make use of certain single descriptor indices that are able to capture relevant information about the events in a simple way. Although a high level of confidence has been shown, in recent years, in the capabilities of ENSO data to predict droughts, many uncertainties still remain. One of the main challenges is the attribution problem. Are all the droughts due to ENSO? How does it work? Why do some ENSO events produce drought and while others do not in the same region? The link between ENSO and drought is still an open book. Much more has to be done before the global community is able to fully understand and obtain reliable prediction methods. Meanwhile, ENSO data seem to be satisfactory for predicting droughts, at least in some major regions of the world.

References

Alexander, M.A. 1992. Midlatitude atmosphere-ocean interaction during El Niño. Part II: The Northern Hemisphere atmosphere. J. Climate 5:959–972 .

Barr, J., B. Choulai, S. Maiha, B. Wayi, and K. Kamnanaya. 2001. Papua New Guinea country case study: Impacts and response to the 1997–98 El Niño event. In: Once Burned Twice Shy? Lessons Learned from the 1997–98 El Niño. United Nations University Press, New York, pp. 159–175.

Bjerknes, J. 1969. Atmospheric teleconnections from the equatorial Pacific. Monthly Weath. Rev. 97:163–172.

Bratton, M. 1987. Drought, food and the social organization of small farmers in Zimbabwe. In: M. Glantz (ed.), Drought and Hunger in Africa: Denying Famine a Future. Cambridge University Press, Cambridge.

Cane, M., G. Eshel, and R. Buckland. 1994. Forecasting Zimbabwean maize yield using eastern equatorial Pacific sea surface temperature. Nature 370: 204–205.

Cardenas, P., and L. Naranjo. 2000. El Niño, La Oscilación del Sur y El ENOS. Papel en la predecibilidad de elementos climáticos. Anal. Fís. 95:203–212.

Cole, J.E., and E. R. Cook. 1998. The changing relationship between ENSO variability and moisture balance in the continental United States. Geophys. Res. Lett. 25:4529–4532.

Glantz, M. 1996. Currents of Change. El Niño Impact on Climate and Society. Cambridge University Press, Cambridge.

Glantz, M.H., M. Betsill, and K. Crandall. 1997. Food security in southern Africa: Assessing the use and value of ENSO information. NOAA project report. Boulder, CO: ESIG/NCAR.

Grimm, A.M., V.R. Barros, and M.E. Doyle. 2000. Climate variability in southern South America associated with El Niño and La Niña events. J. Clim. 13:35–58.

Kaloumaira, A. 2001. Fiji country case study: Impacts and response to the 1997–98 El Niño event. In: Once Burned Twice Shy? Lessons Learned from the 1997–98 El Niño. United Nations University Press, New York, pp. 101–114.

Magaña, V., and A. Quintanar. 1997. On the use of a general circulation model to study regional climate. In: Proceedings of the Second UNAM-Cray Supercomputing Conference on Earth Sciences, Mexico City. Cambridge University Press, Cambridge, UK, pp. 39–48.

Maunder, W.J. 1992. Dictionary of Global Climate Change. Chapman and Hall, New York.

Nicholson, S.E., and Kim, E. 1997. The relationship of the El-Niño Southern Oscillation to African rainfall. Intl. J. Climatol. 17:117–135.

Ogallo, L.A. 1994. Validity of the ENSO-related impacts in eastern and southern Africa. In: Usable Science: Food Security, Early Warning, and El Niño. United Nations Environment Program, Nairobi, Kenya.

Philander, S.G. 1989. El Niño, La Nina, and the Southern Oscillation. Academic Press, New York.

Poveda, G., and O.J. Mesa. 1997. Feedbacks between hydrological processes in tropical South America and large scale oceanic-atmospheric phenomena. J. Climate 10:2690–2702.

Rasmussen, E.M. 1987. Global climate change and variability: Effects on drought and desertification in Africa. In M.H. Glantz (ed.), Drought and Hunger in Africa: Denying Famine a Future. Cambridge University Press, Cambridge.

Rebello E., 2000. Anomalías climáticas e seus impactos no brasil durante o evento "El Niño" de 1982–83 e previsao para o evento "El Niño" de 1997–98 [in Portuguese]. Consecuencias climáticas e hidrológicas del evento El Niño a escala regional y local. Incidencia en América del Sur. UNESCO.

Riebsame, W.E., S.A. Changnon, and T.R. Karl. 1990. Drought and Natural Resource Management in the United States: Impacts and Implications of the 1987–1989 Drought. Westview Press, Boulder, CO.

Ropelewski, C.F., and M.S. Halpert. 1986. North American precipitation and temperature patterns associated with the El Niño-Southern Oscillation (ENSO). Monthly Weath. Rev, 114:2352–2362.

Ropelewski, C.F., and M.S. Halpert. 1987. Global and regional scale precipitation patterns associated with the El Niño-Southern Oscillation. Monthly Weath. Rev. 115:1606–1626.

Sirabaha, S., and J. Caesar. 2000. Significance of the El Niño-Southern Oscillation for Southeast Asia. In Proceedings of the Workshop on the Impacts of El Niño and La Niña on Southeast Asia, Hanoi, Vietnam, February 21–23, 2000.

Stone, R., and A. Auliciems. 1992. SOI phase relationships with rainfall in eastern Australia. Intl. J. Climatol. 12:625–636.

Trenberth, K.E. 1995. Atmospheric circulation climate changes. Climat. Change 31:427–453.

Tsegay, W.G., A. Demlew, and H. Yibrah. 2001. Ethiopia country case study: Impacts and response to the 1997–98 El Niño event. In: Once Burned Twice Shy? Lessons Learned from the 1997–98. El Niño. United Nations University Press, Tokyo, pp. 88–100.

U.S. Department of Agriculture, 1999. El Niño and rice production. World Geogr. News.

Walker, G.T. 1924. Correlation in seasonal variations of weather, IX: A further study of world weather. Mem. Ind. Meteorol. Dept. 24:275–332.

Wolter, K., and M.S. Timlin. 1993. Monitoring ENSO in COADS with a seasonally adjusted principal component index. In: Proceedings of the 17th Climate Diagnostics Workshop, Norman, OK. University of Oklahoma, Norman, pp. 52–57.

Wyrtki, K. 1975. El Niño: The dynamic response of the equatorial Pacific Ocean to atmospheric forcing. J. Phys. Oceanogr. 5:572–584.

CHAPTER FOUR

Techniques to Predict Agricultural Droughts

ZEKAI ŞEN AND VIJENDRA K. BOKEN

In general, the techniques to predict drought include statistical regression, time series, stochastic (or probabilistic), and, lately, pattern recognition techniques. All of these techniques require that a quantitative variable be identified to define drought, with which to begin the process of prediction. In the case of agricultural drought, such a variable can be the yield (production per unit area) of the major crop in a region (Kumar, 1998; Boken, 2000). The crop yield in a year can be compared with its long-term average, and drought intensity can be classified as nil, mild, moderate, severe, or disastrous, based on the difference between the current yield and the average yield.

Statistical Regression

Regression techniques estimate crop yields using yield-affecting variables. A comprehensive list of possible variables that affect yield is provided in chapter 1. Usually, the weather variables routinely available for a historical period that significantly affect the yield are included in a regression analysis. Regression techniques using weather data during a growing season produce short-term estimates (e.g., Sakamoto, 1978; Idso et al., 1979; Slabbers and Dunin, 1981; Diaz et al., 1983; Cordery and Graham, 1989; Walker, 1989; Toure et al., 1995; Kumar, 1998). Various researchers in different parts of the world (see other chapters) have developed drought indices that can also be included along with the weather variables to estimate crop yield. For example, Boken and Shaykewich (2002) modfied the Western Canada Wheat Yield Model (Walker, 1989) drought index using daily temperature and precipitation data and advanced very high resolution radiometer (AVHRR) satellite data. The modified model improved the predictive power of the wheat yield model significantly. Some satellite data-based variables that can be used to predict crop yield are described in chapters 5, 6, 9, 13, 19, and 28.

The short-term estimates are available just before or around harvest time. But many times long-term estimates are required to predict drought for next year, so that long-term planning for dealing with the effects of drought can be initiated in time. Long-term predictions are especially useful for food-grain exporting countries, such the United States and Canada, because grain-exporting agencies are interested in knowing the amount of food grains available for export next year to help set the export targets. For such prediction, time series analysis can be performed.

Time Series Analysis

Weather data during a growing season cannot be used for obtaining the long-term estimates simply because the long-term estimates are required before crops are even sown. As the yield is known to be influenced most by weather conditions during the growing season, it is a common practice to estimate yield using weather data. Attempts to obtain long-term estimates that do not employ weather data are limited.

As an alternative to weather data, annual time series of yield data is used to obtain the long-term yield estimates by modeling the series. In a time series analysis, a variable to be forecasted (yield, in the present case) is modeled as a function of time:

$$Y_t = f(t) + \varepsilon_t \qquad [4.1]$$

where Y_t is yield for year t, $f(t)$ is a function of time t, and ε_t refers to error (i.e., the difference between observed yield and forecasted yield for year t). Once a functional relationship between yield and time is developed, yield for the year ahead can be forecasted. A few techniques (e.g., linear trend, quadratic trend, simple exponential smoothing, double exponential smoothing, simple moving averaging, and double moving averaging) can be used to model an yield series (Boken, 2000).

Pattern Recognition

Pattern recognition (PR) is a process to classify an object by analyzing the numerical data that characterize the object. Various academic fields, such as image processing, medical engineering, criminology, speech recognition, and signature identification, have applied PR to classify objects of interest (Duda et al., 2001). However, PR techniques have not been exploited for drought prediction. The process of PR begins with the identification of a variable that can be used to define the object under study. In the case of agricultural drought, the object can be yield of a major crop in the region.

Various pattern recognition techniques are available in the literature (Jain and Flynn, 1993; Duda et al., 2001), but only a few techniques are relevant to the case of agricultural drought. First, potential variables that affect drought are derived from weather data and satellite data. For example, these variables can be average monthly temperature, total monthly precipitation, and variables based on satellite data during the growing season. V. K. Boken, in a just concluded analysis, derived 32 variables to develop pattern recognition models to predict drought for selected crop districts in Saskatchewan, Canada. Both two-variable and multiple-variable cases were considered. In the two-variable case, an error-correction (EC) procedure (Kumar et al., 1998; Duda et al., 2001) was applied and, in the multiple-variable case, both linear (linear discriminant analysis) and nonlinear (nearest neighbor analysis) techniques were attempted using SAS software (SAS Institute Inc., Cary, North Carolina, United States). In the case of the EC procedure, two variables were selected at a time to examine the presence of a solution vector to linearly separate drought and nondrought events. An iterative procedure was applied using a computer program, but no solution vector was found. This reiterates the complexity involved in the analysis of agricultural drought. To proceed further, the multiple-variable case was investigated and a subset of significant or most suitable variables was determined. To find the subset of significant variables, the STEPDISC procedure of SAS software was used. Using these significant variables for each crop district, the linear and nonlinear techniques were applied to develop models for classifying an event as drought or nondrought.

Linear Discriminant and Nearest Neighbor Analysis

A function was obtained by applying the linear discriminant technique on the subset of variables, which can be used to classify a subset of variables as drought or nondrought. To develop such a function, the whole data set for multiple years was used. The training set was used to develop the linear discriminant function (LDF), and the testing set was used to test the classification performance of the LDF. For applying linear discriminant analysis, the within-category distribution must be normal. A nonparametric technique (nearest neighbor) was also attempted. Using this technique one can classify a subset of variables as drought or nondrought based on the category of the neighboring subsets, and the normality assumption is not required. Up to 83% of accuracy to classify or predict drought was obtained.

Stochastic or Proababilistic Analysis

In the mathematical modeling of agricultural droughts, most often soil moisture records are taken as the basis where a time series of the soil moisture contents, $X_1, X_2, X_3, \ldots, X_n$ is truncated at a threshold soil moisture value, X_0, as shown in figure 4.1. An agricultural drought can be defined on the basis of some objective, random, probabilistic, or statistical properties or features:

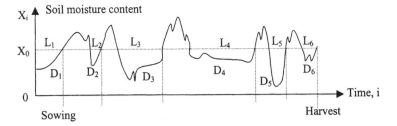

Figure 4.1 Variation in soil moisture content during a crop-growing season and its comparison with a threshold value, X_0. L and D refer to length and dry period, respectively.

1. A wet spell occurs when any time series value at ith instant is greater than the threshold level. The difference $(X_i - X_0)$ is named as the soil moisture surplus (SMS) when it exceeds zero and is called a dry spell or soil moisture deficit (SMD) when it is less than zero.
2. A sequence of wet spells preceded and suceeded by a dry spell is referred to as the duration of wet period, during which there is no moisture stress and plant growth is normal. If the two successive dry spells that separate a wet period are X_i and X_j, then the duration of this wet period is equal to $(j - i)$.
3. Similarly, if a sequence of dry spells is preceded and suceeded by a wet spell, it is then referred to as the duration of dry period, which might restrict the crop development during vegetative or reproductive phenological phases. If the two successive wet spells that separate a dry period are X_k and X_l, then the duration of this dry period is equal to $(l - k)$.
4. If a dry spell (continuation of SMD) is followed by a wet spell, then there is a transition from the drought period to wet period (continuation of SMS) (i.e., $X_i < X_j$).
5. Similarly, if a wet spell is followed by a dry spell, then there is a transition from the wet period to drought period (i.e., $X_i > X_j$).
6. The maximum dry duration in the record of past soil moisture observations corresponds to the most critical agricultural drought period that has occurred in the history of the record site. Such a critical period is directly related to the critical phenological phases and is important to crop yield estimation.
7. The summation of water deficits during the whole drought period gives the total drought severity. This is equivalent to accumulation of soil moisture needed to offset the agricultural drought, which in turn is directly related to rainfall surplus.
8. Finally, the division of the accumulated soil moisture by drought duration can determine the average severity of agricultural drought.

It is possible to calculate almost all of the above objective drought features, provided measured soil moisture records are available. Accordingly, their statistical average, standard deviation, correlation coefficient, skewness, and probability distribution function (PDF) can also be determined.

In contrast, if the interest lies in drought frequency, then the probability statements can also be calculated from the same record. For instance, $P(X_i > X_0)$ and $P(X_i < X_0)$ express simply the SMS and SMD probabilities, respectively. These basic probabilities help construct a probabilistic model that can be used to predict agricultural drought durations (Şen, 1976).

There is no procedure so far for accurately predicting the time of drought occurrence and durations or areal extent of drought. Although various subjective approaches were used in the past, they all failed. In modern times, drought estimations are sought on the basis of objective and systematic scientific procedures, and along this line the probability theory provides a convenient procedure for drought predictions. These techniques, in general, are used for depicting the quantitative relationships between the weather variables and the drought characteristics. For instance, multiple regression analysis or Monte Carlo simulation techniques are used to answer questions concerning regional and temporal drought frequencies.

The majority of drought analysis has concentrated on temporal assessments. The first classical approach to statistically analyzing droughts was evaluating the instantaneously smallest value in a measured sequence of basic variables such as soil moisture recorded at a single site (Gumbel, 1963). This method gives information on the maximum value of drought duration magnitude with a prescribed period of time such as 10, 25, 50, or 100 years. Yevjevich (1967) presented the first objective definition of temporal droughts. Applications of the above method have been performed by Downer et al. (1967), Llamas and Siddiqui (1969), Saldarriaga and Yevjevich (1970), Millan and Yevjevich (1971), Guerrero-Salazar (1973), Guerrero-Salazar and Yevjevich (1975), Şen (1976, 1977, 1980a) and brief descriptions have been presented by Dracup et al. (1980). Due to the analytical difficulties, regional droughts have been studied less. The first study of regional drought was by Tase (1976), who performed many computer simulations to explore various drought properties. Different analytical solutions of drought occurrences have been proposed by Şen (1980b) through random field concept. However, these studies are limited in the sense that they investigate regional drought patterns without temporal considerations.

Below, a systematic approach is presented for the calculation of temporal and regional drought occurrences by simple probability procedures. Recent improvements in statistical methods have tended to place a new emphasis on rainfall studies, particularly with respect to a better understanding of persistence (continuity of dry spells) effects (Şen, 1989, 1990).

Temporal Drought Models

Statistical theory of runs provides a common basis for objectively defining and modeling critical drought given a time series (Feller, 1967). A constant soil moisture truncation level divides the whole series into two complementary parts: those greater than the truncation level, which are referred to as the positive run in statistics, a SMS period in the agricultural sense, and, similarly, a negative run or SMD period. Feller (1967) also gave a definition of runs based on recurrence theory and Bernoulli trials as follows.

A sequence of n events, S (success, SMS) and F (failure, SMD), contains as many S runs of length r as there are non-overlapping, uninterrupted blocks containing exactly r events S each. This definition is not convenient practically because it does not say anything about the start and end of the run (i.e., drought). In contrast, a definition of runs given by Feller (1967) seems to be most revealing for the analysis of various drought features because a run is defined as a succession of similar events preceded and succeeded by different events with the number of similar events in the run referred to as its length (figure 4.1).

Independent Bernoulli Model Truncation of a soil moisture series X_i ($i = 1, 2, \ldots, n$) at a constant level yields two complementary and mutual distinct events—namely, SMS and SMD—with respective probabilities p and q (i.e., $1 - p$). If the probability of the longest run-length, L (i.e., critical agricultural drought duration) in a sample size of i is equal to 1 and is denoted by $P_i\{L = 1\}$, then for sample size $i = 1$, one can simply deduce,

$$P_1\{L = 0\} = q \quad [4.2]$$

$$P_1\{L = 1\} = p \quad [4.3]$$

Since the occurrences of the elementary events are assumed independent from each other, the combined probabilities for $i = 2$ can be written as

$$P_2\{L = 0\} = P_1\{L = 0\} q \quad [4.4]$$

$$P_2\{L = 1\} = P_1\{L = 1\} q + P_1\{L = 0\} p \quad [4.5]$$

$$P_2\{L = 2\} = P_1\{L = 1\} p \quad [4.6]$$

Simply, $P_2\{L = 0\}$ indicates an SMD followed by another SMD. The first term on the right-hand side in $P_2\{L = 1\}$ represents the SMS followed by SMD, and the second term represents an SMD followed by an SMS. Finally, $P_2\{L = 2\}$ is the combination of SMS followed by another SMS event. It is possible to develop the same probability concepts for a soil moisture time series of length n (Şen, 1980a).

Dependent Bernoulli Model In the derivation of drought probabilities above, the occurrence of successive SMD and SMS is considered as independent from each other. However, in nature, there is a tendency of SMD to follow SMD, which implies dependence between successive occurrences. The simplest representation of dependence can be achieved by considering the relative situation of two successive time intervals. This leads to four possible outcomes as transitional probabilities, which are referred to also as conditional probability statements in the probability theory. For instance,

$P(-/+)$ implies the probability of SMD $(-)$ at current time interval on the condition (given) that there is SMS $(+)$ in the following time interval. In contrast, according to the joint probability definition, it is possible to state (1) transition from an SMS to an SMS with probability $P(+/+)$; (2) transition from an SMD to an SMS with probability $P(-/+)$; (3) transition from an SMS to an SMD with probability $P(+/-)$; and, finally, (4) transition from a soil moisture to an SMD with probability $P(-/-)$. Four soil moisture joint probability statements are $P(+,+) = P(+/+)P(+)$; $P(-,+) = P(-/+)P(+)$; $P(+,-) = P(+/-)P(-)$ and $P(-,-) = P(-/-)P(-)$. Similar to the independent Bernoulli case, there are two state probabilities: SMS $P(+)$ and SMD $P(-)$ probabilities. Because transition and state probabilities are independent of each other, the relationships between them can be written as:

$$P(+) = P(+/+)P(+) + P(+/-)P(-) \quad [4.7]$$

$$P(-) = P(-/+)P(+) + P(-/-)P(-) \quad [4.8]$$

where equation 4.7 expresses the probability of an SMS in the current time interval with its first right-hand side term as the probability of SMS $P(+)$, in the previous time interval with its transition $P(+/+)$ from SMS to SMS, and the second term on the right-hand side representing the transition $P(+/-)$ from SMD in the previous time interval to SMS in the current time interval. Equation 4.8 has similar interpretations. Furthermore, due to the mutual exclusiveness of probabilities, the following sequences of probability statements are also valid. Any time interval may have either SMS or SMD cases with state probabilities $P(+)$ or $P(-)$, respectively, whereas transitional probabilities are valid between two successive time intervals, given that the state is in SMS in the previous interval. The derivation mechanism of agricultural drought probabilities are the same as independent Bernoulli case, but a slight change of notation is necessary due to the dependent nature of the successive events. Hence, the probability of the longest critical drought duration, L, being equal to an integer value, j, in a sample size of i with a surplus state at the final stage will be denoted by $P_i^+ \{L = j\}$. Accordingly one can write

$$P_1^- \{L = 0\} = P(-) = q \quad [4.9]$$

$$P_1^+ \{L = 1\} = P(+) = p \quad [4.10]$$

If two successive time intervals ($i = 2$) are considered, it is possible to obtain the following by enumeration:

$$P_2^- \{L = 0\} = P_1^- \{L = 0\} P(-/-) \quad [4.11]$$

$$P_2^+ \{L = 1\} = P_1^- \{L = 0\} P(+/-) \quad [4.12]$$

$$P_2^- \{L = 1\} = P_1^+ \{L = 1\} P(-/+) \quad [4.13]$$

$$P_2^+ \{L = 2\} = P_1^+ \{L = 1\} P(+/+) \quad [4.14]$$

Equation 4.11 refers to the transition from SMD to SMD, whereas Equation 4.13 shows transition from SMS to SMD. Similarly, probabilities can be determined for sample size i. The numerical solutions of these equations for different sample sizes can be obtained using computer porgrams and are presented in figure 4.2, on the basis of a given SMS probability ($p = 0.7$). Using this graph, it is possible to read probability of critical agricultural drought of a given duration for given number of samples.

Markov Model

Although dependence between successive SMS or SMD events is accounted for simply by the dependent Bernoulli model, in nature dependences are more persistent. To model critical agricultural droughts more realistically, the second-order Markov process is presented. This process requires three-interval basic transitional probabilities in addition to two-interval probabilities. The SMS and SMD probabilities remain as they were in the previous models. The complete description of the second-order Markov model for critical drought probability predictions is presented by Şen (1990). The new set of transitional probabilities can be defined as 8 ($i = 2^3 = 8$), which are mutually exclusive and collectively exhaustive alternatives. For example, one of the the eight alternatives will be:

$$P(+/+-) = P(X_i > X_0, X_{i-1} > X_0, X_{i-2} > X_0) \qquad [4.15]$$

Here, $P(+/+-)$ refers to the probability of an SMS at current interval, given that two successive intervals included an SMD and SMS, respectively. In contrast, mutual exclusiveness implies that $P(+/++) + P(-/++) = 1$; $P(-/-+) + P(-/+-) = 1$; $P(+/-+) + P(-/-+) = 1$ and $P(+/--) +$

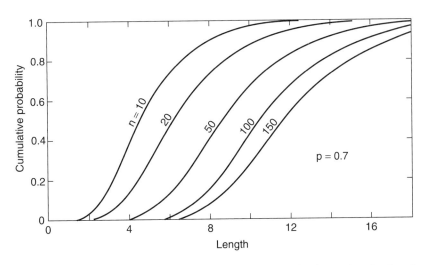

Figure 4.2 Cumulative probabilities of critical agricultural drought of given duration (length in years) and return period n (in years).

48 BASIC CONCEPTS AND DROUGHT ANALYSIS

$P(-/--) = 1$. The critical agricultural drought durations for the first two samples are the same as in the dependent Bernoulli case. However, when the sample size is greater than two, the relevant drought probabilities differ. Numerical solution of these equations are achieved through the use of digital computers, and some of examplary results are shown in figures 4.3 and 4.4.

Spatio-Temporal Drought Models Almost all the studies in the literature are confined to temporal drought assessment, with few studies concerning areal coverage. But there is a significant regional dimension to agricultural drought that occurs at regional scales. Most often, droughts affect not only one country but many countries, in different regional proportions. In this section we include models that represent both spatial (regional) and temporal drought behaviors simultaneously.

In regional studies, clustering of dry spells in a region will be referred to

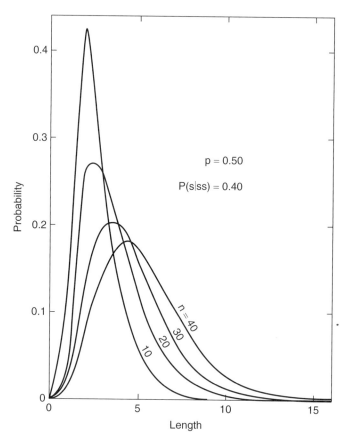

Figure 4.3 Critical drought duration distribution for different sample sizes at $P(+/+\ +) = 0.40$.

TECHNIQUES TO PREDICT AGRICULTURAL DROUGHTS 49

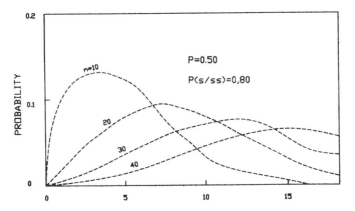

Figure 4.4 Expectation of the critical drought duration for different sample sizes and $P(+/++) = 0.8$.

as "drought area" and clustering of the wet spell as "wet area." Here we explain two different regional drought models. The first model relies on probabilities of dry and wet areas, spell probabilities, p_r and q_r, respectively. Because these two events are mutually exclusive, $p_r + q_r = 1$. This model assumes that once a subarea of agricultural land is hit by a dry spell, it remains under this state in the subsequent time instances. Therefore, as time passes, the number of dry spells hitting the subareas increases steadily until the whole region comes under the influence of drought. Such a regional model has been referred to as regional persistence model (Şen, 1980b). The application of this model is convenient for agricultural droughts in arid and semiarid regions where long drought periods occur.

The second model takes into account the regional as well as the temporal occurrence probabilities of wet and dry spells. The probabilities of temporal SMS (p_t) and SMD (q_t) are mutually exclusive, and therefore, $p_t + q_t = 1$. In this model, in an already drought-stricken area, subareas are subject to temporal drought effects in the next time interval. This model is also known as multiseasonal model because it can be applied for a duration that may include several dry and wet periods. Since agriculture is a seasonal activity, this seasonal model is suitable for agricultural drought modeling.

Regional Drought Modeling Let an agricultural land be divided into m mutually exclusive subareas, each with the same chance of spatial and temporal drought. The Bernoulli distribution theory can be used to find the extent of drought area, A_d, during a time interval, Δt. The probability of n_1 subareas affected by drought can be written according to Bernoulli distribution as (Feller, 1967)

$$P_{\Delta t}(A_d = n_1) = \binom{m}{n_1} p_r^{n_1} q_r^{m-n_1} \qquad p_r + q_r = 1.0 \qquad [4.16]$$

This implies that out of m possible drought-prone subareas, n_1 have SMD, and hence the areal coverage of drought is equal to n_1 or n_1/m. For the subsequent time interval, Δt, there are $(m - n_1)$ drought-prone subareas. Assuming that the evolution of possible SMD and SMS spells along the time axis is independent over mutually exclusive subareas, similar to the concept in equation 4.16, it is possible to write for the second time interval (Tase, 1976; Şen, 1980b). The PDFs of areal agricultural droughts for this model are shown in figure 4.5 with parameters $m = 10$, $p_r = 0.3$, $p_t = 0.2$ and $i = 1, 2, 3, 4,$ and 5. The probability functions exhibit almost symmetrical forms regardless of time intervals, although they have very small positive skewness.

Another version of the multiseasonal model is interesting when the number of continuously SMD subareas appear along the whole observation period. In such a case, the probability of drought area in the first time interval can be calculated using equation 4.16. At the end of the second time interval, the probability of j subareas with two successive SMDs given that already n_1 subareas had SMD in the previous interval can be expressed as

$$P_{2\Delta t}(A_d = j | A_d = n_1) = P_{\Delta t}(A_d = n_1) \binom{n_1}{j} p_t^j q_t^{n_1-j} \quad [4.17]$$

This expression computes the probability of having n_1 subareas to have SMD, out of which j subareas are hit by two SMDs; in other words, there are $(n_1 - j)$ subareas with one SMD. Hence, the marginal probability of continuous SMD subarea numbers is

$$P_{2\Delta t}(A_d = j) = \sum_{k=0}^{m-j} P_{\Delta t}(A_d = k + j) \binom{k+j}{j} p_t^j q_t^k \quad [4.18]$$

In general, for the ith time interval, it is possible to write

$$P_{i\Delta t}(A_d = j) = \sum_{k=0}^{m-j} P_{(i-1)\Delta t}(A_d = k + j) \binom{k+j}{j} p_t^j q_t^k \quad [4.19]$$

The numerical solutions of this expression are presented in figure 4.6 for $m = 10$, $p_r = 0.3$, and $p_t = 0.5$. The probability distribution function is positively skewed.

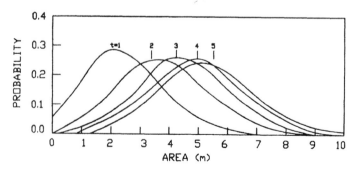

Figure 4.5 Probability of drought area for multiseasonal model ($m = 10$; $p_r = 0.3$; $p_t = 0.2$).

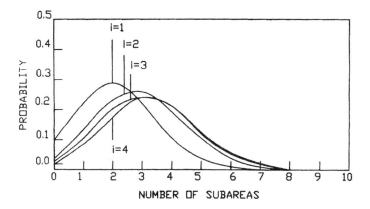

Figure 4.6 The variation in drought probabilities according to the number of subareas.

Drought Parameters The global assessment of model performances can be achieved on the basis of drought parameters such as averages (i.e., expectations and variances), but for drought predictions, the PDF expressions as derived above are significant. The expected (i.e., average) number of SMDs, $E_i(A_d)$, over a region of m subareas during time interval, $i\Delta t$, is defined as

$$E_i(A_d) = \sum_{k=0}^{m} k P_{i\Delta t}(A_d = k) \qquad [4.20]$$

Similarly, the variance, $V_i(A_d)$, of a drought-affected area is given by definition as

$$V_i(A_d) = \sum_{k=0}^{m} k^2 P_{i\Delta t}(A_d = k) - E_i^2(A_d) \qquad [4.21]$$

The drought-stricken average area within the whole region can be determined as:

$$E_i(A_d) = m p_r \sum_{k=0}^{i-1} q_r^k \qquad [4.22]$$

or, succinctly,

$$E_i(A_d) = m(1 - q_r^i) \qquad [4.23]$$

Furthermore, the percentage of agricultural drought area, $P^i{}_A$, can be calculated by dividing both sides by the total number of subareas, m, leading to

$$P^i_A = (1 - q_r^i) \qquad [4.24]$$

Figure 4.7 shows the variation in a drought-affected area with the number of SMD subareas, i, for given SMD probability, q_r.

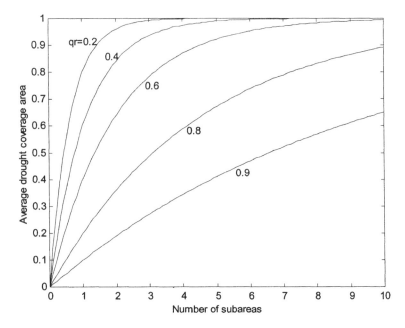

Figure 4.7 Estimation of drought-affected area.

Spatio-temporal drought behaviors were investigated by Sirdas and Şen (2003) for Turkey, where the drought period, magnitude, and standardized precipitation index (SPI) values were presented to depict the relationships between drought duration and magnitude.

Conclusions

This chapter has focused on some of techniques that can be used to predict agricultural drought. Whereas statistical regression, time series, and pattern recognition techniques were directly based on the crop yield variation to define drought, the probabilistic technique was centered on soil moisture surplus or deficit concept. It must be noted, however, that an intricate relationship exists between soil moisture deficit and crop yield or agricultural drought. Therefore, the probabilistic method described in this chapter can be indirectly linked to crop yield to predict agricultural drought.

References

Boken, V.K. 2000. Forecasting spring wheat yield using time series analysis: A case study for the Canadian Prairies. Agron. J. 92:1047–1053.

Boken, V.K., and C.F. Shaykewich. 2002. Improving an operational wheat yield model for the Canadian Prairies using phenological-stage-based normalized difference vegetation index. Intl. J. Remote Sensing 23:4157–4170.

Cordery, I., and A.G. Graham. 1989. Forecasting wheat yields using water budgeting model. Aust. J. Agric. Res. 40:715–728.

Diaz, R.A., A.D. Matthias, and R.J. Hanks. 1983. Evapotranspiration and yield estimation of spring wheat from canopy temperature. Agron. J. 75:805–810.

Downer, R., M.M. Siddiqui, and V. Yevjevich. 1967. Applications of runs to hydrologic droughts. Hydrology Paper 23. Colorado State University, Fort Collins.

Dracup, A.J., K.S. Lee, and E.G. Paulson. 1980. On the statistical characteristics of drought events. Water Resources Res. 16:289–196.

Duda, R.O., P.E. Hart, and D.G. Stork. 2001. Pattern Classification, 2nd ed., John Wiley and Sons, New York.

Feller, W. 1967. An Introduction to Probability Theory and Its Application. John Wiley and Sons, New York.

Guerrero-Salazar, P.L.A., and V. Yevjevich. 1975. Analysis of drought characteristics by the theory of runs. Hydrology Paper 80. Colorado State University, Fort Collins.

Gumbel, E.J. 1963. Statistical forecast of droughts. Bull. Intl. Assoc. Sci. Hydrol. 8:5–23.

Idso, S.B., J.L. Hatfield, R.D. Jackson, and R.J. Reginato. 1979. Grain yield prediction: extending the stress-degree-day approach to accommodate climatic variability. Remote Sens. Environ. 8:267–272.

Jain, A.K., and P.J. Flynn. (eds.). 1993. Three-Dimensional Object Recognition Systems. Elsevier, New York.

(Boken) Kumar, V. 1998. An early warning system for agricultural drought in an arid region using limited data. J. Arid Environ. 40:199–209.

(Boken) Kumar, V., C.E. Haque, and M. Pawlak. 1998. Determining agricultural drought using pattern recognition. In: J.C. Lehr and H.J. Selwood (eds.), Prairie Perspectives—Geographical Essays. Department of Geography, University of Winnipeg, Winnipeg, Manitoba, Canada, pp. 74–80.

Llamas, J., and M.M. Siddiqui. 1969. Runs of precipitation series. Hydrology Paper 33. Colorado State University, Fort Collins.

Millan, J., and V. Yevjevich. 1971. Probabilities of observed droughts. Hydrology Paper 50. Colorado State University, Fort Collins.

Sakamoto, C.M. 1978. The Z-index as a variable for crop yield estimation. Agric. Meteorol. 19:305–313.

Saldarriaga, J., and V. Yevjevich. 1970. Application of run-lengths to hydrologic time series. Hydrology Paper 40. Colorado State University, Fort Collins.

Şen, Z. 1976. Wet and dry periods of annual flow series. J. Hydraul. Div. ASCE Proc. Pap. 12497. 102:1503–1514.

Şen, Z. 1977. Run-sums of annual flow series. J. Hydrol. 35:311–324.

Şen, Z. 1980a. Statistical analysis of hydrologic critical droughts. J. Hydraul. Div. ASCE Proc. Pap. 14134. 106:99–115.

Şen, Z. 1980b. Regional drought and flood frequency analysis: Theoretical considerations. J. Hydrol. 46:265–279.

Şen, Z. 1989. The theory of runs with applications to drought prediction—Comment. J. Hydrol. 110:383–391.

Şen, Z. 1990. Critical drought analysis by second order Markov chain. J. Hydrol. 120:183–203.

Sirdas, S., and Z. Şen. 2003. Spatio-temporal drought analysis in the Trakya region, Turkey. J. Hydrol. Sci. 48:809–819.

Slabbers, P.J., and F.X. Dunin. 1981. Wheat yield estimation in northwest Iran. Agric. Water Manag. 3:291–304.

Tase, N. 1976. Area-deficit-intensity characteristics of droughts. Hydrology Paper 87. Colorado State University, Fort Collins.

Toure, A., D.J. Major, and C.W. Lindwall. 1995. Comparison of five wheat simulation models in Southern Alberta. Can. J. Plant Sci. 75:61–68.

Walker, G.K. 1989. Model for operational forecasting of western Canada wheat yield. Agric. Forest Meteorol. 44:339–351.

Yevjevich, V. 1967. An objective approach to definition and investigation of continental hydrologic droughts. Hydrology Paper 23. Colorado State University, Fort Collins.

PART II

REMOTE SENSING

CHAPTER FIVE

Monitoring Drought Using Coarse-Resolution Polar-Orbiting Satellite Data

ASSAF ANYAMBA, COMPTON J. TUCKER,

ALFREDO R. HUETE, AND VIJENDRA K. BOKEN

There are two distinct categories of remotely sensed data: satellite data and aerial data or photographs. Unlike aerial photographs, satellite data have been routinely available for most of the earth's land areas for more than two decades and therefore are preferred for reliably monitoring global vegetation conditions.

Satellite data are the result of reflectance, emission, and/or back scattering of electromagnetic energy (figure 5.1) from earth objects (e.g., vegetation, soil, and water). The electromagnetic spectrum is very broad, and only a limited range of wavelengths is suitable for earth resource monitoring and applications. The gaseous composition (O_2, O_3, CO_2, H_2O, etc.) of the atmosphere, along with particulates and aerosols, cause significant absorption and scattering of electromagnetic energy over some regions of the spectrum. This restricts remote sensing of the earth's surface to certain "atmospheric windows," or regions in which electromagnetic energy can pass through the atmosphere with minimal interference. Some such windows include visible, infrared, shortwave, thermal, and microwave ranges of the spectrum.

The shortwave-infrared (SWIR) wavelengths are sensitive to moisture content of vegetation, whereas the thermal-infrared region is useful for monitoring and detecting plant canopy stress and for modeling latent and sensible heat fluxes. Thermal remote sensing imagery is acquired both during the day and night, and it measures the emitted energy from the surface, which is related to surface temperatures and the emissivity of surface materials.

This chapter focuses on the contribution of visible and infrared wavelengths to global drought monitoring, and chapter 6 discusses visible, infrared, and thermal wave contributions. Under microwave windows, the satellite data can be divided into two categories: active microwave and

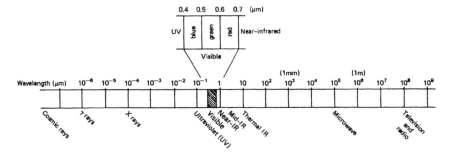

Figure 5.1 The electromagnetic spectrum [from Lillesand and Keifer, 2000].

passive microwave. Chapters 7 and 8 describe applications of passive and active microwave remote sensing to drought monitoring, respectively.

Some Remote Sensing Systems for Monitoring Drought

Landsat

Early use of satellite data was pioneered by the Landsat series originally known as the Earth Resource Technology Satellite (ERTS; http://landsat7.usgs.gov/index.php). Landsat was the first satellite specifically designed for broad-scale observation of the earth's land surface. A series of Landsat satellites (Landsats 1–5 and 7; 6 failed at launch) have provided visible and near-infrared data since 1972, with additional bands in the shortwave infrared and thermal regions for Landsat 4, 5, and 7. The Landsat satellites orbit the earth in sun-synchronous mode with a repeat cycle over any given location of 16 days. The earlier Landsat satellites provided imagery with a multispectral scanning instrument (MSS) at a resolution or pixel size of 80 m. Landsats 4 and 5 carried an additional sensor known as the thematic mapper (TM) that provided image data at a 30-m resolution over the visible, near-, and shortwave-infrared wavelengths with a 120-m thermal band. Currently, Landsat 7 carries an enhanced thematic mapper (ETM+) sensor that provides improved radiometric calibration, a panchromatic band at 15 m, visible through shortwave infrared bands at 30 m, and a thermal band at 60 m spatial resolution.

SPOT

The Systeme Probatoire pour l'Observation de la Terre (SPOT) satellite sensor series (http://www.spotimage.fr/home/), developed by Center National d'Etudes Spatiales (CNES) in France, was first launched in 1986 and provides data at a spatial resolution of 10 m (panchromatic mode) and 20 m (multispectral mode) using what is known as the high-resolution visible

(HRV) sensor. The HRV instruments have been in service on SPOT 1, 2, and 3. Although the SPOT HRV repeat cycle of 26 days is longer and has only 3 spectral bands as opposed to 7 on Landsat TM and ETM+, the spatial resolution is an improvement, and the SPOT sensors can also acquire stereoscopic pairs of images for a given location, which is useful in topographic mapping. SPOT 4 was launched in May 1998 with an additional sensor named VEGETATION (VGT), designed specifically to monitor land surface parameters on a global basis at 1 km spatial resolution. VGT monitors vegetation dynamics on a daily basis and globally in four spectral bands, from visible to shortwave infrared.

NOAA-AVHRR

The advanced very high resolution radiometer (AVHRR) instrument on board the NOAA series of satellites (http://www.nesdis.noaa.gov/) has provided daily data on the global environment for the last 20 years. Although the NOAA series of satellites was designed for weather and climate observations, its broad bands in the visible through thermal-infrared portions of the spectrum have been used effectively for large-scale monitoring of vegetation dynamics, especially in arid and semiarid areas (Tucker et al., 1983; Justice, 1986). Since 1981, the compilation of vegetation measurements from NOAA-7, -9, -11, -14, and currently NOAA-16 has provided a continuous stream of high temporal resolution data. Although the nominal spatial resolution for NOAA data is 1 km, satellite storage limitations enforced degrading of 1-km data to 4-km data to achieve global coverage on a daily basis. Data at about 1 km resolution can only be gathered if there is a local receiving station when the satellite passes overhead.

IRS

The Indian Space Research Organization (ISRO; http://www.isro.org/programmes.htm) currently has several satellites under the Indian Remote Sensing Satellite (IRS) system for natural resource monitoring and management. These data are distributed by ANTRIX Corporation Ltd., the commercial arm of the ISRO, and also by Space Imaging Corporation in the United States. The IRS-1C and IRS-1D satellites together provide continuous global coverage with many advanced capabilities. The IRS-P5 and IRS-P6 offer a very high-resolution panchromatic camera for cartographic applications as well as specific capabilities for agricultural applications.

MODIS

One of the most advanced remote sensing systems for land surface studies is NASA's Moderate Resolution Imaging Spectroradiometer (MODIS; http://modis.gsfc.nasa.gov/) sensor on board the Earth Observing System

(EOS) Terra and Aqua (http://aqua.gsfc.nasa.gov/) platforms, launched in December 1999 and May 2002, respectively. The MODIS instruments provide global coverage in 36 spectral bands with a spatial resolution ranging from 250 m to 1 km. A number of spectral bands and band combinations on this instrument are invaluable to drought, vegetation, and climate studies. Bands 1 and 2 provide 250-m resolution imagery in the red and near-infrared (NIR) regions, and bands 3–7 provide 500-m resolution imagery in the visible, NIR, and SWIR regions. This includes bands 5 (1230–1250 nm) and 6 (1628–1652 nm). In these bands, leaf water content influences the canopy reflectance response. The remaining bands are at 1 km spatial resolution and include measurements in the thermal infrared as well as optical data useful for ocean, atmosphere, and cryosphere applications. This sensor therefore fills the spatial resolution data gap between the Landsat and SPOT satellite series and the coarser NOAA-AVHRR data. Already a number of valuable land products have been assembled and are publicly available for use through various distribution centers (http://edcdaac.usgs.gov/modis/dataprod.html). The products include the traditional normalized difference vegetation index (NDVI) data, as well as the enhanced vegetation index (EVI) data, which offer improvements over the NDVI by minimizing saturation problems at high biomass conditions and by reducing atmosphere and soil background noise (Huete et al., 2002; Justice et al., 2002).

Remote Sensing of Vegetation

Remote sensing of vegetation is accomplished by using the strong coupling between reflected visible and NIR radiation with the physiological condition of leaves and their density. Photosynthesis generally occurs through CO_2 gas exchange between hydrated chloroplast cells that are in direct contact with the intercellular air spaces within the leaves (figure 5.2; Gates et al., 1965).

Chlorophyll is a strong absorber of visible energy, and the interaction of near-infrared wavelengths with vegetation provide important clues as to the structures of plant leaves. A consequence of the internal structure of leaves is that visible and near-infrared solar radiation passing through the leaves are deflected and scattered due to the refractive index, differences between hydrated cells ($\eta \sim 1.3$ μm) and air spaces ($\eta = 1.0$ μm), and the irregular pattern of cell facets in the mesophyll leaf. These scattering effects increase the effective path length of radiation as it passes through the leaves, resulting in increased absorption of visible light by plant pigments and liquid water. In contrast, there is negligible absorption at NIR wavelengths, which results in an enhanced spectral reflectance response (figure 5.3; Tucker and Sellers, 1986).

In the natural world, plant leaves are organized into canopies of varying structures and leaf orientations. The spectral reflectance and absorptance of the leaf elements are modified by the canopy structure and the amount

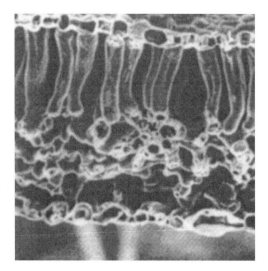

Figure 5.2 A scanning electron micrograph of transverse view of a mature broad-bean leaf magnified 420×. The upper half of the leaf interior is more disorganized, and extensive intercellular air spaces expose the surfaces of mesophyll cells directly to air contact (from Troughton and Donaldson, 1972).

of green leaf density (figure 5.4). Thus, the determinants of the spectral reflectance of plant canopies include physiological condition of leaves, their structural arrangement and density within the plant canopy, and the underlying substrate, when the plant canopy is open (Tucker, 1980a). Sensitivity of satellite data to these characteristics of leaves becomes the basis for monitoring drought conditions using satellite data.

Vegetation Indices

The spectral properties of vegetation canopies, mentioned above, make it possible to monitor vegetation dynamics and their spatial and temporal variability using various remote-sensing platforms. Satellite data-based detection of vegetation health and stress depends on the strong relationship between simple transforms of reflected red and near-infrared energy and the intercepted or absorbed photosynthetically active radiation (APAR) of the plant canopy. Several such transforms, referred to as vegetation indices (VIs), are based on the unique spectral signature of green vegetation in the red and NIR portions of the spectrum and form the basis for quantitative assessment of vegetation condition using satellite data. According to Jackson and Huete (1991), VIs can be divided into two groups: slope-based and distance-based VIs.

Slope-Based Vegetation Indices

The slope-based VIs are simple arithmetic combinations that exploit the contrast between the spectral response patterns of vegetation in the red and NIR portions of the electromagnetic spectrum. Some of these vegetation indices include the ratio vegetation index (RVI), normalized difference

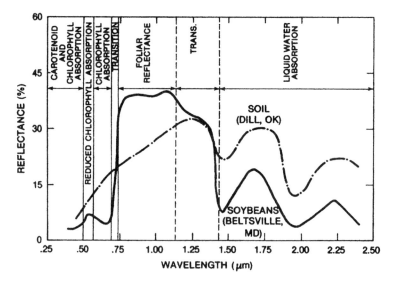

Figure 5.3 Examples of spectral reflectance of soil and vegetation at different wavelengths. Note the high spectral contrasts between green vegetation and soil in the red and near infrared spectral regions (from Tucker and Sellers, 1986).

vegetation index (NDVI), and transformed vegetation index (TVI). The RVI, proposed by Rouse et al. (1974) using Landsat multispectral scanner (MSS) imagery, is a simple division of the reflectance values in the NIR band by those in the red band.

Normalized Difference Vegetation Index The NDVI was also proposed by Rouse et al. (1974) as a spectral VI that isolates green vegetation from its background using Landsat MSS digital data. It is expressed as the difference between the NIR and red (RED) bands normalized by their sum:

$$\text{NDVI} = \frac{\text{NIR} - \text{RED}}{\text{NIR} + \text{RED}} \qquad [5.1]$$

The NDVI is the most commonly used VI because it has a desirable measurement scale ranging from -1 to 1 with zero as an approximate value of no vegetation. Negative values represent nonvegetated surfaces, whereas values close to 1 have very dense vegetation. The NDVI and RVI have the ability to reduce external noise factors such as topographic effects and sun-angle variations.

Transformed Vegetation Index The TVI, proposed by Deering et al. (1975), modifies the NDVI by taking its square root and adding a constant of 0.50. This creates a VI scale consisting of mainly positive values approximating a more normal distribution. There are no theoretical differences between the NDVI and TVI in terms of quantitative values or

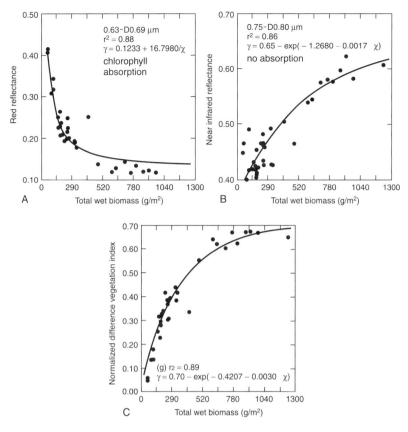

Figure 5.4 The relationships between total wet biomass of grass canopy (>90 green) and (A) the red (0.63–0.69 μm) spectral radiance, (B) near-infrared (0.75–0.80 μm) spectral radiance, and (C) the normalized difference vegetation index (from Tucker, 1977).

vegetation identification. In fact, all of the slope-based VIs can be shown to be functionally equivalent to each other (Perry and Lautenschlager, 1984).

Distance-Based Vegetation Indices

Perpendicular Vegetation Index The family of distance-based vegetation indices were originally derived from the perpendicular vegetation index (PVI) formulated by Richardson and Wiegand (1977). The principal objective of these VIs is to eliminate the effect of soil brightness over surfaces of incomplete vegetation cover where a mixture of green vegetation and soil background dominates the surface. This is particularly important in detecting the presence of vegetation in arid, semiarid, and subhumid environments. The procedure for deriving the PVI is based on the soil line concept, which describes the typical range of soil signatures in red/near-infrared bispectral plots. The soil line is computed by linear regression of

NIR against red band measurements for a sample of bare soil pixels. Pixels falling near the soil line are assumed to be sparsely vegetated, whereas those farther away, in the direction of increasing NIR and decreasing red, represent increasing amounts of vegetation. Soil lines may be specific to one soil type or more general to a variety of soils within an image or satellite data set. The complexity of the derivation of these distance-based indices has resulted in inconsistencies in their formulation for assessing vegetation status and condition, limiting their applications to regional studies where soil-vegetation characteristics can be clearly segregated. Improvements to the PVI have yielded three other PVIs suggested by Perry and Lautenschlager (1984), and Qi et al. (1994) and referred to as PVI1, PVI2, and PVI3.

Vegetation Indices Based on Orthogonal Transformation Vegetation indices based on orthogonal transformation include the difference vegetation index (DVI) also suggested by Richardson and Wiegand (1977), the green vegetation index (GVI) of the tasseled cap transformation (Kauth and Thomas, 1976), Misra's green vegetation index (MGVI) based on the Wheeler-Misra transformation (Wheeler et al., 1976; Misra et al., 1977), and Principal Components Analysis (PCA) (Singh and Harrison, 1985; Fung and LeDrew, 1988). These indices involve decorrelation of the original bands to extract a new set of components that separate vegetation from other surface materials.

Optimized Indices There is a class of VIs based on semiempirical radiative transfer theory that use both slope- and distance-based properties of spectral data in red and NIR plots. These indices are referred to as optimized indices and include the soil-adjusted vegetation index (SAVI) proposed by Huete (1988). The SAVI aims to minimize the effects of soil background on the vegetation signal by incorporating a soil adjustment factor into the denominator of the NDVI equation:

$$\text{SAVI} = \frac{\text{NIR} - \text{RED}}{(\text{NIR} + \text{RED} + L)} \cdot (1 + L) \qquad [5.2]$$

where L is the soil adjustment factor that takes into account first-order, differential penetration of red and NIR energy through a canopy in accordance with Beer's law. There are modified forms of the SAVI that include the transformed soil-adjusted vegetation index (TSAVI) by Baret and Guyot (1991), and the modified soil-adjusted vegetation index (MSAVI) suggested by Qi et al. (1994), based on a modification of the L factor of the SAVI. All of these modifications are intended to improve correction to the soil background brightness for different conditions of surface vegetation cover.

Optimized indices also include the atmosphere resistant vegetation index (ARVI) by Kaufman and Tanré (1992), the enhanced vegetation index (EVI) by Huete et al. (1994), and the aerosol-free vegetation index (AFRI) by Karnieli et al. (2001). These indices incorporate atmosphere and canopy radiative transfer theory for optimized retrieval of vegetation properties from satellite data (Verstraete and Pinty, 1996).

Ideally, optimized and distance-based VIs are superior to slope-based indices because they attempt to minimize or remove atmosphere and soil brightness noise that may limit a quantitative assessment of green vegetation; however, their robustness over global vegetation conditions remains to be tested. The simplicity of sloped-based indices both in terms of numerical results and interpretation has meant that such indices, such as the well-known NDVI, can be used for vegetation monitoring and drought detection at regional as well as global scales.

Relationship between NDVI and Crop Yield

Total dry matter of plant or crop yield are related to APAR (Kumar and Monteith, 1982; Monteith, 1977), and the NDVI is highly correlated with APAR (Daughtry et al. 1983; Hatfield et al., 1984; Wiegand and Richardson, 1984; Asrar et al., 1985). Thus, the use of NDVI data for crop condition assessment, yield estimation, and hence drought monitoring has been intensively analyzed (Aase and Siddoway, 1980; Tucker, 1980b; Tucker et al., 1981; Weigand and Richardson, 1984; Boken and Shaykewich, 2002). The NDVI has also been related to many vegetation canopy characteristics, including leaf area index (LAI), green biomass, and percent cover (Wiegand and Richardson, 1987).

Droughts reduce photosynthesis on account of low rainfall, which reduces total dry matter accumulation and yields and results in lower NDVI values (figure 5.5). In arid and semiarid areas, the rainfall is the principal determinant of primary production and has been found to be highly correlated with the NDVI, although this correlation differs slightly across various climatic regimes (figure 5.6; Malo and Nicholson, 1990; Nicholson et al., 1990; Tucker and Nicholson, 1999).

Droughts usually begin unnoticed and develop cumulatively with their impacts that are not immediately observable by ground data (Kogan, 1997; Kogan, 2002). The key to using the NDVI to monitor and assess droughts is thus to have accurate time series satellite data over long periods. This has been achieved with intercalibrated data from the multiple series of AVHRR sensors on board the polar-orbiting meteorological satellites of the U.S. National Oceanic and Atmospheric Administration (NOAA). Below-normal NDVI values would indicate the occurrence of droughts. Below are some examples of using NDVI time series data derived from NOAA-AVHRR to study drought patterns and their impacts on agricultural production for Africa.

Drought Monitoring Applications for Africa

Monitoring drought and crop production in Africa is not straightforward in part because of the large footprint of the NOAA-NDVI data (~4–8 km

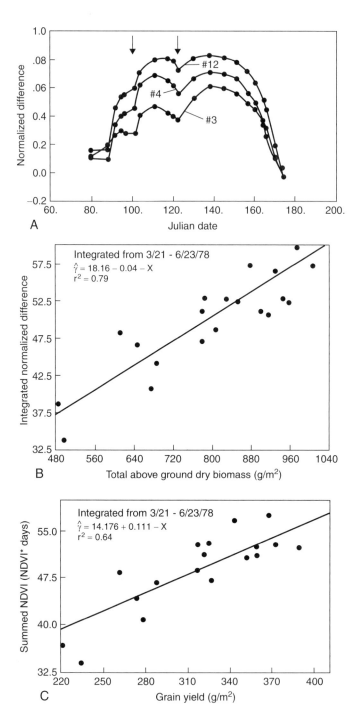

Figure 5.5 (A) Temporal evolution of NDVI (normalized difference vegetation index) of wheat during the growing season. (B) The relationship between NDVI and total above-ground dry matter accumulation. (C) The relationship between wheat yield and NDVI during the growing season (from Tucker et al., 1980, 1981).

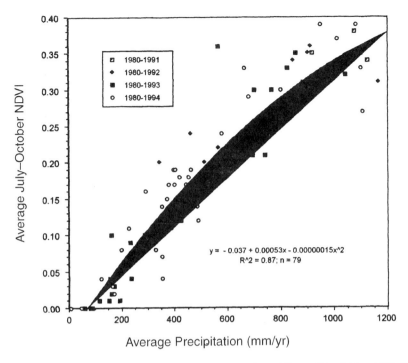

Figure 5.6 Relationship between July–October NDVI and rainfall over West Africa, 1980–97 (from Tucker and Nicholson, 1999).

spatial resolution), the complexity of crop mixtures with surrounding vegetation types, and the small farm sizes that dominate the African landscape. Many major organizations, however, operationally use NDVI data to monitor drought and famine in Africa, such as the Food and Agricultural Organization's Global Information and Early Warning System (GIEWS; http://www.fao.org/giews/), the U.S. Agency for International Development's Famine Early Warning System Network (FEWSNET; http://www.fews.net/; chapter 19), and the U.S. Department of Agriculture's Foreign Agricultural Service (USDA/FAS; http://www.fas.usda.gov/pecad/pecad.html). In all of the following examples for the Sahel, East Africa, and southern Africa, the monthly NDVI data for 20+ years were used (Tucker, 1996; Anyamba et al., 2001, 2002; Los et al., 2001).

The Sahel

Prolonged and severe droughts occurred in the Sahel region of Africa in the 1970s and 1980s (Tucker et al., 1983, 1986; Hielkema et al., 1986; Justice et al., 1986). These droughts were monitored by comparing monthly NDVI values with the long-term means. Figure 5.7 shows reduction in NDVI values on account of reduced precipitation for Mali in the Sahelian region of Africa.

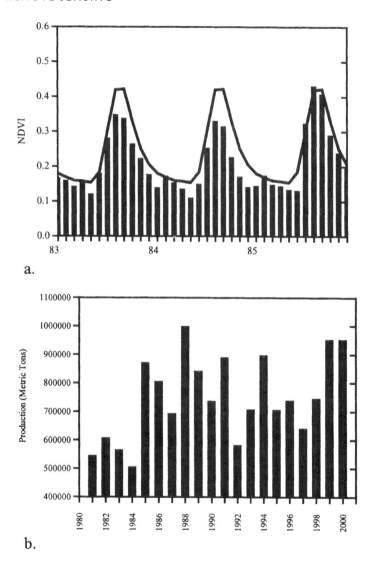

Figure 5.7 (a) The NDVI monthly values (bars) and long-term (1982–2000) means (thick line) for Mali during 1983–85, and (b) the annual variation in millet production for Mali during 1981–2000.

During the period from 1983 to 1985, Mali, like other Sahelian countries, experienced one of the most severe droughts in recent history. This region has only one growing season from June to October, with a peak rainy season in July–August. The mean NDVI vegetation profile shows this growing season pattern, with a peak in NDVI during August–September every year. During the 1983–84 period, rainfall was below normal over most of the Sahel region, resulting in below-normal NDVI values. This was followed by a recovery in NDVI values to more normal levels in 1985. Drought evolved slowly in early 1983 before reaching a peak in 1984 when NDVI was 35% below normal.

Millet and sorghum are the staple crops in this semiarid region of Africa. Millet production for Mali, covering the period 1981–2000, is shown in figure 5.7b. The NDVI data showed that 1984 was the year of the most severe drought, resulting in the lowest production in 20 years, about 500,000 metric tons of millet. This was the lowest point over several years of progressively decreased production from 1980 to 1983, following years of persistent drought during the 1970s.

The spatial pattern of NDVI anomalies for 1984 (figure 5.8) shows a region-wide drought across the Sahel region, from Senegal to Sudan in the east, with NDVI anomalies ranging well below normal: -20% to as low as -80% in Chad. This drought led to large-scale famine, starvation, and loss of human life and livestock.

East Africa

Most of East Africa has a bimodal rainfall distribution and hence two agricultural growing seasons. The short growing season typically begins in late September and ends in November. The long growing season typically begins in February or March and continues through May or June. Figure 5.9 shows the evolution of monthly NDVI vis-à-vis the long-term monthly NDVI mean for Kenya (East Africa) for 1983–86 and 1996–99.

Like rainfall, NDVI values over East Africa exhibit a bimodal pattern, with maximum values in April and November. This pattern follows the evolution of both the long and the short rainy seasons that are governed by the Inter-Tropical Convergence Zone (ITCZ), following the twice-yearly solar passage over the equatorial zone. For 1983 and 1984, the monthly NDVI values for both short and long rainy seasons were lower than the

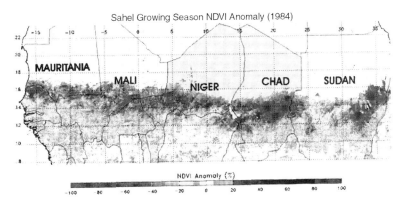

Figure 5.8 Spatial variation in NDVI anomaly across the Sahel region during the growing season of 1984.

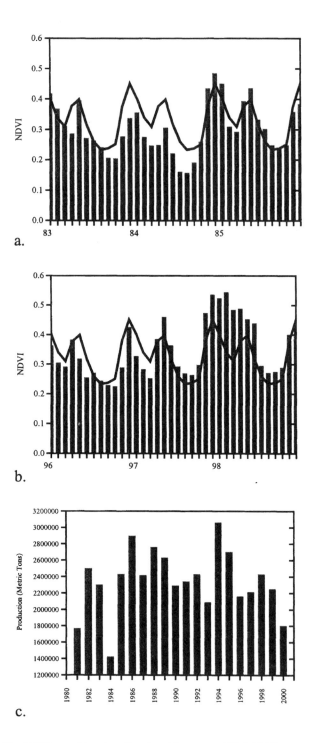

Figure 5.9 The NDVI monthly values (bars) and long-term (1982–2000) means (thick line) for Kenya during (a) 1983–85 and (b) 1996–98. (c) The annual variation in maize production for Kenya during 1981–2000.

long-term mean values, especially in 1984, and only slightly recovered to near their long-term average values in 1985. This deficit is a reflection of the large-scale rainfall deficit that affected the entire Sahelian region and extended into East Africa. The NDVI declined by 10–40% of normal NDVI during the period from the end of 1983 to the end of 1984 only slightly recovering to near normal in 1985. This indicates that the persistence of earlier drought conditions may have dampened the expected improvement in vegetation conditions with the return of near-normal rainfall conditions in 1985.

The NDVI patterns in 1996 to mid-1997 were associated with drought conditions that prevailed over most of East Africa during this period. NDVI declined by 20% for most of 1996 and early 1997 but was above normal for the period from late 1997 through 1998.

The drought of 1984 caused significant reduction in maize production (figure 5.9c). In 1984, only 1.4 million tons of maize was produced, compared to 3.0 million tons in 1994 and 2.4 million tons in 1998. Periods of reduced maize production (1984, 1992, 1996, 2000) corresponded to the occurrence of El Niño/Southern Oscillation (ENSO) cold events (which typically resulted in below-normal rainfall) and below-normal NDVI values (figures 5.9a,b).

The spatial anomaly patterns of drought conditions during March—May (critical growing period) 1984 are shown in figure 5.10. The drought pattern in 1984 was most pronounced in Kenya, especially the eastern part, and most of Ethiopia and Somalia. A large part of East Africa had NDVI values that were 10–60% below normal. Monthly NDVI in 1992 was well below normal (figure 5.11a). Figure 5.11b shows the interannual variability in maize production for Zimbabwe from 1981 to 2000.

Southern Africa

In southern Africa, the worst drought of the century occurred during 1991–1992 and affected nearly 100 million people. The most affected countries were Zimbabwe and Botswana.

The years 1983, 1987, 1992, 1995, and 1998 showed large deficits in production, with the lowest production in 1992 (figure 5.11b). All of these years represented periods of ENSO warm events, which were associated with below-normal rainfall and the prevalence of drought conditions over most of the region (Cane et al., 1994; Phillips et al., 1998). Drought years also led to inadequate forage and poor nutrition for animals, resulting in reduced carcass weight and increased cattle mortality (figure 5.12).

Unlike East Africa, southern Africa has only one growing season that normally begins in November and ends in April. Figure 5.11a shows the temporal evolution of monthly NDVI values and the respective long-term means for Zimbabwe for 1991–93. The time series data showed a unimodal pattern, with the NDVI reaching a maximum of about 0.4 units in February—March after the maximum rainfall in December—January. The dry season typically includes the period from late May to October,

72 REMOTE SENSING

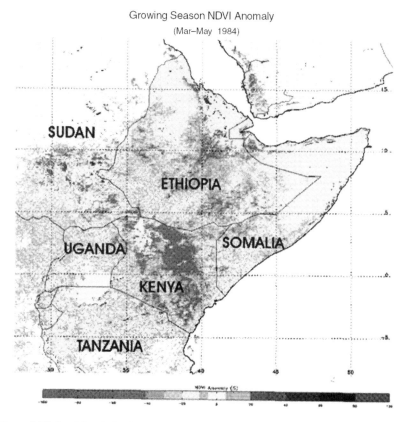

Figure 5.10 Spatial variation in NDVI anomaly across East Africa during the critical growing season of 1984.

with associated low NDVI values (typically ~ 0.2). During 1991–93, NDVI values over the growing season were lower than their long-term means, indicating the prevalence of drought conditions. The NDVI during the 1992–93 period was below the long-term average. In 1992, the region received below-normal rainfall, causing region-wide drought conditions with NDVI anomalies reaching 80% below normal.

Future Research Needs

The SWIR portion of the electromagnetic spectrum is a region over which leaf water content influences the reflectance response from vegetation canopies, enabling one to assess the drought status of crops and vegetation (Tucker, 1980a). However, the operational use of the SWIR region for drought studies has been restricted by problems in separating the effects of canopy structure and geometry from those of water status. Ceccato et al. (2001) have shown that by combining the NIR and SWIR data into

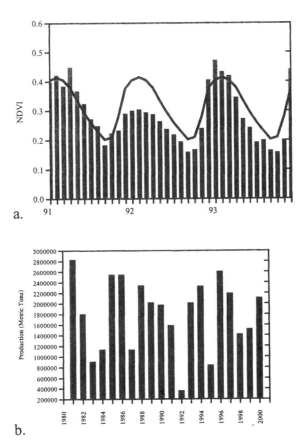

Figure 5.11 (a) The NDVI monthly values (bars) and long-term (1982–2000) means (thick line) for Zimbabwe during 1991–93. (b) The annual variation in maize production for Zimbabwe during 1981–2000.

NDVI-like formulations, one can retrieve variations induced by leaf water contents. Gao (1996) proposed a normalized difference water index (NDWI) for detecting and monitoring vegetation liquid water content using narrow-band (<10 nm) data. With the launch of new satellite sensor systems, more research can be expected on the development of operational and global drought monitoring approaches involving vegetation water status using the SWIR region.

Combining land-surface temperatures (T_s) with vegetation indices (VI) is also of great interest in drought monitoring studies (Nemani and Running, 1989). Under drought conditions, soil moisture is reduced, and hence evapotranspiration declines, causing leaf temperature to rise. Therefore, thermal infrared energy emitted from the vegetation canopy can be used to detect increased temperature of leaves to monitor drought conditions. Vegetation and crop stress indices, such as the surface moisture index (SMI;

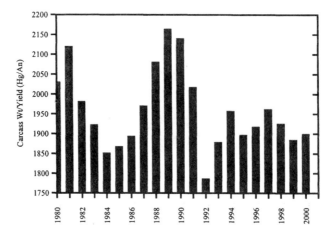

Figure 5.12 The variation in beef and veal production in Botswana during 1981–2000.

Nemani et al., 1993) and the water deficit index (WDI; Moran et al., 1994) use thermal and vegetation index measurements of vegetated surfaces to estimate the soil moisture status through the rate of evaporative water loss.

The SMI is based on the slope of the AVHRR-NDVI and maximum temperature values. Gillies et al. (1997) later applied inversion techniques to improve the estimates of available soil moisture from VI–T_s scatterplots. These concepts, however, need to be incorporated into vegetation indices to be suitable for global monitoring of droughts.

Conclusions

Since the 1980s, NDVI time series data derived from the series of polar-orbiting NOAA-AVHHR instruments have established a baseline (or reference) for monitoring global terrestrial vegetation. Early applications were primarily aimed at validating the utility of the NDVI data by comparing field measurements to satellite measurements of vegetation characteristics. Perhaps one of the most critical findings from these early field studies was that in semiarid areas, NDVI was linearly related to rainfall and thus could be used as an indicator of past rainfall or vegetation conditions. Field studies of primary production in conjunction with analysis of satellite vegetation indices indicated a strong linear relationship between NDVI and seasonal primary production, particularly over semiarid and arid environments. These are areas that are subject to lowest rainfall leading to droughts that result in food shortages and famine. Although AVHRR data are at coarse resolution, crop yield estimates can be satisfactorily made through comparisons of the time series data with historical production statistics at national and regional levels and through comparisons between current crop-growing conditions with those of previous years or their historical mean.

However, there are some problems associated with using the data set in semiarid areas. Increases in atmospheric water vapor have the tendency to reduce near-infrared reflectance and therefore decrease NDVI values, resulting in an underestimation of primary production. Aerosols from fires and volcanoes may further attenuate NDVI values in various ways depending on the underlying surface. At the regional scale, other factors such as vegetation type, soil type, and topographic variations may alter NDVI values and may have to be taken into consideration for improved crop yield estimates and assessments. In general the NDVI data presented here are more than adequate for monitoring vegetation conditions over fairly large areas, in the order of 100 km^2. The data, however, remain inadequate for monitoring and evaluating the status of crops at the field or individual plot level. Most of the regions where these data are used comprise complex mixtures of cropland, natural vegetation, and nonvegetated areas. This therefore adds complexity to the interpretation of AVHRR-NDVI data at local scales.

Alternative vegetation indices such as SAVI, MSAVI, PVI, GVI, and other advanced satellite systems (such as SPOT VGT, SeaWiFS, and Terra/Aqua-MODIS instruments) may offer improvements in drought impact studies on crop yield assessments, but they have not yet been validated or implemented globally, other than in specific case studies at regional scales. The 4+ years of global MODIS EVI data (2000–2003) that are now available may still be insufficient for constructing a baseline or reference database from which one could measure anomalies in drought conditions. Effective translation coefficients would be needed to bridge any improved indices with the NDVI, as well as to extend the various indices across different sensors and platforms. Despite the shortcomings, the NDVI data set has become a principal layer of information for drought monitoring at regional or global scales.

References

Aase, J. K, and F.H. Siddoway. 1980. Spring wheat yield estimates from spectral reflectance measurements. Agron. J. 72:149–154.

Anyamba, A., C.J. Tucker, and J.R. Eastman. 2001. NDVI anomaly patterns over Africa during the 1997/98 ENSO warm event. Intl. J. Remote Sens. 22:1847–1859.

Anyamba, A., C.J. Tucker, and R. Mahoney. 2002. El Niño to La Niña: Vegetation response patterns over East and southern Africa during 1997–2000 period. J. Climate 15:3096–3103.

Asrar, G., E.T. Kanemasu, R.D. Jackson, and P.J. Pinter. 1985. Estimation of total above-ground phytomass production using remotely sensed data. Remote Sens. Environ. 17:211–220.

Baret, F., and G. Guyot. 1991. Potentials and limits of vegetation indices for LAI and APAR assessments. Rem. Sens. Environ. 35:161–173.

Boken, V.K., and C.F. Shaykewich. 2002. Improving an operational wheat yield model for the Canadian Prairies using phenological-stage-based normalized difference vegetation index. Int. J. Remote Sens. 23:4157–4170.

Cane, M.A., G. Eshel, and R.W. Buckland. 1994. Forecasting Zimbabwean maize yields using eastern equatorial Pacific sea-surface temperature. Nature 370: 204–205.

Cecatto, P., S. Flasse, S. Tarantola, S. Jacquemoud, and J.M. Gregoire. 2001. Detecting vegetation leaf water content using reflectance in the optical domain. Remote Sens. Environ. 77:22–33.

Daughtry, C.S.T., K.P. Galio, and M.E. Bauer. 1983. Spectral estimates of solar radiation intercepted by corn canopies. Agron. J. 75:527–531.

Deering, D.W., J.W. Rouse, R.H. Haas, and J.A. Schell. 1975. Measuring "forage production" of grazing units from Landsat MSS data. In: Proceedings of the 10th International Symposium on Remote Sensing of Environment, II, University of Michigan, pp. 1169–1178.

Fung, T., and E. LeDrew. 1988. The determination of optimal threshold levels for change detection using various accuracy indices. Photogram. Eng. Remote Sens. 54:1449–1454.

Gao, B.C. 1996. NDWI—A normalized difference water index for remote sensing of vegetation liquid water from space. Remote Sens. Environ. 58:257–266.

Gates, D.M., H.J. Keegan, J.C. Schleter, and V.P. Weidner. 1965. Spectral properties of plants. Appl. Optics 4:11.

Gillies, R.R., T.N. Carlson, J. Cui, W.P. Kustas, and K.S. Humes. 1997. A verification of the "triangle" method for obtaining surface soil water content and energy fluxes from remote measurements of the Normalized Difference Vegetation Index (NDVI) and surface radiant temperature. Intl. J. Remote Sens. 18:3145–3166.

Hatfield, J.L., G. Asrar, and E.T. Kanemasu. 1984. Intercepted photosynthetically active radiation in wheat canopies estimated by spectral reflectance. Remote Sens. Environ. 14:65–76.

Hielkema, J.U., S.D. Prince, and W.L. Astle. 1986. Rainfall and vegetation monitoring in the Savanna Zone of the Democratic Republic of Sudan using NOAA advanced very high-resolution radiometer. Intl. J. Remote Sens. 7:1499–1514.

Huete, A.R. 1988. A soil-adjusted vegetation index (SAVI). Remote Sens. Environ. 25:53–70.

Huete, A., C. Justice, and H. Liu. 1994. Development of vegetation and soil indices for MODIS-EOS. Remote Sens. Environ. 49:224–234.

Huete, A., K. Didan, T. Miura, E.P. Rodriguez, X. Gao, and L.G. Ferreira. 2002. Overview of the radiometric and biophysical performance of the MODIS vegetation indices. Remote Sens. Environ. 83:195–213.

Jackson, R.D., and A.R. Huete. 1991. Interpreting vegetation indices. Prev. Vet. Med. 11:185–200.

Justice, C.O. 1986. Monitoring the grasslands of semi-arid Africa using NOAA-AVHRR data. Intl. J. Remote Sens. 7.

Justice, C.O., B.N. Holben, and M.D. Gwynne. 1986. Monitoring East African vegetation using AVHRR data. Intl. J. Remote Sens. 7:1453–1474.

Justice, C.O., J.R.G. Townshend, E.F. Vermote, E. Masuoka, R.E. Wolfe, N. Saleous, D.P. Roy, and J.T. Morissette. 2002. An overview of MODIS Land data processing and products. Remote Sens. Environ. 83:3–15.

Karnieli, A., Y.J. Kaufman, L. Remer, and A. Wald. 2001. AFRI—aerosol free vegetation index. Remote Sens. Environ. 77:10–21.

Kaufman, Y.J., and D. Tanré. 1992. Atmospherically resistant vegetation index (ARVI) for EOS-MODIS. IEEE Trans. Geosci. Remote Sens. 30:261–270.

Kauth, R.J., and G.S. Thomas. 1976. The tasseled cap—a graphic description of the spectral temporal development of agricultural crops as seen by Landsat.

In: Proceedings of the Symposium on Machine Processing of Remotely Sensed Data. Perdue University, West Lafayette, IN, pp. 41–51.

Kogan, F.N. 1997. Global drought watch from space. Bull. Am. Meteorol. Soc. 78:621–636.

Kogan, F.N. 2002. World droughts in the new millennium from AVHRR-based vegetation health indices. EOS Trans. Am. Geophys. Union 83:557.

Kumar, M. and J.L. Monteith. 1982. Remote sensing of plant growth. In: H. Smith (ed.), Plants and the Daylight Spectrum. Pitman, London, pp. 133–144.

Lillesand, T.M., and R.W. Keifer. 2000. Remote Sensing and Image Interpretation. John Wiley & Sons, New York.

Los, S.O., G.J. Collatz, L. Bounoua, P.J. Sellers, and C.J. Tucker. 2001. Global interannual variations in sea surface temperature and land surface vegetation, air temperature and precipitation. J. Climate 14:1535–1549.

Malo, A.R., and S.N. Nicholson. 1990. A study of rainfall and vegetation dynamics in the African Sahel using normalized difference vegetation index. J. Arid Environ. 19:1–24.

Misra, P.N., S.G. Wheeler, and R.E. Oliver. 1977. Kauth-Thomas brightness and greenness axes (IBM personal communication). Contract NAS-9-14350, RES 23-46.

Monteith, J.L. 1977. Climate and the efficiency of crop production in Britain. Phil. Trans. R. Soc. B 271:277–282.

Moran, M.S., T.R. Clarke, Y. Inoue, and A. Vidal. 1994. Estimating crop water deficits using the relation between surface-air temperature and spectral vegetation index. Remote Sens. Environ. 46:246–263.

Nemani, R.R., L.L. Pierce, and S.W. Running. 1993. Developing satellite-derived estimates of surface moisture status. J. Appl. Meteorol. 32:548–557.

Nemani, R.R., and S.W. Running. 1989. Estimation of regional surface resistance to evapotranspiration from NDVI and thermal-IR AVHRR data. J. Appl. Meteorol. 28:276–284.

Nicholson, S.E., M.L. Davenport, and A.R. Malo. 1990. A comparison of the vegetation response to rainfall in the Sahel and East Africa, using normalized difference vegetation index from NOAA AVHRR. Clim. Change 17:209–242.

Perry, C. Jr., and L.F. Lautenschlager. 1984. Functional equivalence of spectral vegetation indices. Remote Sens. Environ. 14:169–182.

Phillips, J., M.A. Cane, and C. Rosenzweig. 1998. ENSO, seasonal rainfall patterns and maize yield variability in Zimbabwe. Agric. Forest. Meteor. 90:39–50.

Qi, J., A. Chehbouni, A.R. Huete, Y.H. Kerr, and S. Sorooshian. 1994. A modified soil adjusted vegetation index. Remote Sens. Environ. 48:119–126.

Richardson, A.J., and C.L. Wiegand. 1977. Distinguishing vegetation from soil background information. Photogram. Eng. Remote Sens. 43:1541–1552.

Rouse, J.W. Jr., R.H. Haas, D.W. Deering, J.A. Schell, and J.C. Harlan. 1974. Monitoring the vernal advancement and retrogradation (green wave effect) of natural vegetation. NASA/GSFC type III final report. National Aeronautics and Space Administration, Greenbelt, MD.

Singh, A., and A. Harrison. 1985. Standardized principal components. Intl. J. Remote Sens. 6:883–896.

Troughton, J.H., and L.A. Donaldson. 1972. Probing Plant Structure. Chapman and Hall, London.

Tucker, C.J. 1980a. Leaf water content in the infrared. Remote Sens. Environ. 10:23.

Tucker, C.J. 1980b. A critical review of remote sensing and other methods for nondestructive estimation of standing crop biomass. Grass Forage Sci. 35:177–182.

Tucker, C.J. 1996. History of the use of AVHRR data for land applications. In: G. D'Souza, A.S. Belward, and J.P. Malingreau (eds.), Advances in the use of NOAA AVHRR data for land applications. Kluwer, Boston, pp. 1–20.

Tucker, C.J., B.N. Holben, J.H. Elgin, and J.E. McMurtrey. 1981. Remote sensing of total dry matter accumulation in winter wheat. Remote Sens. Environ. 11:171–189.

Tucker, C.J., C.L. Vanpreat, E. Boerwinkel, and A. Gaston. 1983. Satellite remote sensing of total dry matter in the Senegalese Sahel: 1980–1984. Remote Sens. Environ. 13:461–474.

Tucker, C.J., C.O. Justice, and S.D. Prince. 1986. Monitoring the grasslands of the Sahel 1984–1985. Intl. J. Remote Sens. 7:1571–1582.

Tucker, C.J., and S.E. Nicholson. 1999. Variation in the size of the Sahara Desert from 1980 to 1997. Ambio 28:587–591.

Tucker, C.J., and P.J. Sellers. 1986. Satellite remote sensing of primary production. Intl. J. Remote Sens. 7:1395–1416.

Verstraete, M.M., and B. Pinty. 1996. Designing optimal spectral indices for remote sensing applications. IEEE Trans. Geosci. Remote Sens. 34:1254–1265.

Wheeler, S.G., P.N. Misra, and A.Q. Holmes. 1976. Linear dimensionality of Landsat agricutural data with implications for classification. In: Proceedings of the Symposium on Machine Processing of Remotely Sensed Data. Perdue University, West Lafayette, IN.

Wiegand, C.L., and A.J. Richardson. 1984. Leaf area, light interception, and yield estimates from spectral components analysis. Agron. J. 76:543–548.

Wiegand, C.L., and A.J. Richardson. 1987. Spectral components analysis: Rationale, and results for three crops. Intl. J. Remote Sens. 8:1011–1032.

CHAPTER SIX

NOAA/AVHRR Satellite Data-Based Indices for Monitoring Agricultural Droughts

FELIX N. KOGAN

Operational polar-orbiting environmental satellites launched in the early 1960s were designed for daily weather monitoring around the world. In the early years, they were mostly applied for cloud monitoring and for advancing skills in satellite data applications. The new era was opened with the series of TIROS-N launched in 1978, which has continued until present. These satellites have such instruments as the advanced very high resolution radiometer (AVHRR) and the TIROS operational vertical sounder (TOVS), which included a microwave sounding unit (MSU), a stratospheric sounding unit (SSU), and high-resolution infrared radiation sounder/2 (HIRS/2). These instruments helped weather forecasters improve their skills. AVHRR instruments were also useful for observing and monitoring earth surface. Specific advances were achieved in understanding vegetation distribution. Since the late 1980s, experience gained in interpreting vegetation conditions from satellite images has helped develop new applications for detecting phenomenon such as drought and its impacts on agriculture. The objective of this chapter is to introduce AVHRR indices that have been useful for detecting most unusual droughts in the world during 1990–2000, a decade identified by the United Nations as the International Decade for Natural Disasters Reduction.

AVHRR-Based Vegetation Indices

Radiances measured by the AVHRR instrument onboard National Oceanic Atmospheric Administration (NOAA) polar-orbiting satellites can be used to monitor drought conditions because of their sensitivity to changes in leaf chlorophyll, moisture content, and thermal conditions (Gates, 1970; Myers, 1970). Over the last 20 years, these radiances were converted into indices that were used as proxies for estimating various vegetation conditions

(Kogan, 1997, 2001, 2002). The indices became indispensable sources of information in the absence of in situ data, whose measurements and delivery are affected by telecommunication problems, difficult access to environmentally marginal areas, economic disturbances, and political or military conflicts. In addition, indices have advantage over in situ data in terms of better spatial and temporal coverage and faster data availability.

The AVHRR-based indices used for monitoring vegetation can be divided into two groups: two-channel indices, and three-channel indices. The normalized differences vegetation index (NDVI) is derived from two-channel data, the visible (VIS, 0.58–0.68 μm), and near infrared (NIR, 0.72–1.1 μm) and is defined as NDVI = (NIR − VIS)/(NIR + VIS). The NDVI has been widely used for characterizing distribution of vegetation (Tarpley et al., 1984; Justice et al., 1985; Tucker and Sellers, 1986; Boken and Shaykewich, 2002) and for monitoring vegetation conditions (Kogan, 1995). However, this approach is insufficient for monitoring crops specifically during drought periods because crop health depends not only on the water stress but also on thermal conditions. Therefore, three-channel indices were introduced to monitor impacts of moisture and thermal stresses on the vegetation conditions.

New Method and Data

The new numerical method, introduced in the late 1980s, is based on the combination of VIS, NIR, and thermal (10.3–11.3 μm) channels (Kogan, 1997, 2001). This method is built on three basic environmental laws: law of minimum (LOM), law of tolerance (LOT), and the principle of carrying capacity (CC). The Leibig's LOM postulates that primary production is proportional to the amount of the most limiting resource contributing to growth and is at its lowest when one of the factors affecting it is at the extreme minimum. The Shelford's LOT states that the effect of each environmental factor on an organism or ecosystem is maximum or minimum when the environmental factor ranges between the limits of tolerance. With regard to these laws, the CC is defined as the maximal population size of a given species that resources of a habitat can support (Ehrlich et al., 1977; Orians, 1990).

The new method was applied to the NOAA global vegetation index (GVI) data set issued routinely since 1985 (Kidwell,1997). The GVI is produced by sampling the AVHRR-based 4-km (global area coverage format; GAC) daily radiances in the VIS, NIR, and IR (10.3–11.3 and 11.5–12.5 μm), which were truncated to 8-bit precision and mapped to a 16-km^2 latitude/longitude grid. To minimize the cloud effects, these maps were composited over a seven-day period by saving radiances for the day that had the largest difference between NIR and VIS channels.

Because AVHRR-based radiances have both interannual and intraannual noise (varying illumination and viewing, sensor degradation, satellite navigation and orbital drift, atmospheric and surface conditions, methods of data sampling and processing, communication and random errors), their removal is crucial for improving data interpretation. Therefore, the initial processing included postlaunch calibration of VIS and NIR, calculation of NDVI, and converting IR radiance to brightness temperature (BT), which was corrected for nonlinear behavior of the sensor (Rao and Chen, 1995, 1999). The three-channel algorithm routines included a complete removal of high-frequency noise from NDVI and BT values, stratification of world ecosystems, and detection of medium-to-low frequency fluctuations in vegetation condition associated with weather variations (Kogan, 1997). These steps were crucial in order to use AVHRR-based indices as a proxy for temporal and spatial analysis and interpretation of weather-related vegetation condition and health.

Finally, three indices characterizing moisture (VCI), thermal (TCI), and vegetation health (VT) conditions were constructed following the principle of comparing a particular year NDVI and BT with the entire range of their variation during the extreme (favorable/unfavorable) conditions. Based on the LOM, LOT, and CC, the extreme conditions were derived by calculating the maximum (max) and minimum (min) NDVI and BT values using 14-year satellite data. The maximum/minimum criteria were used to classify carrying capacity of ecosystems in response to climate and weather variations. The VCI, TCI, and VT were formulated as:

$$\text{VCI} = [(\text{NDVI} - \text{NDVI}_{min})/(\text{NDVI}_{max} - \text{NDVI}_{min})]100 \qquad [1]$$

$$\text{TCI} = [(\text{BT}_{max} - \text{BT})/(\text{BT}_{max} - \text{BT}_{min})]100 \qquad [2]$$

$$\text{VT} = a(\text{VCI}) + (1 - a)\text{TCI} \qquad [3]$$

where NDVI, NDVI_{max}, and NDVI_{min} are the smoothed weekly NDVI, its multiyear absolute maximum and minimum, respectively; BT, BT_{max}, and BT_{min} are similar values for brightness temperature; and a is a coefficient that quantifies a share of VCI and TCI contribution to the vegetation condition. For example, if other conditions are near normal, vegetation is more sensitive to moisture during canopy formation (leaf appearance) and to temperature during flowering. Therefore, the share of moisture contribution into the total vegetation condition is higher than temperature during leaf canopy formation and lower during flowering. Because moisture and temperature contributions during a vegetation cycle are currently not known, the share of weekly VCI and TCI can be assumed to be equal.

Table 6.1 explains the algorithm development. As seen in the first row, both NDVI and BT data fluctuate considerably, primarily due to clouds, sun-sensor position, bidirectional reflectance, and random noise. In June, for example, clouds triggered considerable reduction in NDVI and BT; but such a reduction in August was smaller. The smoothing procedure (second row) eliminates outliers and emphasizes seasonal cycle. Since the smoothed NDVI was close to multiyear maximum (third row) values and BT was close to minimum (fourth row) values, VCI, TCI, and VT are >60,

Table 6.1 Coefficient used for algorithm development for various vegetation indices

Parameters	Normalized difference vegetation index				Brightness temperature			
	May	June	July	August	May	June	July	August
Raw	0.60	0.22	0.45	0.39	35.3	9.9	31.7	15.4
Smooth	0.39	0.41	0.44	0.45	27.8	28.5	29.3	26.7
Max	0.42	0.44	0.45	0.47	31.0	31.3	31.7	31.1
Min	0.31	0.32	0.34	0.35	27.0	27.5	27.8	26.4
V(T)CI	73	75	91	83	80	74	62	94
VT	77	74	76	88				

indicating good vegetation health. In contrast, during drought years, these indices will be <40, indicating vegetation stress.

Monitoring Major Droughts

United States

The United States is the world largest producer and leading exporter of agricultural products, including grains. Drought occurs almost every year somewhere in the nation, affecting agriculture (see chapter 9). Severe droughts occurred in the United States in 1988, 1989, 1996, and 2000. The 1988 drought cost around $40 billion in damages to the U.S. economy in human health, environment, and wildlife. Grain production fell below domestic consumption probably for the first time in the second half of last century (Reibsame et al., 1990; Kogan, 1995). AVHRR-based estimates show that by the end of June 1988, vegetation experienced stress in the most productive areas of the Great Plains, the U.S. grain basket (figure 6.1). Total world grain production in 1988 dropped by 3% (FAO, 2000).

Drought in 1989 and 1996 began early and by the end of April affected the primary winter wheat areas. Compared to earlier droughts, the 1988 vegetation stress (black color in figure 6.1) was not seen so early. By July the 1988 drought turned into a national disaster, affecting vegetation during the most critical mid-season period. The crop yield anomalies during the years of major droughts are well related to vegetation health indices (Kogan, 1995, 1997, 2000).

Former Soviet Union

If the United States is the largest seller of grains, the former Soviet Union (FSU) has been and will likely be the largest U.S. grain buyer. Since the breakup of the USSR in 1991, stagnation in technology-related grain growth in combination with frequent droughts led to serious grain shortages. Therefore, monitoring of the FSU grain production is very important for U.S. grain growers and traders.

Since 1991, FSU experienced droughts during different years, as identified by the AVHRR-based indices (figure 6.1, right panel) and affected 100–150 million acres of crops and rangeland, reducing the total FSU grain production by 10–15% (20–30 million metric tons). Independent countries incurred up to 30% of grain losses. The worse economic problems occurred when major drought affected both winter (mostly Ukraine and southern Russia) and spring grain crops (eastern regions in 1991, 1996, and 1998). These droughts were confirmed by climatological observations. In these cases, crop yield anomalies showed strong correlation with vegetation health indices (Kogan, 1997, 2001).

Argentina

Argentina is the second largest exporter (next to the United States) of corn and coarse grains, and the third largest exporter of wheat (FAO, 2000). Droughts and dry spells are frequent and devastating in this country. Since 1985, Argentina experienced two major and several minor droughts. By all standards, the most damaging droughts occurred during 1988–89 and 1989–90 crop seasons (figure 6.2), when the country lost between 15 and 20% of the total grain production. The minor droughts were less intensive and affected smaller areas, causing 5–10% reduction in crop yields. These results were validated by the ground measurements of wheat yield in Cordoba province (Seiler et al., 2000).

China

China produces grains and cotton, mostly for domestic consumption. From time to time, China also imports small amounts of agricultural commodities. However, in 1994, China unexpectedly imported a huge volume of cotton, exceeding by almost twofold the largest purchases since 1981. These imports were preceded by a cotton yield reduction three years in a row: 22% in 1992–93, 11% in 1993–94, and 7% in 1994–95. Our investigation indicated that this reduction can be attributed to unfavorable growing conditions (vegetation stress) across the main cotton-growing areas (figure 6.3).

Of all three years, the most severe vegetation stress (both moisture and thermal) occurred in 1992, which also showed the largest yield reduction. Some deterioration of vegetation conditions was also observed in 1994, but the drought-related stress was partially offset by a almost normal summer rainfall. Unlike the other two drought years, the 1993 drought was due to excessive moisture as revealed by the AVHRR-derived data.

Conclusions

This chapter has described a drought monitoring technique based on the estimation of green canopy stress from AVHRR-derived indices that characterize vegetation health, moisture, and thermal conditions. The products

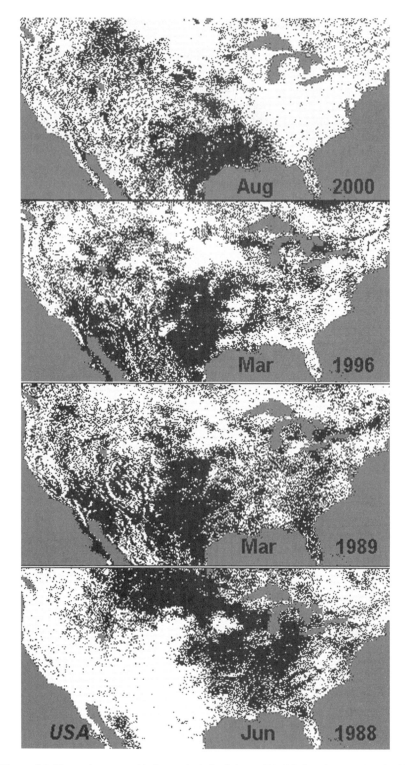

Figure 6.1 Vegetation stress (dark areas) derived from AVHRR-based vegetation health index for monitoring drought in the United States and the former Soviet Union (FSU).

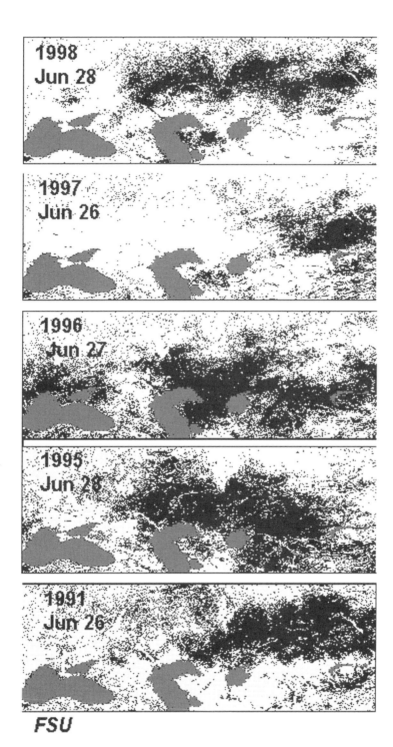

Figure 6.1 *continued*

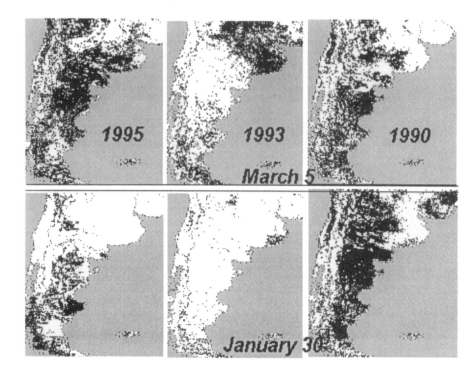

and data discussed in this chapter are delivered on real-time basis (every Monday) and are available on the Web (http://orbitnet.nesdi.noaa.gov/crad/sat/vci). The data can detect global and regional vegetation health, moisture and thermal conditions, and fire risk potential, which are all related to droughts. An important part of this process is to develop a close cooperation with users and to receive their feedback on the performance of these products. Encouraging comments from various users from different countries have been received.

The National Environmental Satellite, Data and Information Service (NESDIS; www.nesdis.noaa.gov) provides two- to four-month-long hands-on (on-site) training on using this technology. This includes access to satellite data, hardware, and software. The users are required to match their country's conventional data with satellite-based products for validating the products and to develop new applications based on mutual interests. Another way of interaction with users is a long-term cooperation program. Among recent projects, the most successful cooperation, which led to the development of new PC-based data processing system and AVHRR-based crop yield models, was with China, Kazakhstan and Israel (supported by the U.S. Agency for International Development), and with Poland (supported by U.S.–Poland binational fund).

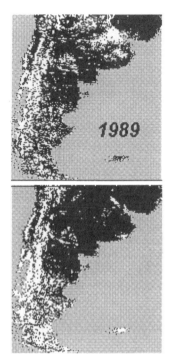

Figure 6.2 Vegetation stress (dark areas) derived from AVHRR-based vegetation health index for monitoring drought in Argentina.

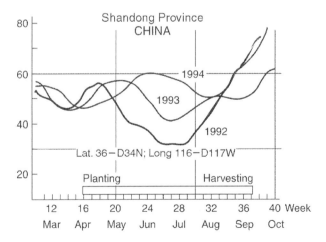

Figure 6.3 Vegetation health dynamics during the cotton-growing season in Shandong Province of China.

References

Boken, V.K., and C.F. Shaykewich. 2002. Improving an operational wheat yield model for the Canadian prairies using phenological-stage-based normalized difference vegetation index. Intl. J. Remote Sens. 23:4157–4170.

Ehrlich, P.R., A.H. Ehrlich, and J.P. Holdren. 1977. Ecoscience: Population, Resources, Environment. W.H. Freeman, San Francisco.

FAO. 2000. Crop production. Food and Agriculture Organization, Rome. Available http://fao.org.

Gates, D.M. 1970. Physical and physiological properties of plants. In: Remote Sensing with Specific Reference to Agriculture and Forestry. National academy of Sciences, Washington, DC, pp. 224–252.

Justice, C.O., J.R.G. Townshend, B.N. Holben, and C.J. Tucker. 1985. Analysis of the phenology of global vegetation using meteorological satellite data. Intl. J. Remote Sens. 6:1271–1283.

Kidwell, K.B. (ed.). 1997. NOAA polar orbiting data user's guide. Technical Report. U.S. Department of Commerce, Washington, DC.

Kogan, F.N. 1995. Droughts of the late 1980s in the United States as derived from NOAA polar orbiting satellite data. Bull. Am. Meteorol. Soc. 7:655–668.

Kogan, F.N. 1997. Global drought watch from space. Bull. Am. Meteorol. Soc. 78:621–636.

Kogan, F.N. 2000. Global drought detection and impact assessment from space. In: D.A. Wilhite (ed.), Drought: A Global Assessment, vol. 1. Hazard and Disaster Series. Routledge, London, pp. 196–210.

Kogan, F.N. 2001. Operational space technology for global vegetation assessment. Bull. Am. Meteorol. Soc. 82:1949–1964.

Kogan, F.N. 2002. World droughts in the new millennium from AVHRR-based vegetation health indices. Eos 83:557–564.

Myers, V.I. 1970. Soil, water, and plant relations. In: Remote Sensing with Specific Reference to Agriculture and Forestry. National Academy of Sciences, Washington, DC, pp. 253–267.

Orians, G.H. 1990. Ecological sustainability. Environment 32:10.

Rao, C.R.N., and J. Chen. 1995. Inter-satellite calibration linkages for the visible and near-infrared channels of the Advanced Very High Resolution Radiometer on the NOAA-7, -9, and -11 spacecrafts. Intl. J. Remote Sens. 16:1931–1942.

Rao, C.R.N., and J. Chen. 1999. Revised post-launch calibration of the visible and near-infrared channels of the Advanced Very High Resolution Radiometer on the NOAA-14 spacecraft. Intl. J. Remote Sens. 20:3485–3491.

Riebsame, W.E., S.A. Changnon, and T.R. Karl. 1990. Drought and Natural Resource Management in the United States: Impacts and Implications of the 1987–1989 Drought. Westview Press, Boulder, CO.

Seiler, R.A., F. Kogan, and G. Wei. 2000. Monitoring weather impact and crop yield from NOAA AVHRR data in Argentina. Adv. Space Res. 26:1177–1185.

Tarpley, J.P., S.R. Schnieder, and R.L. Money. 1984. Global vegetation indices from NOAA-7 Meteorological satellite. J. Appl. Meteorol. 23:491–494.

Tucker, C.J., and P.J. Sellers. 1986. Satellite remote sensing of primary production. Intl. J. Remote Sens. 7:1395–1416.

CHAPTER SEVEN

Passive Microwave Remote Sensing of Soil Moisture and Regional Drought Monitoring

THOMAS J. JACKSON

Mitigating the effects of drought can be improved through better information on the current status, the prediction of occurrence, and the extent of drought. Soil moisture can now be measured using a new generation of microwave remote sensing satellites. These measurements can be used to monitor drought conditions on a daily basis over the entire earth. The quality of these products will continue to improve over time as new sensors are launched. These satellite products, combined with existing in situ observations and models, should be exploited in drought monitoring, assessment, and prediction.

Measuring soil moisture on a routine basis has the potential to significantly improve our understanding of climatic processes and strengthen our ability to model and forecast these processes. Leese et al. (2001) concluded that the optimal approach to monitoring soil moisture would be a combination of model-derived estimates using in situ and remotely sensed measurements. In this regard, each method produces soil moisture values that are both unique and complementary. This concept is essentially the process of data assimilation described by Houser et al. (1998). In situ measurements of soil moisture have been made in a few countries over the past 70 years (Robock et al., 2000). However, due to cost and sensor limitations, there are few soil moisture sensor systems available today, especially for automated measurements. A lack of routine observations of soil moisture has led to the use of surrogate measurements (i.e., antecedent precipitation index) and modeled estimates, which limits the possibility of physically based model validation and acceptance.

Current tools to predict drought, such as drought indices and Global Climate Models (GCMs), do not include any direct observations of the soil condition, which is critical for agriculture. Passive microwave remote sensing instruments respond to the amount of moisture in the soil. Several methods have the potential to provide both soil moisture and drought information. In the past, the options have been limited by the availability of satellite systems. Even with these limitations, investigators have explored the potential of these data in soil moisture studies with some success. Within the next few years, a wide range of new and significantly improved satellites will be launched that will offer new opportunities. In this chapter, the basis of passive microwave remote sensing is presented with a description of alternative techniques for retrieving soil moisture. A review of current and future satellite systems is presented, along with examples of soil moisture studies that illustrate how this information can be used for drought monitoring and assessment.

Passive Microwave Remote Sensing of Surface Soil Moisture

Physical Basis

Microwave remote sensing provides a direct measurement of the surface soil moisture for a range of vegetation cover conditions. Two basic approaches are used, passive and active. In passive methods, the natural thermal emission of the land surface (or brightness temperature) is measured at microwave wavelengths using very sensitive detectors. Only passive microwave methods are treated in this chapter. Chapter 8 provides details on active microwave systems.

The microwave region of the electromagnetic spectrum consists of frequencies between 0.3 and 30 GHz. This region is subdivided into bands, which are often referred to by a lettering system. Some of the relevant bands that are used in earth remote sensing are K (18–27 GHz), X (8–12 GHz), C (4–8 GHz), and L (1–2 GHz). Frequency and wavelength are used interchangeably. Wavelength (in centimeters) approximately equals 30 times frequency (in GHz). Within these bands only small ranges exist that are protected for scientific applications, such as radio astronomy and passive sensing of the earth's surface. A general advantage of microwave sensors, in contrast to visible and infrared, is that observations can be made through cloud cover because the atmosphere is nearly transparent, particularly at frequencies <10 GHz. In addition, these measurements are not dependent on solar illumination and can be made at any time of the day or night.

Microwave sensors operating at very low frequencies (<6 GHz) provide the best soil moisture information. At low frequencies, attenuation and scattering problems associated with the atmosphere and vegetation are less significant, the instruments respond to a deeper soil layer, and a higher sensitivity to soil water content is present.

Low-frequency passive sensors provide information on the surface reflectivity. An examination of relationship between reflectivity and soil moisture is essential to estimate soil moisture from microwave remote sensing data.

Soil Moisture and Reflectivity Assuming that the earth is a plane surface with surface geometric variations and volume discontinuities much less than the wavelength, only refraction and absorption of the media need to be considered. This situation permits the use of the Fresnel reflection equations as a model of the system (Ulaby et al., 1986). These equations predict the surface reflectivity as a function of dielectric constant (k) and the viewing angle (θ) based on the polarization of the sensor (H = horizontal or V = vertical).

$$r^H = \left| \frac{\cos\theta - \sqrt{k - \sin^2\theta}}{\cos\theta + \sqrt{k - \sin^2\theta}} \right|^2 \quad [7.1]$$

$$r^V = \left| \frac{k\cos\theta - \sqrt{k - \sin^2\theta}}{k\cos\theta + \sqrt{k - \sin^2\theta}} \right|^2 \quad [7.2]$$

Polarization refers to the orientation of the electromagnetic waves with respect to the surface. The dielectric constant of soil is a composite of the values of its components (air, soil, and water). The basic reason that microwave remote sensing can provide soil moisture information is that there is a large difference between the dielectric constants of water ($\sim$80) and the other components ($<$5).

Based on an estimate of the mixture dielectric constant derived from the Fresnel equations and soil texture information, volumetric soil moisture can be estimated using an inversion of the dielectric mixing model (i.e., Hallikainen et al., 1985). The depth of soil contributing to the measurement is about one-quarter the wavelength (based on a wavelength range of 2–21 cm). As noted above, it is desirable to use low frequencies because the measurement at these frequencies provides more information on the soil column.

Soil Moisture and Brightness Temperature Passive microwave remote sensing uses radiometers that measure the natural thermal microwave emission within a narrow band centered on a particular frequency. The measurement provided is the brightness temperature in degrees Kelvin, T_B, which includes contributions from the atmosphere, reflected sky radiation, and the land surface. Atmospheric contributions are negligible at frequencies $<$10 GHz, and the cosmic radiation contribution to sky radiation has known values that vary only slightly in the frequency range used for observations of soil water content.

The brightness temperature of a surface is equal to its emissivity (e) multiplied by its physical temperature (T).

$$T_B = eT \quad [7.3]$$

The emissivity is equal to 1 minus the reflectivity, which provides the link to the Fresnel equations and soil moisture for passive microwave remote sensing. Figure 7.1 illustrates the relationships between emissivity and soil moisture that can be expected at a high and a low microwave frequency. If the physical temperature is determined independently, the emissivity can be determined from T_B. The physical temperature can be estimated using surrogates based on satellite surface temperature, air temperature observations, or model predictions (i.e., Owe and van de Griend, 2001).

Microwave Measurement and Vegetation For natural conditions, a variation in vegetation type and density is likely to be encountered. The presence of vegetation has a major impact on the microwave measurement. Vegetation reduces the sensitivity of the relationship to changes in soil water content by attenuating the soil signal and by adding its own microwave emission to the measurement. This attenuation increases with increasing microwave frequency, which is another important reason for using lower frequencies. Attenuation is characterized by the optical depth of the vegetation canopy. Jackson and Schmugge (1991) presented a method for estimating optical depth that used information on the vegetation type (typically derived from land cover) and vegetation water content, which is estimated using visible or near infrared remote sensing.

Soil Moisture Retrieval Algorithms

Recent efforts to develop research and operational methods for estimating soil moisture retrieval algorithms for the advanced microwave scanning radiometer (AMSR) instruments onboard the NASA Aqua and NASDA ADEOS-II satellites (Njoku et al., 2000) have resulted in the formalization of several alternative approaches. For the most part, all these methods are based on the same basic relationships but are implemented differently. Series of equations used to estimate soil moisture involve many variables related to frequency, polarization, and viewing angle of the sensor, describing physical temperature and atmospheric profile (Njoku and Li, 1999). These equations are solved using forward calculations of T_B or inversions for soil moisture.

Most research and applications involving passive microwave remote sensing of soil moisture have emphasized low frequencies (L band). In this range, it is possible to develop soil moisture retrievals based on a single H polarization observation (Jackson, 1993). It is well known that with H polarization T_B is more sensitive to soil moisture than V polarization. This approach relies on ancillary data on temperature, vegetation, land cover, and soils. Atmospheric corrections are assumed to be negligible at these frequencies. The single channel/ancillary data approach has been tested and calibrated using aircraft L-band observations (Jackson et al., 1999) and higher frequency satellite measurements (Jackson, 1997; Jackson and Hsu, 2001). Standard error of estimated values in L-band aircraft experiments were on the order of 3% volumetric soil moisture. The higher frequency satellite studies had larger errors (>5%). As noted previously, the optical depth computation approach used in Jackson and Schmugge (1991) has

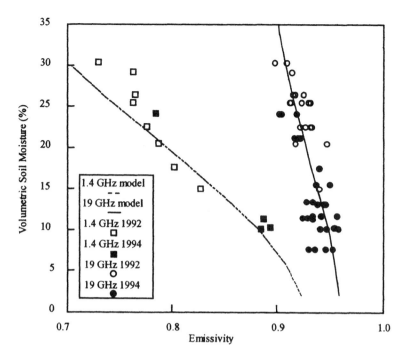

Figure 7.1 Observed and modeled relationships between microwave emissivity and soil moisture (Jackson, 1997).

a high degree of uncertainty at higher frequencies. In addition, at high frequencies other investigators have found that the single scattering albedo must be considered (Owe et al., 1992; Njoku and Li, 1999).

The weakness of the vegetation correction has led to a reconsideration of the application of the single channel/ancillary data algorithm when using high-frequency microwave data for soil moisture estimation. As a result, several investigators have examined a variation of the approach proposed in Njoku and Li (1999). In this approach, a series of equations for predicting the brightness temperature for several channels is solved iteratively. Several channels (polarizations and/or frequencies) can be used subject to the constraint that the number of unknowns is less than or equal to the number of equations. Based on practical considerations and existing data sources, a soil moisture retrieval algorithm uses the lowest microwave frequency available for both polarizations. Therefore, there will be two independent observations. By controlling and specifying parameters, the unknowns in the radiative transfer equations can be reduced to the dielectric constant and another free variable, such as the vegetation optical depth. A more exhaustive validation of this approach is needed.

Another multiple-channel approach to soil moisture retrieval was proposed by Wigneron et al. (2000). They suggested using measurements made at several viewing angles. A challenge in this approach is obtaining simultaneous, consistent multiple angle measurements. Finally, there have been attempts to use the polarization difference (PD; the difference between the vertical and the horizontal brightness temperature i.e., $T_B^V - T_B^H$) for soil moisture retrieval. As described by Njoku et al. (2000), an advantage of this approach is that it normalizes out the physical temperature. However, because it does not use physically based equations, the retrieval ultimately relies on calibration. The polarization difference approach has been explored in a number of studies. These investigations did not involve actual soil moisture but rather related indices.

Microwave Polarization Difference Index and Vegetation Parameters

The microwave polarization difference index (MPDI), an alternative to the polarization difference, is expressed as:

$$\text{MPDI} = c(T_B^V - T_B^H)/(T_B^V + T_B^H) \qquad [7.4]$$

where c is a scaling factor. Tucker (1989) and Tucker and Choudhury (1987) have attempted to exploit high-frequency passive microwave remote sensing in monitoring drought. They used 37-GHz polarization difference data collected by the scanning multifrequency microwave radiometer (SMMR). At this frequency the brightness temperature is dominated by vegetation effects and, in particular, vegetation water content. It was hypothesized that these observations could enhance products based on the normalized difference vegetation index (NDVI; chapters 5 and 6). Comparisons were made between the PD and NDVI for several study areas during periods of drought and nondrought. It was observed that the PD was very sensitive to changes in the NDVI at lower NDVI levels. However, as NDVI increased, the PD saturated at a low value. The polarization difference approaches zero as vegetation level increases.

Teng et al. (1995) summarized much of the work that has been done to relate microwave polarization information to the NDVI. They found the PD to be more sensitive to vegetation cover in regions of sparse vegetation and to NDVI in densely vegetated regions. These authors further used satellite 37-GHz PD data over the U.S. Midwest and found that drought years could be differentiated from normal years during the preplanting and early stages of crop growth. At a point in the growth cycle through harvest there was little information contained in the PD between years or months. The type of crop also had an influence on the potential information available using the PD. They concluded that NDVI provides more information on vegetation conditions; however, the PD provides unique information during periods when the NDVI is not useful.

Teng et al. (1995) concluded that PD and NDVI data are complementary for drought analyses. They noted that in some situations (the 1988 drought in the U.S. Midwest in particular) the NDVI can detect drought early in the season for severely affected regions. However, other regions that are not as severely affected are often not detected until later in the growth cycle. The PD information would be of particular use in these situations.

Satellite Passive Systems

A passive microwave instrument of particular note is the Special Satellite Microwave/Imager (SSM/I) onboard the Defense Meteorological Satellite Platforms. These polar orbiting satellites have been in operation since 1987 and provide high frequencies and two polarizations (table 7.1) except for 22 GHz (V only). Spatial resolution of the SSM/I is very coarse, as shown in table 7.1.

The SSM/I utilizes conical scanning, which provides measurements at the same viewing angle at all beam positions on a swath of 1200 km. This makes data interpretation more straightforward and simplifies image comparisons. There have been as many as four SSM/I satellites in operation at any given period. Therefore, frequent and even multiple daily passes are typical for most regions of the earth. Data from the SSM/I are publicly available (http://www.saa.noaa.gov/).

Interpreting data from the SSM/I to extract surface information requires accounting for atmospheric effects on the measurement. When one considers the atmospheric correction, the significance of vegetation attenuation, and the shallow contributing depth of soil for these high frequencies, it becomes apparent that the data are of limited value for estimating soil water content. However, data from the SSM/I can be used under some circumstances, such as in arid and semiarid areas with low amounts of vegetation. Figure 7.1 includes the results obtained over the southern Great Plains of the United States, from Jackson (1997), using SSM/I data. Value-added products from the SSM/I sensors include a wide range of atmospheric and oceanic variables. However, for reasons noted above, there have been few attempts to generate standard land surface products.

Another current satellite option is the Tropical Rainfall Measurement Mission (TRMM) Microwave Imager (TMI). It is a five-channel, dual-polarized passive microwave radiometer with a constant viewing angle. The lowest TMI frequency is 10 GHz (table 7.1), about half that of the SSM/I. The TMI has higher spatial resolution as compared to the SSM/I. TRMM is not a polar orbiting satellite and only provides coverage of the tropics, which includes latitudes between 38°N and 38°S for the TMI instrument. However, a unique capability of the TMI is its ability to collect data daily, and in many cases more often, within certain latitude ranges. This could facilitate multitemporal and diurnal analyses. These data are publicly available (http://trmm.gsfc.nasa.gov/data_dir/data.html). Jackson and Hsu (2001) have retrieved soil moisture from TMI observations. These studies showed the potential of the improved spatial resolution, higher temporal repeat coverage, and lower frequency as compared to the SSM/I.

Although no longer in operation, another satellite instrument of interest is the scanning multifrequency microwave radiometer (SMMR). This

Table 7.1 Characteristics of passive microwave satellites

Satellite	Frequency (GHz)	Polarization[a]	Horizontal resolution (km × km)	Equatorial crossing time (local)
SSM/I	19.4	H and V	69 × 43	0543 h
	22.2	V	60 × 40	0922 h
	37.0	H and V	37 × 28	1000 h
	85.5	H and V	15 × 13	
TMI	10.7	V, H	59 × 36	
	19.4	V, H	31 × 18	
	21.3	H	27 × 17	Changes
	37.0	V, H	16 × 10	
	85.5	V, H	7 × 4	
AMSR (NASA)	6.9	H and V	75 × 43	
	10.7	H and V	48 × 27	
	18.7	H and V	27 × 16	
	23.8	H and V	31 × 18	1330 h
	36.5	H and V	16 × 9	
	89.0	H and V	7 × 4	
AMSR (Japan)	6.9	H and V	71 × 41	
	10.7	H and V	46 × 26	
	18.7	H and V	25 × 15	
	23.8	H and V	23 × 14	1030 h
	36.5	H and V	14 × 8	
	89.0	H and V	6 × 4	

[a]H = horizontal; V = vertical.

instrument operated on the Nimbus-7 satellite between 1978 and 1987. It was a polar orbiting satellite and SMMR had a constant view angle of 50.3° and a swath of 780 km. Dual polarization T_B was measured at frequencies of 6.6, 10.7, 18, 21, and 37 GHz. Spatial resolution for C band was very coarse (∼150 km), and peculiarities of operation resulted in a long repeat cycle. SMMR data are also available (http://nsidc.org/data/nsidc-0036.html).

Recently several multifrequency passive microwave satellite systems have been launched, and more systems are planned. These systems offer lower frequency channel operating at C or L band, which should provide a more robust soil moisture measurement, and better spatial resolution. These satellites include, for example, the National Aeronautics and Space Administration (NASA) Aqua, the Japanese Advanced Earth Observation Satellite (ADEOS-II), the Naval Research Lab Windsat, the European Space Agency Soil Moisture Ocean Salinity Mission (SMOS), and the NASA Hydrosphere States Mission (Hydros).

Aqua was launched in May 2002 (http://aqua.nasa.gov/AMSRE3.html), and ADEOS-II was launched in the same year. Each includes an instrument called the advanced microwave scanning radiometer (AMSR). As shown in table 7.1, these are multifrequency systems that include a 6.9-GHz channel with 60-km spatial resolution. AMSR holds great promise for estimating soil water content in sparsely vegetated regions and is the best possibility in the near term for mapping soil water. Based on published results and supporting theory (Owe et al., 1992; Njoku and Li, 1999), this instrument should be able to provide information about soil water content in regions of low vegetation cover, with less than 1 kg/m^2 vegetation water content. Aqua and ADEOS-II can provide observations with nominal equatorial crossing times of 1330 and 1030, respectively.

As opposed to previous passive microwave satellite missions, Aqua and ADEOS-II include soil moisture as a product. On Aqua it is a standard product, and on ADEOS-II it is a research product. The algorithm planned for use with Aqua is a variation of the multichannel approach described in Njoku and Li (1999). Several types of soil moisture products are to be produced. These include a daily swath product and a global composite. The swath products include a retrieval of soil moisture for each pixel observed. Results will be composited to a standard grid to generate a global map of surface soil moisture with a nominal spatial resolution of 25 km. Following a period of calibration/validation, the soil moisture products should be available on a daily basis. Examples of the types of products can be found on the Web (http://sharaku.eorc.nasda.go.jp/AMSR/index_e.htm).

Windsat was launched in 2003 and includes a multifrequency passive microwave radiometer system with a C-band channel. This system includes the AMSR and other frequencies and offers additional polarization options. The equatorial crossing time is 0630 h. It is a prototype of one component of the next generation of operational polar orbiting satellites that the United States will be implementing by 2010. Experience gained by using these science missions will provide the basis for future operational products. Research programs are underway to develop and implement space-based systems with a 1.4-GHz channel that would provide improved global soil moisture information. The challenge with low-frequency passive microwave remote sensing from satellite platforms has been to achieve a useful spatial resolution subject to the constraints of antenna size. Toward that goal, the European Space Agency is developing a sensor system called the Soil Moisture Ocean Salinity (SMOS) mission (Wigneron et al., 2000). SMOS will use synthetic aperture radiometry techniques to overcome the resolution–antenna-size problem. It is scheduled for launch after 2007.

Drought-Related Investigations

As described in previous sections, passive microwave remote sensing of soil moisture and associated applications such as drought monitoring have been limited by the availability of reliable synoptic daily products. This is the result of sensor system limitations, which are expected to be overcome in the near future. The following examples illustrate the types of information that can be expected from these new sensors and exploratory studies using previous sensor systems.

Regional-Scale Soil Moisture Dynamics Using Passive Microwave Sensors

Washita '92 was a large-scale study of remote sensing and hydrology conducted using an aircraft-based L-band mapping radiometer over the Little Washita watershed in southwestern Oklahoma, United States (Jackson et al., 1995). Passive microwave observations were made over a nine-day period in June 1992. The watershed was saturated with a great deal of standing water at the outset of the study. During the experiment, no rainfall occurred, and observations of surface soil water content exhibited a drydown pattern. Observations of surface soil water content were made at sites distributed over the area. Significant variations in the level and rate of change in surface soil water content were noted over areas dominated by different soil textures.

Passive microwave observations were made on eight of the nine days of the study period. The radiometer data were processed to produce brightness temperature maps of a 740-km^2 area at a 200-m resolution on each of the eight days. Using the single-channel soil water content retrieval algorithm described in previous sections, these brightness temperature data were converted to soil water content images. Grayscale images for each day are shown in figure 7.2. These data exhibited significant spatial and temporal patterns. Spatial patterns are associated with soil textures, and temporal patterns are associated with drainage and evaporative processes. These results clearly show that consistent information can be extracted from low-frequency passive microwave data. They also illustrate the existence of the spatial and temporal variability that cannot be captured by point observations. The basic concepts developed in the Washita '92 experiment were evaluated in a follow-up to the study that expanded the spatial domain to 10,000 km^2 and expanded the time period to one month. Results presented by Jackson et al. (1999) verified that the watershed scale of the Washita '92 experiment could be extrapolated in both space and time scales compatible with satellite observation systems.

Surface and Profile Relationships

Information on the spatial and temporal variation of the unsaturated zone of the soil can be used to estimate recharge. This information can be of significant value if the data are provided as a spatially distributed product. Remote sensing satisfies the spatial and temporal needs; however, it cannot be used to directly assess the entire depth of the unsaturated zone without some additional information. Establishing a link between the easily accessible surface layer and the full soil profile has long been a research goal. A foundation for this endeavor is described by Jackson (1980). In that study, under the assumption of hydraulic equilibrium within a soil

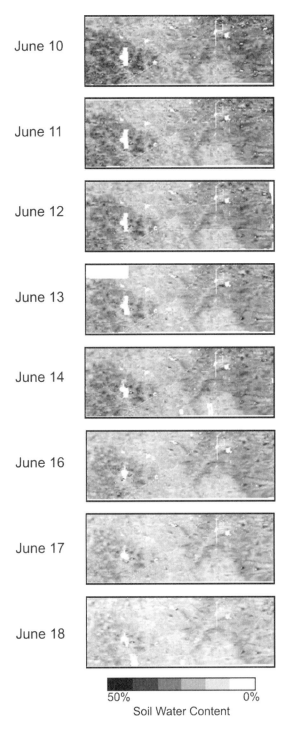

Figure 7.2 Washita '92 temporal soil moisture products (Jackson et al., 1995). The upper left corner is 563844mE, 3872666mN and the lower right is 609444mE, 3854066mN (UTM Zone 14N). Pixel resolution is 200 m.

profile of known properties, it was shown that a theoretical basis exists for surface–profile relationships and that the chances of success can be improved with additional observations at greater soil depths at particular times of the day (early morning/predawn).

Arya et al. (1983) examined the correlation between surface observations and the soil moisture profile. They observed that the correlation decreases as the depth of the soil profile increases. Better results would be expected for vegetated fields than for bare soil because vegetation tends to make the surface soil moisture profile more homogeneous with depth. The authors also compared the differences in the profile water determined using this approach and using the measured net surface flux. In this study, the two approaches were nearly equal, which could indicate that no recharge or flux across the lower boundary was occurring.

Jackson et al. (1987) combined spatially distributed remotely sensed surface observations of soil moisture over a large area in the Texas High Plains region, United States, of the Ogallala Aquifer with limited ground profile observations to produce preplanting profile soil moisture maps. The conventional approach to generating the soil moisture product involved sampling the profile at selected locations and then developing a contour map. The accuracy of this product depended on the number of points and how well they represented the local conditions at the field scale. In the remote sensing approach, a correlation was established between (1) the surface observation determined using 1.4-GHz passive microwave data and (2) the profile soil moisture at the observation points. Using this relationship at each remote sensing data point, an estimate of profile soil moisture was produced. If repeated on a temporal basis, this technique could provide spatial information on the flux of the soil water profile.

Soil Moisture-Related Indices

Radiation Aridity Index Reutov and Shutko (1987) established a linkage between microwave brightness temperature and an integrated climate parameter called the radiation aridity index (S). This is computed as follows:

$$S = \frac{\text{Annual radiation balance}}{(\text{Latent heat of vaporization})(\text{Annual precipitation})} \quad [7.5]$$

The authors cite numerous studies that link this climate variable to runoff, biological activity, and economic productivity. It is essentially the ratio between incident energy and the energy used to evaporate moisture from the soil.

Using extensive records from regions in Russia and surrounding states, they assembled soil moisture and temperature data as well as the data to compute S. Brightness temperature values were simulated using the observed soil moisture and temperature. Average brightness temperatures during growing season were computed. Individual regions were then classified into one of several landscape types and a range of S and T_B for each was determined. Figure 7.3 shows a clear correlation. It is important to note that at low values of the seasonal T_B, there is a lower sensitivity in S than at higher values of T_B. When error and variance are considered, this result suggests that the T_B is useful for the lower range and that this corresponds to the vegetated as opposed to semiarid conditions.

Antecedent Precipitation Index The antecedent precipitation index (API) is based on the summation of the precipitation for the current day and the API for the previous day reduced by a moisture depletion coefficient. Blanchard et al. (1981) and McFarland and Harder (1982) performed some of the first analysis involving API and satellite-based microwave brightness temperature. Using higher frequency microwave data (19 GHz) collected by the electronically scanned microwave radiometer (ESMR), they examined relationships in the Southern Great Plains (SGP) region of the United States. They found, for major wheat-producing regions, that drought conditions and, to a degree, soil moisture conditions, could be detected before the full canopy development and after the harvest. At full canopy stage, the microwave measurement at this frequency was related to the moisture condition of the canopy. The key result was that a correlation existed between T_B and API for low levels of vegetation.

There have been attempts to relate API to T_B using data from nearly all the passive microwave satellite systems that have flown in space. Choudhury and Golus (1988) used SMMR data. They broadened the ESMR analysis to longer wavelengths and an extended period of time. They also looked at the SGP but over a larger region with varying vegetation levels. These authors noticed a vegetation effect on the relationship between T_B and API. The next step in much of the research was to develop regression

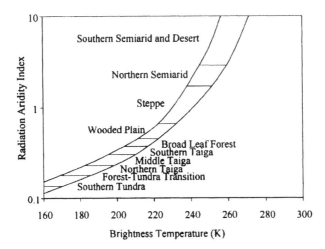

Figure 7.3 Relationship between the radiation aridity index and microwave brightness temperature for different climatic zones (Reutov and Shutko, 1987).

relationships between T_B and API that included a quantitative vegetation correction. This was accomplished using NDVI with a variety of satellites. One example is the study by Ahmed (1995). In that study, a long record of SMMR C-band data in the SGP was analyzed. Individual vegetation regions (as defined by NDVI) were identified, and a regression function was established for each. The performance of the regression varied with the NDVI level. Sensitivity depended on the slope, and the slope depended on NDVI. Equations were developed to predict the regression slope based on NDVI. Teng et al. (1993) performed a similar analysis using SSM/I data.

Wang (1985) examined SMMR C and X bands and Skylab S-194 L-band observations over the SGP. Two different regions were defined by vegetation level, and a regression was developed for each. The results showed that the vegetation level affected the sensitivity and that the sensitivity for a given region decreased as frequency increased. It was concluded that at 10.7 GHz it would be difficult to monitor soil moisture under even light vegetation. A wide variety of studies have been conducted that established regional and seasonal relationships between T_B and API. None of these efforts produced a robust and transferable approach or product.

Conclusions

Soil moisture maps, an improved index of drought and inputs to GCMs, are possible through the use of passive microwave remote sensing of soil moisture. These products could provide information useful for drought monitoring and prediction. Because drought indices and GCMs do not consider actual soil moisture conditions, this approach could have great value for agriculture. Results could improve the accuracy, timing, and reliability of the indices and predictions. Global soil moisture data will be a reality as new and improved satellite sensor systems are implemented. It is also possible that these products could be integrated with vegetation index data and existing drought indices to develop improved methods for monitoring and predicting drought.

References

Ahmed, N.U. 1995. Estimating soil moisture from 6.6-GHz dual polarization, and/or satellite derived vegetation index. Intl. J. Remote Sens. 16:687–708.

Arya, L.M., J.C. Richter, and J.F. Paris. 1983. Estimating profile water storage from surface zone soil moisture measurements under bare field conditions. Water Resources Res. 19:403–412.

Blanchard, B.J., M.J. McFarland, T.J. Schmugge, and E. Rhodes. 1981. Estimation of soil moisture with API algorithms and microwave emission. Water Resources Bull. 17:767–774.

Choudhury, B.J., and R.E. Golus. 1988. Estimating soil wetness using satellite data. Intl. J. Remote Sens. 9:1251–1257.

Hallikainen, M.T., F.T. Ulaby, M.C. Dobson, M.A. El-Rayes, and L. Wu. 1985.

Microwave dielectric behavior of wet soil, I, Empirical models and experimental observations. IEEE Trans. Geosc. Remote Sens. 23:25–34.

Houser, P.R., W.J. Shuttleworth, J.S. Famiglietti, H.V. Gupta, K. Syed, and D.C. Goodrich. 1998. Integration of soil moisture remote sensing and hydrologic modeling using data assimilation. Water Resources Res. 34:3405–3420.

Jackson, T.J. 1980. Profile soil moisture from surface measurements. J. Irrigat. Drain. Div. ASCE 106:81–92.

Jackson, T.J. 1993. Measuring surface soil moisture using passive microwave remote sensing. Hydrol. Process. 7:139–152.

Jackson, T.J. 1997. Soil moisture estimation using special satellite microwave/imager satellite data over a grassland region. Water Resources Res. 33:1475–1484.

Jackson, T.J., M.E. Hawley, and P.E. O'Neill. 1987. Preplanting soil moisture using passive microwave sensors. Water Resources Bull. 23:11–19.

Jackson, T.J., and A.Y. Hsu. 2001. Soil moisture and TRMM microwave imager relationships in the Southern Great Plains 1999 (SGP99) experiment. IEEE Trans. Geosc. Remote Sens. 39:1632–1642.

Jackson, T.J., D.M. Le Vine, A.Y. Hsu, A. Oldak, P.J. Starks, C.T. Swift, J.D. Isham, and M. Haken. 1999. Soil moisture mapping at regional scales using microwave radiometry: The Southern Great Plains hydrology experiment. IEEE Trans. Geosc. Remote Sens. 37:2136–2151.

Jackson, T.J., D.M. Le Vine, C.T. Swift, T.J. Schmugge, and F.R. Schiebe. 1995. Large area mapping of soil moisture using the ESTAR passive microwave radiometer in Washita '92. Remote Sens. Environ. 53:27–37.

Jackson, T.J., and T.J. Schmugge. 1991. Vegetation effects on the microwave emission from soils. Remote Sens. Environ. 36:203–212.

Leese, J., T. Jackson, A. Pitman, and P. Dirmeyer. 2001. GEWEX/BAHC international workshop on soil moisture monitoring, analysis and prediction for hydrometeorological and hydroclimatological applications. Bull. Am. Meterol. Soc. 82:1423–1430.

McFarland, M.J., and P.H. Harder. 1982. Crop moisture condition assessment with passive microwave radiometry. In: Proceedings of the 3rd Conference on Hydrometeorology. American Meteorological Society, Boston, MA, pp. 4.1–4.5.

Njoku, E., T. Koike, T. Jackson, and S. Paloscia. 2000. Retrieval of soil moisture from AMSR data. pp. 525–533. In: P. Pampaloni and S. Paloscia (eds.), Microwave Radiometry and Remote Sensing of the Earth's Surface and Atmosphere. VSP Publications, Zeist, The Netherlands, pp. 525–533.

Njoku, E.G., and L. Li. 1999. Retrieval of land surface parameters using passive microwave measurements at 6 to 18 GHz. IEEE Trans. Geosc. Remote Sens. 37:79–93.

Owe, M., and A.A. van de Griend. 2001. On the relationship between thermodynamic surface temperature and high-frequency (37 GHz) vertically polarized brightness temperature under semi-arid conditions. Intl. J. Remote Sens. 22:3521–3532.

Owe, M., A.A. van de Griend, and A.T. Chang. 1992. Surface soil moisture and satellite microwave observations in semiarid southern Africa. Water Resources Res. 28:829–839.

Reutov, Y.A., and A.M. Shutko. 1987. Interconnection of the brightness temperature in radio-frequency range with the radiation aridity index. Issledov. Zemli Kosmos. 6:42–48.

Robock, A., K.Y. Vinnikov, G. Srinivasan, J. Entin, S. Hollinger, N. Speranskaya, S. Liu, and A. Namkai. 2000. The global soil moisture data bank. Bull. Am. Meteorol. Soc. 81:1281–1299.

Teng, W.L., P.C. Doraiswamy, and J.R. Wang. 1995. Temporal variations of the microwave polarization difference index and its relationship to the normalized difference vegetation index in a densely cropped area. Photogram. Eng. Remote Sens. 61:1033–1040.

Teng, W.L., J.R. Wang, and P.C. Doraiswamy. 1993. Relationship between satellite microwave radiometric data, antecedent precipitation index, and regional soil moisture. Intl. J. Remote Sens. 14:2483–2500.

Tucker, C.J. 1989. Comparing SMMR and AVHRR data for drought monitoring. Intl. J. Remote Sens. 10:1663–1672.

Tucker, C.J., and B.J. Choudhury. 1987. Satellite remote sensing of drought conditions. Remote Sens. Environ. 23:243–251.

Ulaby, F.T., R.K. Moore, and A.K. Fung. 1986. Microwave Remote Sensing: Active and Passive, vol. III. From Theory to Application. Artech House, Dedham, MA.

Wang, J.R. 1985. Effect of vegetation on soil moisture sensing observed from orbiting microwave radiometers. Remote Sens. Environ. 17:141–151.

Wigneron, J.P., P. Waldteufel, A. Chanzy, J.C. Calvet, and Y. Kerr. 2000. Two-dimensional microwave interferometer retrieval capabilities over land surfaces (SMOS mission). Remote Sens. Environ. 73:270–282.

CHAPTER EIGHT

Active Microwave Systems for Monitoring Drought Stress

ANNE M. SMITH, KLAUS SCIPAL,

AND WOLFGANG WAGNER

Remote sensing can provide timely and economical monitoring of large areas. It provides the ability to generate information on a variety of spatial and temporal scales. Generally, remote sensing is divided into passive and active depending on the sensor system. The majority of remote-sensing studies concerned with drought monitoring have involved visible–infrared sensor systems, which are passive and depend on the sun's illumination. Radar (radio detection and ranging) is an active sensor system that transmits energy in the microwave region of the electromagnetic spectrum and measures the energy reflected back from the landscape target. The energy reflected back is called backscatter. The attraction of radar over visible–infrared remote sensing (chapters 5 and 6) is its independence from the sun, enabling day/night operations, as well as its ability to penetrate cloud and obtain data under most weather conditions. Thus, unlike visible–infrared sensors, radar offers the opportunity to acquire uninterrupted information relevant to drought such as soil moisture and vegetation stress.

Drought conditions manifest in multiple and complex ways. Accordingly, a large number of drought indices have been defined to signal abnormally dry conditions and their effects on crop growth, river flow, groundwater, and so on (Tate and Gustard, 2000).

In the field of radar remote sensing, much work has been devoted to developing algorithms to retrieve geophysical parameters such as soil moisture, crop biomass, and vegetation water content. In principle, these parameters would be highly relevant for monitoring agricultural drought. However, despite the existence of a number of radar satellite systems, progress in the use of radar in environmental monitoring, particularly in respect to agriculture, has been slower than anticipated. This may be attributed to the complex nature of radar interactions with agricultural targets and the suboptimal configuration of the satellite sensors available in the 1990s

(Ulaby, 1998; Bouman et al., 1999). Because most attention is still devoted to the problem of deriving high-quality soil moisture and vegetation products, there have been few investigations on how to combine such radar products with other data and models to obtain value-added agricultural drought products. This chapter provides a brief overview of radar sensor systems and the principles involved in the interaction of microwave energy with agricultural targets.

The two main radar systems with potential for agricultural monitoring are synthetic aperture radars (SARs) and scatterometers. While SARs offer high ground resolution suitable for providing information on a farm level, scatterometers allow frequent sampling (daily to weekly) at a regional scale. Scatterometers have similar spatial and temporal sampling characteristics as spaceborne radiometers, which are discussed in chapter 7. Progress with the use of these two radar systems is discussed, with emphasis on how the information could be used to monitor agricultural droughts.

Sensor Systems

Investigations into the potential for radar in geosciences began in the 1960s and gained momentum with the launch of the SEASAT satellite in 1978. In the 1980s and early 1990s, given the impending launch of a number of satellite systems, a variety of aircraft and NASA space shuttle missions were conducted in support of radar applications in land, sea, and ice monitoring. Radar satellite systems were launched by Europe (ERS-1 and ERS-2), Japan (JERS-1), and Canada (RadarSat) in the early 1990s. Recently, the European Space Agency launched ENVISAT, which has an Advanced Synthetic Aperture Radar (ASAR) on board. More satellites carrying radar instruments are scheduled for launch in the next few years (ALOS in 2004; Radarsat-2, METOP and TerraSAR in 2005).

The two broad categories of radar instruments, SAR and scatterometer, provide data on different spatial and temporal scales. The SAR systems provide data with much higher spatial but poorer temporal resolution than scatterometer systems. For example, the ERS-1 and ERS-2 SAR provide 25-m spatial resolution data over an area of 100×100 km with a 35-day repeat cycle. However, due to data costs and operational constraints, SAR acquisitions of an area are much less frequent in practice. In contrast, the scatterometer onboard ERS-1 and ERS-2 provides 50-km spatial resolution data with a repeat cycle of every 3–4 days. In the case of RadarSat, the SAR can be configured in a number of modes with differing spatial and temporal resolutions. The ScanSAR mode provides information over a 500-km swath at 100-m spatial and 3- to 4-day temporal resolution compared to the fine or standard modes that provide information over a 50- to 100-km swath at high spatial (9–25 m) but low temporal (24-day) resolution. In deciding the value of each of these different radar systems, one should critically assess the information requirements of a particular application. It is not only important to know what parameter is of interest, but also at what spatial and temporal scale the information is required.

The radar satellite systems of the 1990s were primarily designed for sea and ice monitoring rather than for applications on land. Thus, the frequency (or wavelength), polarization, incidence angle, and sampling characteristics (both in space and time) of these satellite radar systems were not always ideal for monitoring land-surface processes. The latest generation of spaceborne radars have more advanced technical capabilities such as multiple polarization imaging and higher spatial resolution. It is expected that these new features will considerably enhance the usefulness of these sensors for agricultural applications. To facilitate a better understanding of the capabilities and limitations of past and new radar systems, the effects of system and target parameters on radar backscatter are discussed in the next section.

Physical Fundamentals

The interaction of radar energy with agricultural targets is different from that with visible–infrared energy. Interpretation of radar data requires an understanding of the physical processes involved in the scattering of electromagnetic waves by objects that are smaller, comparable to, and larger than the wavelength. Therefore modeling these processes is relatively difficult because simplifying mathematical assumptions often results in a lack of correspondence with physical reality. In an agricultural context, the radar signal can interact with vegetation or soil only, but more likely there is scattering within the canopy and the return signal is from multiple sources.

The attenuation of the radar signal and the interaction with vegetation and soil depends on a number of system and target parameters. The frequency (or wavelength), polarization, incidence angle, look direction, and resolution of the sensor all influence the sensor capabilities; important target parameters are surface roughness, complex dielectric constant, structural geometry, slope angle, and orientation of the target. The distinction between system and target parameters is convenient for understanding radar interactions. However, all parameters interact to influence each other. For example, the roughness of the target as seen by the sensor is dependent on the wavelength, the local angle of the sensor, which is a function of both the look angle of the sensor and the slope angle of the target, and the look direction, which affects the geometry of the target in question.

System Parameters

Polarization Radar instruments distinguish the polarization of the transmitted and the received electromagnetic waves. In microwave remote sensing, "horizontal" polarization (H) means that the electric field vector of the electromagnetic wave is oriented parallel to the earth's surface, while "vertical" polarization (V) means that a wave arrives at the interface with a vertical component (but a horizontal component also exists). When the polarization of the transmitting antenna is the same as for reception of the signal, one speaks of like-polarization (HH or VV), and otherwise of cross-polarization (HV or VH). Until the launch of ENVISAT, all satellite systems used linear combinations in either the horizontal (HH in the case of RadarSat and JERS-1) or vertical directions (VV in the case of ERS-1 and ERS-2).

Crop structure has a strong effect on the polarization of the backscattered waves. Generally, if the target has a strong vertical structure (e.g., grain crops), scattering is dominated by the returns from the underlying ground surface. Modified by vegetation density and radar frequency, HH polarization tends to interact to a lesser extent with the crop than VV and penetrates more effectively to the underlying soil. Cross-polarized signals (HV or VH) depend on multiple reflections between the canopy and the soil to depolarize the transmitted signal. Experimental and modeling studies suggest that the ability to estimate soil and vegetation parameters will be enhanced significantly with the launch of fully-polarized SAR systems such as RadarSat-2 and ALOS (Bindlish and Barros, 2000).

Frequency The effect of radar frequency on backscatter is largely a function of the size and dielectric properties of the target. The dielectric constant of water, which constitutes 80–90% of plant matter and is present in the soil, differs markedly depending on radar frequency. As frequency decreases and wavelength increases, the size of the target components become smaller relative to the sensor, and the scattering surfaces appear "smoother," making scattering by vegetation elements less efficient.

In the presence of dense vegetation, at high frequencies or smaller wavelengths, radar backscatter is primarily a function of canopy scattering, and the soil has a minor impact. At lower frequencies or longer wavelengths, the impact of soil becomes greater as the radar signal penetrates farther into the canopy. It is usually believed that radars operating in X or K band (table 8.1) are not sensitive to soil moisture when crops are fully established, whereas L- and P-band radars penetrate crop and grass canopies even at the height of the growing season.

As a result of differential backscatter from cotton and alfalfa in Ku and C band, Moran et al. (1998) suggested that dual frequency would be useful in estimating vegetation status and soil moisture. Unfortunately, no spaceborne SAR system has yet been approved that would allow instantaneous measurements at two different frequencies.

Incident Angle The incident angle is defined as the angle between the incident radar beam and the vertical to the intercepting surface. In early studies using single-channel C-band radar, steeper incident angles (15–25°) were generally found to be more useful in estimating soil moisture in presence or absence of vegetation. In absence of vegetation and at steeper angles, the radar backscatter is affected to a lesser extent by surface roughness, but in presence of vegetation, the path length through and attenuation by the

Table 8.1 Radar frequencies and corresponding wavelengths

Band	Frequency (GHz)	Wavelength (cm)
Ka	27–40	0.8–1.1
K	18–27	1.1–1.7
Ku	12–18	1.7–2.4
X	8–12.5	2.4–3.8
C	4–8	3.8–7.5
L	1–2	15–30
P	0.3–1	30–100

vegetation decrease. Conversely, at shallow angles (40–45°) the path length through the canopy increases, and so does the vegetation response. Consequently, backscatter measurements at different incident angles provide useful information about the relative contributions of soil and vegetation to total backscatter. Although SAR sensors have only one antenna that looks at the earth's surface from one direction, the scatterometers onboard ERS-1/2 and METOP have three and six antennas, respectively, that acquire instantaneous measurements at different incidence angles. This multiangle capability of scatterometers enables soil moisture and vegetation effects to be separated quite efficiently (Wagner et al., 1999a).

Target Parameters

The important target characteristics influencing radar backscatter are geometry and dielectric properties. The dielectric constant, which is a measure of the relative effectiveness of a substance as an electrical insulator, ranges from as high as 80 for liquid water at low microwave frequencies to <4 for dry matter. Thus, in an agricultural environment, radar backscatter is a function of soil surface roughness and moisture content and, where vegetation exists, of crop type, phenological state, and vegetation water content.

Soil Radar measurements of bare soil surfaces are very sensitive to the water content in the soil surface layer due to the pronounced increase in the soil dielectric constant with increasing water content. For longer wavelengths, the backscatter coefficient may increase up to about 10-fold from dry to wet soil conditions. Although there is a very strong relationship between the backscatter measurements and the soil moisture content, the retrieval of soil moisture is a challenging task due to the confounding influence of surface roughness.

There are several empirical and theoretical models that describe backscatter from bare soil in terms of the soil moisture and surface roughness (Fung, 1994; Dubois et al., 1995). Unfortunately, many studies found relatively poor agreement between modeled and experimentally observed radar responses. It has been suggested that the inadequate representation of surface roughness variations of real surfaces may be one of the reasons for this failure. Surface roughness is usually expressed in terms of the root mean square (r.m.s.) of height or, in the case of the more complex models, in terms of the autocorrelation function of the surface height variations. It appears that agricultural and other natural soil surfaces are relatively complex and often have a multiscale structure. That is why simple parameters such as the r.m.s. of height fail to adequately describe radar backscatter (Davidson et al., 2000). One way of working around this problem is to use change detection techniques that only interpret changes in soil moisture under stable surface roughness conditions.

Vegetation Crop type, crop phenology, and crop condition influence the architecture and dielectric constant of the canopy and thus radar backscatter. There are numerous studies related to crop discrimination using radar (Bouman and Uenk, 1992; Brisco et al., 1992; Ban and Howarth, 1999; Treitz et al., 2000). The ability to discriminate crop type and phenological stage with radars may be attributed to changes in both canopy geometry and plant biophysical parameters. The radar frequency and incident angle influence the relationship.

Radar backscatter from an agricultural field is normally not influenced just by the plants, but also by the underlying soil and by soil–plant interactions. As indicated in figure 8.1, there are three main contributions from a vegetation canopy: direct backscatter from the vegetation elements, multiple scattering from plants and soil, and the contribution from the soil surface. Accordingly, total canopy backscatter, σ^0, is formulated as the sum of three parts:

$$\sigma^0 = \sigma^0_{\text{veg}} + a_t^2 \sigma^0_{\text{soil}} + \sigma^0_{\text{int}} \qquad [8.1]$$

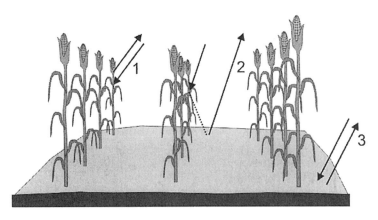

Figure 8.1 The interaction of radar with an agricultural target results in backscatter (1) directly from the crop, (2) from a combination of crop and soil, and (3) directly from the soil [from Brisco and Brown, 1998].

where σ^0_{veg} is the backscatter attributable to vegetation, σ^0_{soil} is the direct backscatter from the underlying soil surface, a_t is the one-way attenuation factor, and σ^0_{int} represents multiple scattering in soil and vegetation.

In many models the interaction term σ^0_{int} is neglected because multiple scattering by vegetation and soil is generally smaller than the direct contributions from the canopy and the soil. One such model is the Cloud Model proposed by Attema and Ulaby (1978), which assumes that the vegetation volume can be represented by a cloud of water droplets that are uniformly distributed throughout the volume. Although the Cloud Model does not take structural effects into account, it has been used with some success to empirically relate ground measurements of vegetation water content, vegetation height, and other crop parameters to backscatter measurements acquired with ground-based, airborne, or spaceborne radar sensors (Bouman, 1991; Taconet et al., 1994; Xu et al., 1996).

Monitoring Drought Stress

High-Resolution SAR Imaging

For assessing drought conditions at a local scale, such as a municipal or watershed level, high-resolution imagery from visible–infrared scanners and SAR systems can be used. The preceding discussion illustrated the potential of SARs for mapping drought-relevant parameters including soil moisture or plant stress. The latter may be manifest in reduced canopy moisture and ultimately in plant growth.

The use of radar for directly inferring drought stress conditions of crops has been limited. In a ground-based scatterometer study, radar backscatter differed between stressed and nonstressed crops, but backscatter was dependent on crop type and crop density (Brisco and Brown, 1990). The separability of the stressed crops was attributed to their stunted growth and lower biomass. Steven et al. (2000) concluded that Radarsat SAR data can be used to detect canopy dehydration under conditions of moisture stress.

The greatest potential for SAR data for drought monitoring is through the estimation of soil moisture. Dobson and Ulaby (1998) provided an overview of common modeling approaches on how to derive soil moisture from SAR data. At the scale of SARs (tens of meters), spatial patterns of surface roughness and both vegetation type and density leave a strong imprint on SAR imagery. Therefore it is not possible to infer soil moisture patterns by simple visual analysis of SAR images. More sophisticated retrieval approaches are required that account for surface roughness and, where vegetation is present, for the effect of vegetation on the SAR signal. Unfortunately, as previously discussed, models often fail to accurately describe backscatter from bare soil surfaces due to the complex and multiscale structure of agricultural and other natural soil surfaces and the limited validity range of the models. Nevertheless, it has been demonstrated that relatively simple change detection approaches, which depend on the availability of a number of consecutive SAR acquisitions, may be successful in tracking soil moisture conditions within a watershed.

A number of approaches use the single-channel SAR data provided by ERS-1/2 (Crevier et al., 1996; Rotunno Filho et al., 1996; Moran et al., 2000; Le Hégarat-Mascle et al., 2002). Bare-soil fields are ideal targets for tracking soil moisture through time, but in an agricultural environment a field does not remain bare throughout the season. The approach of Le-Hégarat-Mascle et al. (2002), for a particular watershed, uses a time series of images from one year to develop an empirical relationship between the mean radar backscatter and mean soil moisture from a number of target sites. In subsequent years, through inversion of the empirical relationship, soil moisture can be directly estimated from the imagery rather than by time-consuming ground-based measurements. The target sites are fields of bare soil or fields with very little vegetation, which may change from image to image in the time series. However, on any one date sufficient sites are selected such that the mean value is representative of the watershed. As an alternative to bare soil, Crevier et al. (1996) and Rotunno Filho et al. (1996) proposed temporal measurements of semipermanent grass fields to monitor soil moisture status of a watershed. The semipermanent grass fields, due to their relative stability in surface characteristics both within and across years, offer a means to normalize the surface roughness and vegetation amounts. This idea is being further examined by Sokol et al. (2002).

The technique of Moran et al. (2000), developed in a rangeland system using ERS-1 data, may be useful in tracking a potential drought. A single image acquired under dry conditions was used to normalize for surface roughness and standing brown litter content in all other images in a time series. The simple subtraction of the dry image backscatter values from the backscatter values in every other image in the time series increased the relationship with soil moisture ($r^2 = 0.93$) compared to using the values extracted directly from each image ($r^2 = 0.27$). The green leaf area index (GLAI), the one-sided green leaf area per unit ground, within the rangeland system was <0.35 and thus could be ignored in terms of influencing soil moisture determination using ERS-1. However, a method was presented to correct for GLAI using a vegetation component derived from optical data.

Further progress in the use of radar can be expected with the launch of technically more advanced multipolar, multifrequency radar satellites. Moran et al. (1998) found, in alfalfa and cotton, that up to a GLAI value of 4, high-frequency Ku (VV) radar was sensitive to increases in GLAI, whereas the lower frequency C band (VV) was sensitive to soil moisture. Above a GLAI value of 4, the Ku band saturated and the sensitivity of C band to soil moisture decreased due to attenuation of the signal by the vegetation. Similarly, Prevot et al. (1993) found that the simultaneous use of X band (VV) which is adapted to biomass estimation and C-band

(HH) radar which is adapted to soil moisture estimation, enabled both soil moisture and LAI of wheat canopies to be estimated. The availability of L band would further increase the potential to estimate soil moisture under denser canopies.

Large-Scale Monitoring Using Scatterometers

Compared to SAR, the number of scientific studies investigating the potential of scatterometers for soil moisture and vegetation retrieval is limited. Most likely this is due to the low spatial resolution of scatterometers (tens of kilometers), which restricts their use to regional applications. Still, considerable progress has been made, and scatterometer-derived soil moisture data, which reflect the atmosphere-related, large-scale component of the soil moisture field (Vinnikov et al., 1999), have already been used as input to crop models to assess drought-induced yield reductions. Here we only review work done with the C-band (VV) scatterometer onboard ERS-1/2 but recognize that there have been impressive first studies using the Ku-band scatterometers onboard the QuikScat and Midori-2 satellites launched in 1999 and 2002, respectively.

Many of the initial ERS scatterometer studies focused on the retrieval of vegetation parameters because a substantial agreement between backscatter and global vegetation index maps has been observed (Frison and Mougin, 1996). Consequently, several models capable of separating soil moisture from vegetation effects have been developed and applied to backscatter time series. Surface roughness is less of a problem for the analysis of scatterometer time series given that, at regional scales, it can be considered to be time invariant. In recent years attention has shifted more and more to the retrieval of soil moisture given the greater than anticipated sensitivity of the C-band scatterometer to soil moisture. For example, application of the model developed by Woodhouse and Hoekman (2000) over a Mediterranean region (Spain) did not properly recover the seasonal vegetation signal but provided soil surface reflectivity values in agreement with the monthly precipitation records.

A soil moisture retrieval technique based on a change detection approach has been developed by Wagner et al. (1999b). The method is capable of separating the effects of soil moisture and vegetation phenology by exploiting the information content provided by the multiple-viewing capabilities of the ERS scatterometer. By comparing instantaneous ERS scatterometer measurements to the lowest and highest backscatter values in the ERS scatterometer time series, a relative measure of the moisture content of the surface soil layer (<5 cm) is obtained. The algorithm has been tested over different climatic regions with success, and the multiyear soil moisture data were derived from remotely sensed data (Wagner and Scipal, 2000; Scipal et al., 2002). The data set is available to other research groups on request and can be viewed on a Web site (IPF, 2003).

Microwaves sense only the first few centimeters of the soil, but for agricultural applications mainly the soil moisture content within the reach of the plant roots is of interest (Jones et al., 2000). To estimate the water content at deeper levels using scatterometer data, Wagner et al. (1999b) proposed a two-layer water balance model that transforms the highly variable surface soil moisture series into a red-noise–like profile of soil moisture time series. When soil hydrologic properties (wilting level, field capacity, and total water capacity) are known, the water content available to plants can be estimated. Over the Ukraine, a comparison with gravimetric soil moisture data from the agro-meteorological station network showed that the soil moisture content in the 0–100-cm layer can be estimated with an accuracy of about 5% volumetric soil moisture. A comparison with in situ soil moisture data from a network of 20 TDR (time domain reflectory) probes in the Duero-Basin in Spain yielded an accuracy of better than 3% volumetric soil moisture for the same layer (Scipal et al., 2003).

Timely information about the regional soil moisture conditions from remote-sensing data is useful for assisting agro-meteorological analysis. The information content can be further enhanced by comparing the present year with the long-term mean and modeling the timing of crop water stress (Barron et al., 2003). Given the availability of nine years of ERS-1/2 scatterometer data, soil moisture anomalies can be easily calculated. Figure 8.2 shows how scatterometer-derived soil moisture anomalies for the months February and March 1999 over Southeast Asia (including India) compare to rainfall anomalies derived from a globally gridded precipitation data produced by the Global Precipitation Climatology Center (GPCC, 2003). In both data sets, anomalous dry conditions centered over southern China can be recognized.

To assess the timing of water availability during critical crop-growth stages, remotely sensed soil moisture data must be combined with crop-growth models. In a study conducted over Russia and the Ukraine, scatterometer-derived soil moisture data were used as input to a crop growth model WOFOST (Supit et al., 1994). This model simulates the daily growth of a specific crop using weather and soil data and following the hierarchical distinction between potential and limited production. The ratio between potential and limited production is the indicator of drought stress. The scatterometer-derived soil moisture is used to replace the soil moisture estimates normally derived using water budget models. The results of the study suggested that provincial yield assessment can be improved through the use of remotely sensed soil moisture information (Wagner et al., 2000).

Conclusions

Radars can potentially be used to monitor soil moisture, plant moisture content, and vegetation production. For example, it has been demonstrated that drought conditions can be inferred from scatterometers by calculating soil moisture anomaly indicators or by using the remotely sensed parameters as input to the more complex crop-growth models to capture cumu-

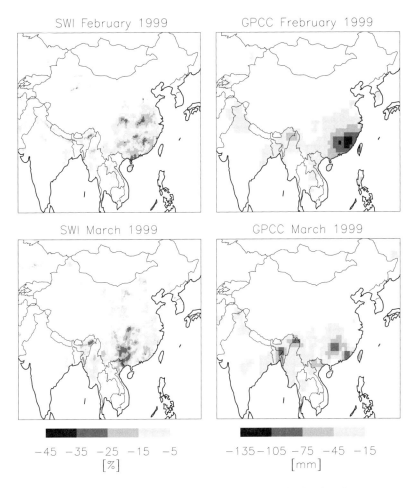

Figure 8.2 Spatial drought pattern of an extreme winter drought that hit China in 1999 as reflected in scatterometer data and gridded precipitation data. Left: difference of February and March 1999 profile soil moisture to the 1992–2000 average derived from scatterometer data; right: difference of February and March 1999 gridded precipitation (GPCC, 1998) to the 1992–2000 average.

lative effects of droughts. However, applications of radars in monitoring drought stress have been limited due to the relative newness of radar technology, the lack of operational radar sensor systems, and the complex interaction of radar waves with natural surfaces. Realistically, much more research and preoperational demonstrations need to be carried out before radars can be accepted as dependable information providers for drought monitoring systems. Nevertheless, a greater understanding of the radar capabilities will lead to the more intelligent use of the SAR systems such as RadarSat-2 (scheduled to be launched in 2005) or METOP (scheduled to be launched in 2005) for operational drought monitoring.

References

Attema, E.P., and F.T. Ulaby. 1978. Vegetation modeled as a water cloud. Radio Sci. 13:357–364.

Ban, Y., and P.J. Howarth. 1999. Multitemporal ERS-1 SAR data for crop classification: A sequential masking approach. Can. J. Remote Sens. 25:438–447.

Barron, J., J. Rockström, F. Gichuki, and N. Hatibu. 2003. Dry spell analysis and maize yields for two semi-arid locations in east Africa. Agric. Forest Meteorol. 117:23–37.

Bindlish R., and A.P. Barros. 2000. Multifrequency soil moisture inversion from SAR measurements with the use of IEM. Remote Sens. Environ. 71:67–88.

Bouman, B.A.M. 1991. Crop parameter estimation from ground-based X-band (3-cm Wave) radar backscattering data. Remote Sens. Environ. 37:193–205.

Bouman, B.A.M., and D. Uenk. 1992. Crop classification possibilities with radar in ERS-1 and JERS-1 configuration. Remote Sens. Environ. 40:1–13.

Bouman, B.A.M., D.W.G. van Kraalingen, W. Stol, and H.J.C. van Leeuwen. 1999. An agroecological modeling approach to explain ERS SAR radar backscatter of agricultural crops. Remote Sens. Environ. 67:137–146.

Brisco, B., and R.J. Brown. 1990. Drought stress evaluation on agricultural crops using C-HH SAR data. Can. J. Remote Sens. 16:39–44.

Brisco, B., and R.J. Brown. 1998. Agricultural applications of Radar. In: Principles and Applications of Imaging Radar. Manual of Remote Sensing. Henderson F.M. and A.J. Lewis (eds), vol. 2. 3rd Edition. John Wiley & Sons, New York.

Brisco, B., R.J. Brown, J.G. Cairns, and B. Snider. 1992. Temporal ground-based scatterometer observations of crops in western Canada. Can. J. Remote Sens. 18:14–21.

Crevier Y., T.J. Pultz, T.I. Lukowski, and T. Toutin. 1996. Temporal analysis of ERS-1 SAR backscatter for hydrology applications. Can. J. Remote Sens. 22:65–76.

Davidson M.W.J., T. Le Toan, F. Mattia, G. Stalino, T. Manninen, and M. Borgeaud. 2000. On the characterization of agricultural soil roughness for radar remote sensing studies. IEEE Trans. Geosci. Remote Sens. 38:630–640.

Dobson, M.C., and F.T. Ulaby. 1998. Mapping soil moisture distributions with imaging radar, In: F.M. Henderson and A.J. Lewis (eds.), Principles and Applications of Imaging Radar, Manual of Remote Sensing, vol. 2. John Wiley and Sons, New York, pp. 435–509.

Dubois, P.C., J. van Zyl, and T. Engman. 1995. Measuring soil moisture with imaging radars. IEEE Trans. Geosci. Remote Sens. 33:915–926.

Frison, P.L., and E. Mougin. 1996. Monitoring global vegetation dynamics with ERS-1 wind scatterometer data. Intl. J. Remote Sens. 17:3201–3218.

Fung, A. 1994. Microwave Scattering and Emission Models and their Applications. Artech House, Norwood, MA.

GPCC. 2003. Global Precipitation Climatology Centre Web site. Available http://gpcc.dwd.de, accessed May 2003.

IPF. 2003. The global soil moisture archive 1992–2000 from ERS scatterometer data. Institute of Photogrammetry and Remote Sensing of the Vienna University of Technology. Available http://www.ipf.tuwien.ac.at/radar/ers-scat/home.htm accessed May 2003.

Jones, R.J.A., P. Zdruli, and L. Montanarella. 2000. The estimation of drought risk in Europe from soil and climate data. In: J.V. Vogt and F. Somma (eds.),

Drought and Drought Mitigation in Europe. Kluwer, Dordrecht, The Netherlands, pp. 133–146.

Le Hégarat-Mascle, S., M. Zribi, F. Alem, A. Weisse, and C. Loumange. 2002. Soil moisture estimation from ERS/SAR: Toward an operational methodology. IEEE Trans. Geosci. Remote Sens. 40:2647–2658.

Moran, M.S., D.C. Hymer, J. Qi, and E.E. Sano. 2000. Soil moisture evaluation using multi-temporal synthetic aperture radar (SAR) in semiarid rangeland. Agric. Forest Meteorol. 105:69–80.

Moran, M.S., A. Vidal, D. Troufleau, Y. Inoue, and T.A. Mitchell. 1998. Ku- and C-band SAR for discriminating agricultural crop and soil conditions. IEEE Trans. Geosci. Remote Sens. 36:265–272.

Prevot L., I. Champion, and G. Guyot. 1993. Estimating surface soil moisture and leaf area index of a wheat canopy using a dual-frequency (C and X bands) scatterometer. Remote Sens. Environ. 46:331–339.

Rotunno Filho, O.C., E.D. Soulis, N. Kouwen, A. Abdeh-Kolahchi, T.J. Pultz, and Y. Crevier. 1996. Soil moisture in pasture fields using ERS-1 SAR data: Preliminary results. Can. J. Remote Sens. 22:95–107.

Scipal, K., W. Wagner, A. Ceballos, J. Martinez, and C. Scheffler. 2003. The potential of scatterometer derived soil moisture for catchment scale modeling. Paper presented at the International Congress on Modelling and Simulation (Modsim 2003), Townsville, Queensland, Australia, July 14–17, 2003.

Scipal, K., W. Wagner, M. Trommler, and K. Naumann. 2002. The global soil moisture archive 1992–2000 from ERS scatterometer data: First results. In: Proceedings of IGARRS 2002, Toronto, Canada, June 24–28, 2002, CDROM. IEEE Inc., Piscataway, NJ.

Sokol, J., T.J. Pultz, A. Deschamps, and D. Jobin. 2002. Polarimetric C-band observations of soil moisture for pasture fields. In: Proceedings of IGARRS 2002, Toronto, Canada, June 24–28, 2002, CDROM. IEEE Inc., Piscataway, NJ.

Steven, M.D., G. Gill, G. Cookmartin, K. Morrison, and K.W. Jaggard. 2000. Radar response to wilting in sugar beet *Beta vulagris*. Aspects Appl. Biol. 60:123–130.

Supit, I., A.A. Hooijer, and C.A. van Diepen. 1994. System description of the WOFOST 6.0 crop simulation model implemented in CGMS, vol. 1. Theory and Algorithms. EUR 15956, Office for Official Publications, European Community, Luxembourg.

Taconet, O., M. Benallegue, D. Vidal-Madjar, L. Prevot, M. Dechambre, and M. Normand. 1994. Estimation of soil and crop parameters for wheat from airborne backscattering data in C and X bands. Remote Sens. Environ. 50: 287–294.

Tate, E.L., and A. Gustard. 2000. Drought definition: A hydrological perspective. In: J.V. Vogt and F. Somma (eds.), Drought and Drought Mitigation in Europe. Kluwer, Dordrecht, The Netherlands, pp. 23–48.

Treitz, P.M., P.J. Howarth, O.R. Filho, and E.D. Soulis. 2000. Agricultural crop classification using SAR tone and texture statistics. Can. J. Remote Sens. 20:18–29.

Ulaby, F.T. 1998. SAR biophysical retrievals: Lessons learned and challenges to overcome. In: Applications Workshop on Retrieval of Bio- and Geo-Physical Parameters from SAR Data for Land, October 21–23, 1998. European Space Research and Technology Center, Nordwijk, The Netherlands.

Vinnikov, K.Y., A. Robock, S. Qiu, J.K. Entin, M. Owe, B.J. Choudhury, S.E. Hollinger, and E.G. Njoku. 1999. Satellite remote sensing of soil moisture in Illinois, United States. J. Geophys. Res. 104:4145–4168.

Wagner, W., G. Lemoine, M. Borgeaud, and H. Rott. 1999a. A study of vegetation cover effects on ERS scatterometer data. IEEE Trans. Geosci. Remote Sens. 37:938–948.

Wagner, W., G. Lemoine, and H. Rott. 1999b. A method for estimating soil moisture from ERS scatterometer and soil data. Remote Sens. Environ. 70:191–207.

Wagner, W., and K. Scipal. 2000. Large-scale soil moisture mapping in western Africa using the ERS scatterometer. IEEE Trans. Geosci. Remote Sens. 38:1777–1782.

Wagner, W., K. Scipal, K. van Deepen, H. Boogaard, R. Beck, and E. Nobbe. 2000. Assessing water-limited crop production with a scatterometer based crop growth monitoring system. In: Proceedings of IGARSS 2000, Honolulu, Hawaii, July 24–28, 2000. IEEE Inc., Piscataway, NJ, pp. 1696–1698.

Woodhouse, I.H., and D.H. Hoekman. 2000. A model-based determination of soil moisture trends in Spain with the ERS-Scatterometer. IEEE Trans. Geosci. Remote Sens. 38:1783–1793.

Xu, H., M.D. Steven, and K.W. Jaggard. 1996. Monitoring leaf area of sugar beet using ERS-1 SAR data. Intl. J. Remote Sens. 17:3401–3410.

PART III

THE AMERICAS

CHAPTER NINE

Monitoring Drought in the United States: Status and Trends

DONALD A. WILHITE, MARK D. SVOBODA,

AND MICHAEL J. HAYES

Drought occurs somewhere in the United States almost every year and results in serious economic, social, and environmental costs and losses. Drought is more commonly associated with the western United States because much of this region is typically arid to semiarid. For example, this region experienced widespread drought conditions from the late 1980s through the early 1990s. The widespread and severe drought that affected large portions of the nation in 1988 resulted in an estimated $39 billion in impacts in sectors ranging from agriculture and forestry to transportation, energy production, water supply, tourism, recreation, and the environment (Riebsame et al., 1991). In the case of agriculture, production losses of more than $15 billion occurred and especially devastated corn and spring wheat belts in addition to reducing exports to other nations. In 1995, the U.S. Federal Emergency Management Agency (FEMA) estimated annual losses attributable to drought at $6–8 billion (FEMA, 1995). Since 1995, drought has occurred in nearly all parts of the country, and many regions have been affected on several occasions and in consecutive years. Most of the eastern United States experienced an extremely severe drought in 1998–99, and in parts of the southeast, drought occurred each year from 1999 through 2002, especially in Florida and Georgia.

Figure 9.1 depicts nonirrigated corn yields for Nebraska for the period from 1950 to 2002. Nebraska is one of the principal agricultural states in the United States, and corn is one of its primary crops. The drought effects on yields are most apparent during the severe droughts of the mid-1950s, mid-1970s, 1980, 1983, 1988–89, and 2000. Extremely wet years, such as 1993 in the eastern part of the state, also depressed corn yields.

Monitoring drought presents some unique challenges because of its distinctive characteristics (Wilhite, 2000). The purpose of this chapter is to document the current status of drought monitoring and assessment in the United States, particularly with regard to the agricultural sector.

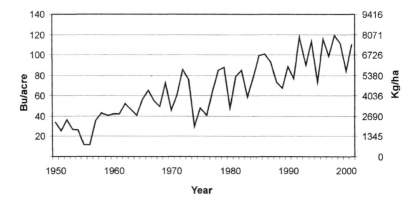

Figure 9.1 Dryland corn yield for Nebraska, 1950–2002 (from U.S. Department of Agriculture, National Agricultural Statistics Service, Washington, DC; www.nass.usda.gov).

Drought Monitoring Program

After the creation of the National Drought Mitigation Center (NDMC) in 1995, one of our first goals was to create a "one-stop shopping" section on our Web site that would provide users with access to all the information for drought monitoring in a timely and reliable fashion. The development of the "drought watch" section was undertaken because no routine national or regional integrated assessment was available from federal agencies. This section has evolved into the "Monitoring Drought" section of the NDMC Web site (drought.unl.edu). Currently, drought monitoring in the United States relies heavily on a product called the U.S. Drought Monitor, which draws on the climate and water supply indicators and indices such as Palmer drought severity index (PDSI) and the standardized precipitation index (SPI).

Palmer Drought Severity Index

The PDSI was developed by W.C. Palmer (1965) for monitoring droughts in terms of their intensity, duration, and spatial extent. Although there are several variations of the index, each variation has characteristics similar to the PDSI (Palmer, 1965; Karl and Knight, 1985; Heddinghaus and Sabol, 1991). Palmer based the PDSI on anomalies in the supply and demand concept of the water balance equation. Inputs into weekly or monthly calculations include precipitation, temperature, and the local antecedent soil moisture conditions. The data are standardized to account for regional differences so that the PDSI values can be compared from one location to another. Therefore, identical PDSI values, in theory, in the Midwest and Texas indicate the same severity of drought, even though actual rainfall deficiencies would be different at the two locations.

Weekly maps of a modified PDSI (Heddinghaus and Sabol, 1991) are produced by the Climate Prediction Center (CPC) of the National Oceanic and Atmospheric Administration (NOAA) and are frequently used in assessments of agricultural conditions around the United States. The PDSI has been used historically in policy decisions by the U.S. Department of Agriculture (USDA) regarding requests for drought relief and by states as triggers for response actions as part of state drought plans. However, the PDSI has limitations that diminish its application, bringing into question the practice of basing agricultural policy decisions solely on the PDSI. These limitations have been well documented (Alley, 1984; Karl and Knight, 1985; Willeke et al., 1994; Kogan, 1995; McKee et al., 1995; Guttman, 1998). The most significant limitations of the PDSI related to monitoring agricultural drought include: (1) an inherent time scale (about 8 months) that makes it difficult to detect emerging drought conditions and shorter-length drought periods; (2) the characteristic that all precipitation is treated as rain, which does not account for snowfall, snow cover, and frozen ground, thus questioning the reliability of the real-time winter PDSI values and preplant soil moisture estimates (one of the PDSI inputs); (3) the characteristic that the natural lag between precipitation and runoff is not considered and that no runoff occurs until the water capacities of the surface and subsurface soil layers are full, which leads to an underestimation of runoff; and (4) wide variations in the "extreme" and "severe" classifications of the PDSI values, depending on location in the country.

If a drought index is going to be spatially comparable and useful for agricultural policy decisions, extreme and severe classifications must occur consistently and with low frequency (Guttman et al., 1992). An additional concern is that the PDSI does not do well in the mountainous western United States, especially since a majority of that region's precipitation falls during the winter as snowfall. The PDSI can also retain values reflecting drought well after a climatological recovery from drought has occurred. All of these limitations reveal the importance of caution when using the PDSI for monitoring agricultural conditions and making related policy decisions.

Standardized Precipitation Index

The SPI was developed to address some of the problems inherent in the PDSI. McKee and his colleagues at Colorado State University (McKee et al., 1993, 1995) designed the SPI to be a relatively simple year-round index for monitoring drought and water supply conditions in Colorado. The SPI supplemented information provided by the PDSI and the surface water supply index (SWSI) developed by Shafer and Dezman (1982). The SPI is based on precipitation alone, whereas the SWSI incorporates snowpack, streamflow, precipitation, and reservoir storage. Calculation of the SPI for the specified time period for any location requires long-term monthly precipitation data for at least 30 years (i.e., the longer the data set, the more reliable the SPI values). The probability distribution function is determined from the long-term record by fitting a function to the data. The cumulative distribution is then transformed using equal probability to a normal distribution with a mean of zero and standard deviation of one so the values of the SPI are expressed in standard deviations (Edwards and McKee, 1997). A particular precipitation total for a specified time period is then identified with a particular SPI value consistent with probability. Positive SPI values indicate greater than median precipitation, whereas negative values indicate less than median precipitation. The magnitude of departure from zero represents a probability of occurrence so that decisions can be made based on this SPI value. An SPI value of less than −1.0 (moderately dry) occurs 16 times in 100 years and an SPI of less than −2.0 (extremely dry) occurs 2–3 times in 100 years.

The fundamental strength of the SPI is that it can be calculated for a variety of time scales. This versatility allows the SPI to be used to monitor short-term water supplies such as soil moisture, which is important for agricultural production, and longer-term water resources such as ground water supplies, streamflow, and lake and reservoir levels, which are important for agriculture and other water users. Colorado uses the SPI information as part of a routine climatic assessment completed by the Water Availability Task Force for Colorado's drought plan. This information is useful for detecting the potential impacts of drought on agriculture and other economic sectors. Determining the linkages between SPI values at different time scales is the subject of considerable research as those involved in monitoring drought seek to identify triggers to initiate various mitigation actions for agriculture and other sectors.

The SPI has a number of advantages over the PDSI. First, it is a simple index and is based only on precipitation. The PDSI calculations are complex because 68 terms are defined as part of the calculation procedure (Soulé, 1992). In spite of the complexity of the PDSI, McKee (personal communication, 1996) believes that the main driving force behind the PDSI is precipitation. Second, the SPI is versatile. It can be calculated on any time scale, which gives the SPI the capability to monitor conditions important for both agricultural and hydrological applications. This versatility is also critical for monitoring the temporal dynamics of a drought, including its onset and termination, which has typically been a difficult task for other indices. Third, because of the normal distribution of SPI values, the frequencies of extreme and severe drought classifications for any location and any time scale are consistent. McKee et al. (1993) suggest an SPI classification scale (table 9.1). Fourth, because it is based only on precipitation and not on estimated soil moisture conditions as is the case with PDSI, the SPI is just as effective during the winter months.

Although developed for use in Colorado, the SPI can be applied to any location with a data set of 30 years or longer. SPI maps for multiple time scales are routinely available on the NDMC Web site (http://drought.unl.edu) in the "drought watch" section and on the Web site of the Western Regional Climate Center (http://www.wrcc.dri.edu/spi/spi.html). The NDMC has disseminated SPI information and software at workshops and through direct e-mail contact with foreign governments, international or-

Table 9.1 Classification of drought categories for the standardized precipitation index (SPI) (according to McKee et al., 1993)

SPI values	Drought category	Time in category
0 to –0.99	Mild drought	34.1%
–1.00 to –1.49	Moderate drought	9.2%
1.50 to –1.99	Severe drought	4.4%
≤–2.00	Extreme drought	2.3%
		~50%

ganizations, and nongovernmental organizations. It is now being used in both operational and research modes in more than 50 countries and has been proven quite effective as part of a comprehensive, integrated early warning system.

The NDMC's experience has been that the SPI detects emerging drought conditions more quickly than the PDSI, a characteristic that is extremely critical in the timely implementation of mitigation and response actions by individuals and governments (Hayes et al., 1999). Many states are using the SPI as part of their efforts to monitor drought and trigger various drought-related mitigation and response actions.

Developing an effective drought monitoring system presents many unique challenges because of the slow onset nature of drought, its spatial extent and duration, and the requirement that multiple indicators and indices be used to properly characterize its severity and potential impacts. In addition to these challenges, the ineffectiveness of drought monitoring systems in the United States and elsewhere is also associated with inadequacies in the systems themselves. First, monitoring systems have often depended on an inadequate network of weather stations. Data from these stations may be reported infrequently (e.g., monthly) so that information is not readily available to decision makers at critical times or decision points. Second, drought monitoring systems are often based on a single parameter or index. Because of the complexities of drought, no single parameter or index can adequately capture the intensity and severity of drought and its potential impacts on a diverse group of users. Each index has strengths and weaknesses, which often vary spatially. Third, the delivery of information products to assess drought severity is often untimely. And fourth, information products are often developed without a clear understanding of user needs, or users are confused about how to apply this information when making critical climate-based decisions.

A comprehensive drought monitoring system has been recommended for many years in the United States (Wilhite et al., 1986; Riebsame et al., 1991; Wilhite and Wood, 1994), but no action on these recommendations had taken place until recently. In 1999 it became apparent that a new approach to drought monitoring was needed to address many of the inadequacies noted above.

U.S. Drought Monitor

The U.S. Drought Monitor was developed through a partnership between the National Drought Mitigation Center at University of Nebraska, the National Centers for Environmental Prediction/Climate Prediction Center (NCEP/CPC) of NOAA, and the USDA's Joint Agricultural Weather Facility (USDA/JAWF). The National Drought Policy Act, passed by the U.S. Congress in the summer of 1998, and the subsequent formation of the National Drought Policy Commission (NDPC) and its working groups in 1999 provided additional momentum to improve drought monitoring efforts in the United States. A working group on monitoring and prediction formed by the NDPC during spring 1999 provided additional opportunities for interactions with a larger group of climatologists throughout the country on drought monitoring issues. The group also helped to form the template for early versions of the U.S. Drought Monitor, first released on an experimental basis on May 20, 1999, as a biweekly product. The Drought Monitor became an operational product in August 1999 when it was officially released at a joint White House press conference conducted by the Department of Commerce and USDA.

The Drought Monitor is maintained on the Web site of the NDMC (http://drought.unl.edu/monitor/_monitor.html). It consists of a map showing which parts of the United States are suffering from various degrees of drought (figure 9.2). The map also accompanies text that describes the drought's current impacts, future threats, and prospects for improvement. The Drought Monitor is derived from several key parameters and ancillary indicators (e.g., fire potential, pasture and range conditions) from different agencies. The six key parameters making up the scheme at this writing are the PDSI, the Climate Prediction Center's Soil Moisture Model (percentiles), the U.S. Geological Survey's daily streamflow (percentiles), percentage of normal precipitation, SPI, and a remotely sensed satellite vegetation health index. Table 9.2 illustrates the drought severity classification system currently used to prepare the map.

The authors of the Drought Monitor rely on different agencies for the inputs required to create the map. The initial draft of the map is produced on Monday and distributed via e-mail to more than 150 climate, water supply, and agricultural specialists throughout the country. These persons are asked to review the map and provide comments. These regional experts often have a better understanding of local situations because of their direct contacts with agricultural, water, and natural resources managers. Based on the comments from these reviewers, the map is revised and a second draft is distributed. The final map is completed by Wednesday night and placed on the Web site at 0730 h (Central Standard Time) each Thursday morning. Previous maps are archived, and users can also see an animation of the past 6- and 12-week periods to better visualize the changing spatial extent and severity of drought conditions across the country. The Drought Monitor map for May 21, 2002, provides an example (figure 9.2).

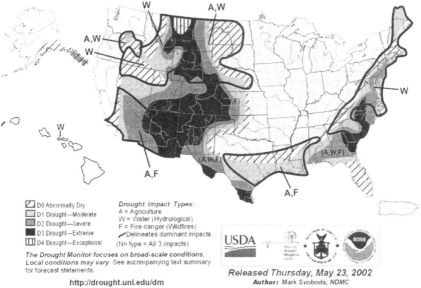

Figure 9.2 U.S. Drought Monitor for May 21, 2002.

The Drought Monitor map classifies droughts on a scale from one to four (D1–D4; i.e., from least to most intense droughts). D4 is a 1-in-50-year event. A fifth category, D0, indicates an abnormally dry area either heading into drought or recovering from it. The Drought Monitor also shows which sectors are experiencing the dominant or primary impacts, using the labels A (agriculture: crops, livestock, range, or pasture), W (water supplies), or F (high risks of fire danger), as shown in figure 9.2. For example, an area shaded and labeled D2 (A) is in general experiencing severe drought conditions that are affecting the agricultural sector more significantly than the water supply sector. The area is not seeing a heightened fire risk in association with this dryness. An area shaded as D2 with no A, W, or F would be experiencing impacts in all three sectors. The final map summarizes all of this information in an easy-to-read format that shows where drought is emerging, lingering, or subsiding.

This map product has been widely accepted and is used by a diverse set of users to track drought conditions across the country. The users include agricultural producers; commodity brokers; water and natural resource managers; congressional delegations; local, state, and federal agencies; and mainstream media. The number of hits on the Drought Monitor Web site increased from 1.75 million in 2000 to more than 5 million in 2002, exceeding our greatest expectations. The Drought Monitor has already become an essential component of the United States's

Table 9.2 Drought Monitor's drought severity classification system

Drought type		Associated ranges of objective indicators					
Category	Description	Palmer drought index	CPC soil moisture model (percentiles)	USGS weekly streamflow (percentiles)	Percent of normal precipitation	Standardized precipitation index	Satellite vegetation health index
D0	Abnormally dry	−1.0 to −1.9	21–30	21–30	<75% for 3 months	−0.5 to −0.7	36–45
D1	Moderate drought	−2.0 to −2.9	11–20	11–20	<70% for 3 months	−0.8 to −1.2	26–35
D2	Severe drought	−3.0 to −3.9	6–10	6–10	<65% for 6 months	−1.3 to −1.5	16–25
D3	Extreme drought	−4.0 to −4.9	3–5	3–5	<60% for 6 months	−1.6 to −1.9	6–15
D4	Exceptional drought	−5.0 or less	0–2	0–2	<65% for 12 months	−2.0 or less	1–5

initiative toward a national drought policy (Wilhite, 2001; Svoboda et al., 2002). For example, the states significantly affected by drought in 2000, 2001, and 2002 used the product to track the severity of drought conditions in their state and make policy decisions on emergency and mitigation actions. Users appreciate a product that simplifies a difficult and complex issue but is based on scientific climatic indices and parameters.

Certainly, the Drought Monitor cannot always capture the local situation accurately. The partners in this activity are striving to improve the science of drought monitoring by improving networks and developing new climatic indices and other assessment tools to make this a better product in the future.

Drought Mitigation

Crop and livestock producers can take numerous actions to reduce the potential impact of drought on their productive capacity. The most obvious of these actions is the introduction of irrigation to provide supplemental water to crops and pastures during drought periods. But irrigation also increases the costs of production through an increase in resource input costs (i.e., energy costs for water applications), and therefore irrigation may not lead to an increase in profit margin. Other mitigation actions that can be used by agricultural producers are the application of climate forecasts and other climate-based information to planting decisions, especially in the preplanting period, leading to changes in crop type/variety, planting date, fertilization practices, and cultivation practices. Livestock producers may also alter the size of herds and grazing plans to reflect shortages in forage and market prices. Increased on-farm storage of forage is an appropriate mitigation strategy to increase the resilience of producers to drought periods.

Summary

The United States continues to experience drought and, as a result, suffers significant economic losses. The U.S. Drought Monitor, developed in 1999 as an experimental product by the NDMC, USDA, and NOAA, has become an operational product that has gained widespread acceptance by scientists, resource managers, policy makers, the business community, and others. The success of this product depends on the collaboration between the principal partners and other organizations, including the user community, and the integration of several climate and water supply indices and other climate parameters into a weekly assessment of drought severity and spatial extent. The Drought Monitor map and the procedures used in its production are transferable to other drought-prone regions, with appropriate modifications.

References

Alley, W.M. 1984. The Palmer drought severity index: Limitations and assumptions. J. Climatol. Appl. Meteorol. 23:1100–1109.

Edwards, D.C., and T.B. McKee. 1997. Characteristics of 20th century drought in the United States at multiple time scales. Climatology Report 97–2. Department of Atmospheric Science, Colorado State University, Fort Collins.

FEMA. 1995. National mitigation strategy. Federal Emergency Management Agency, Washington, DC.

Guttman, N.B. 1998. Comparing the Palmer drought index and the standardized precipitation index. J. Am. Water Resources Assoc. 34:113–121.

Guttman, N.B., J.R. Wallis, and J.R.M. Hosking. 1992. Spatial comparability of the Palmer drought severity index. Water Resources Bull. 28:1111–1119.

Hayes, M.J., M.D. Svoboda, D.A. Wilhite, and O.V. Vanyarkho. 1999. Monitoring the 1996 drought using the standardized precipitation index. Bull. Am. Meteorol. Soc. 80:429–438.

Heddinghaus, T.R., and P. Sabol. 1991. A review of the Palmer drought severity index and where do we go from here? In: Proceedings of the Conference on Applied Climatology 7th, Salt Lake City, Utah, September 10–13, 1991. American Meteorological Society, Boston, MA, pp. 242–246.

Karl, T.R., and R.W. Knight. 1985. Atlas of monthly Palmer hydrological drought indices (1931–1983) for the contiguous United States. Historical Climatology Ser. 3–7. National Climatic Data Center, Asheville, NC.

Kogan, F.N. 1995. Droughts of the late 1980s in the United States as derived from NOAA polar-orbiting satellite data. Bull. Am. Meteorol. Soc. 76:655–668.

McKee, T.B., N.J. Doesken, and J. Kleist. 1993. The relationship of drought frequency and duration to time scales. In: Proceedings of the Conference on Applied Climatology 8th, Anaheim, California, January 17–22, 1993. American Meteorological Society, Boston, MA, pp. 179–184.

McKee, T.B., N.J. Doesken, and J. Kleist. 1995. Drought monitoring with multiple time scales. In: Proceedings of the Conference on Applied Climatology 9th, Dallas, Texas, January 15–20, 1995. American Meteorological Society, Boston, MA, pp. 233–236.

Palmer, W.C. 1965. Meteorological drought. Research Paper no. 45. U.S. Weather Bureau, Washington, DC.

Riebsame, W.E., S.A. Changnon Jr., and T.R. Karl. 1991. Drought and Natural Resources Management in the United States: Impacts and Implications of the 1987–89 Drought. Westview Press, Boulder, CO.

Shafer, B.A., and L.E. Dezman. 1982. Development of a surface water supply index (SWSI) to assess the severity of drought conditions in snowpack runoff areas. Proceedings of the Western Snow Conference, Colorado State University Fort Collins, CO, pp. 164–175.

Soulé, P.T. 1992. Spatial patterns of drought frequency and duration in the contiguous USA based on multiple drought event definitions. Intl. J. Climatol. 12:11–24.

Svoboda, M., D. LeComte, M. Hayes, R. Heim, K. Gleason, J. Angel, B. Rippey, R. Tinker, M. Palecki, D. Stooksbury, D. Miskus, and S. Stephens. 2002. The drought monitor. Bull. Am. Meteorol. Soc. 83:1181–1192.

Wilhite, D.A. 2000. Drought as a natural hazard: Concepts and definitions. In:

D.A. Wilhite (ed.), Drought: A Global Assessment. Hazards and Disasters: A Series of Definitive Major Works (A.Z. Keller, ed.). Routledge, London, pp. 3–18.

Wilhite, D.A. 2001. Moving beyond crisis management. Forum Appl. Res. Public Policy 16:20–28.

Wilhite, D.A., N.J. Rosenberg, and M.H. Glantz. 1986. Improving federal response to drought. J. Climatol. Appl. Meteorol. 25:332–342.

Wilhite, D.A., and D.A. Wood. 1994. Drought management in a changing west: New directions for water policy. IDIC Technical Report Ser. 94–1. International Drought Information Center, University of Nebraska, Lincoln.

Willeke, G., J.R.M. Hosking, J.R. Wallis, and N.B. Guttman. 1994. The national drought atlas. Institute for Water Resources Report 94-NDS-4. U.S. Army Corps of Engineers, Institute for Water Resources, Alexandria, VA.

CHAPTER TEN

Agricultural Drought in North-Central Mexico

JOSÉ ALFREDO RODRÍGUEZ-PINEDA,
LORRAIN GIDDINGS, HÉCTOR GADSDEN,
AND VIJENDRA K. BOKEN

Drought is the most significant natural phenomenon that affects the agriculture of northern Mexico. The more drought-prone areas in Mexico fall in the northern half of the country, in the states of Chihuahua, Coahuila, Durango, Zacatecas, and Aguascalientes (figure 10.1). The north-central states form part of the Altiplanicie Mexicana and account for 30.7% of the national territory of 1,959,248 km². This area is characterized by dry and semidry climates (García, 1981) and recurrent drought periods. The climate of Mexico varies from very dry to subhumid. Very dry climate covers 21%, dry climate covers 28%, and temperate subhumid and hot subhumid climates prevail in 21% and 23% of the national territory, respectively.

About 20 years ago, almost 75% of Mexico's agricultural land was rain-fed, and only 25% irrigated (Toledo et al., 1985), making the ratio of rain-fed to irrigated area equal to 3. However, for the northern states this ratio was 3.5 during the 1990–98 period (table 10.1). Because of higher percentage of rain-fed agriculture, drought is a common phenomenon in this region, which has turned thousands of hectares of land into desert. Though the government has built dams, reservoirs, and other irrigation systems to alleviate drought effects, rain-fed agriculture (or dryland farming) remains the major form of cultivation in Mexico.

Defining a Drought

In Mexico, there is no standard definition for agricultural drought. However, the Comisión Nacional del Agua (CNA; i.e., National Water Commission), which is a federal agency responsible for making water policies, has coined its own definition for drought. This agency determines whether a particular region has been affected by drought, by studying rainfall records

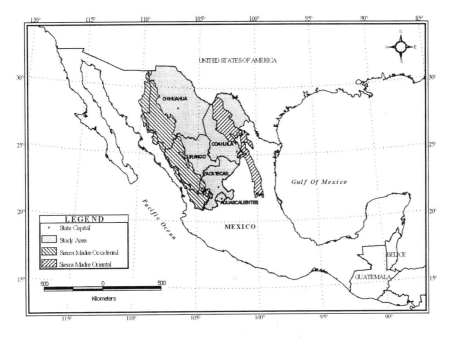

Figure 10.1 Five northern Mexican states selected for the study.

collected from the national climatic network. The national climatic network is spread throughout the country and is managed by the Servicio Meteorológico Nacional (SMN; i.e., National Meteorological Services). The CNA determines, for a municipal region, if the rainfall is equal to or less than one standard deviation from the long-term mean over a time period of two or more consecutive months. If it is, then the secretary of state declares drought for the region. Based on such a definition, drought was declared in 60 municipalities during 2000–01 and in 50 municipalities during 2001–02, out of 67 municipalities in the state of Chihuahua. If a municipality is declared drought-affected, it qualifies for federal funds for drought mitigation.

Table 10.1 The distribution of irrigated and rainfed area in selected states of Mexico

State	Irrigated Area (I)		Rain-fed Area (R)		Area Ratio (R/I)
	(ha)	(%)	(ha)	(%)	
Aguascalientes	35,252	5.5	94,884	4.3	2.7
Coahuila	122,484	19.1	53,765	2.4	0.4
Chihuahua	265,410	41.3	634,129	28.6	2.4
Durango	98,762	15.4	508,880	22.9	5.2
Zacatecas	120,253	18.7	928,559	41.8	7.7
Total/Average	642,161	100	2,220,217	100	3.5

Insurance companies have adopted a different definition of drought for the purpose of payments to farmers during drought periods. These companies define an agricultural drought as a state of insufficient water or rainfall availability such that plants are damaged due to any of the following problems: rickets, dehydration, permanent withering, total or partial desiccation of the reproductive fruit organs or grains, irregular pollination during the embryo formation, or plant death.

Impacts of Droughts

According to the Instituto Nacional de Estadística, Geografía e Informática (INEGI; i.e., National Institute of Statistics, Geography and Informatics; 1999a), maize production in the state of Chihuahua declined from 880,082 tons in 1993 to 303,627 tons in 1995 due to a drought causing $168 million in losses. Likewise, during the 1994–96 period, bean production in Coahuila State decreased from 31,908 tons to 6,035 tons causing an economic loss of about $16 million (INEGI, 2000a). Droughts plagued the state of Chihuahua and other northern states during 1993–2002 and drastically reduced the sowing area, even in irrigated districts. The water level in the 10 largest Chihuahuan reservoirs was reduced to 20% of its total capacity, from 4,259,000 to 851,800 m^3. In Delicias irrigation district, crops were sown only in 25% of the land during spring–summer season due to historically low water levels in the La Boquilla Dam, which permitted the cultivation of only 26,000 ha compared to 104,000 ha during normal years.

Persistent drought also forced a high percentage of productive men of the northwest zone to migrate to the United States in search of jobs. An estimated 16,000 workers moved to the United States, mostly due to lack of agricultural jobs in Mexico (*El Heraldo de Chihuahua*, 2002).

Causative Factors of Drought

Climate in the northern central plateau is mostly arid and semiarid (García, 1981). Orographic barriers and subtropical high-pressure cells are the major factors responsible for formation of dry climate in Mexico (Schmidt, 1989). The two mountain ranges, the Sierra Madre Occidental in the west and the Sierra Madre Oriental in the east, both parallel to the coastlines, are major elongated barriers to moisture flow from the Pacific Ocean and the Gulf of Mexico (figure 10.1). In addition to the orographic barriers, most of the year, the presence of subtropical high-pressure cells over the landmass blocks the progress of a storm and prevents the penetration of moist air (Schmidt, 1986). Because of these conditions that prevail in 70% of the national territory, the northern zone receives less than 40% of the total rainfall, while the southern part with 30% of territory receives the remaining more than 60% of rainfall (Velasco, 1999).

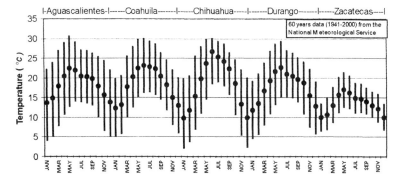

Figure 10.2 Variation in annual temperature for five states of Mexico, for the period from 1941 to 2000. Upper, lower, and middle points of a vertical line represent maximum, mean, and minimum temperatures, respectively.

The five north-central states are typified by hot summers and cold winters (figure 10.2) with moderate to low rainfall (García, 1981; Schmidt, 1986). As a result, the region has a potential annual water loss through evapotranspiration that far exceeds the annual precipitation. The rains occur mostly during the summer months, with extreme annual variability (figure 10.3).

El Niño/Southern Oscillation and Drought

Recent research investigation about the effects of El Niño on the rainfall distribution in northern Mexico shows that this climatic phenomenon has an influence on rainfall patterns. During El Niño years, rainfall was below normal during summers and above normal during winters (Magaña et al.,

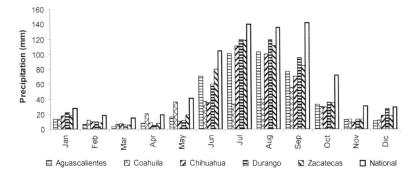

Figure 10.3 Monthly mean rainfall values in the study area and for Mexico showing the main rainfall season from June through September. Coahuila is the state with the smallest value and Durango with the largest one. [Based on 60 years of data (1941–2000) collected from the National Meteorological Service]

2003; Tiscareño et al., 2003). Magaña et al. (2003) also established that during La Niña years, rainfall was normal or little bit above normal during summers, but rainfall showed a declining pattern during winters.

Tiscareño et al. (2003) used variables related to the El Niño/Southern Oscillation (ENSO; sea-surface temperatures and pressure differences in the Pacific Ocean; chapter 3) and correlated them to rainfall data. This allowed the identification of wet and dry periods that coincided with hot and cold periods of the Pacific Ocean waters. Based on this analysis, Tiscareño et al. arrived at the conclusion that the drought of 1982–83, which affected almost the entire country, was a result of the ENSO.

Agricultural production is also affected by the ENSO. This phenomenon has caused a decrease in national grain production since the 1960s (Tiscareño-López, 1999). Tiscareño et al. (2003) also showed considerable interannual yield variations as a result of climate anomalies in Mexico.

Drought Monitoring

Currently drought monitoring in Mexico is in its early stage of development. There are only two institutions that monitor droughts in the country. One of them is the SMN, and the other is the Centro de Investigaciones Sobre la Sequía (CEISS; i.e., Drought Research Center). The SMN monitors droughts by determining precipitation deviations from the normal, and these deviations are regularly published (http://smn.cna.gob.mx/productos/map-lluv/tabla.gif). Essentially, the historical monthly mean and annual mean are compared with the actual monthly and annual precipitation; the deviation is presented as a percentage of mean values. A negative deviation represents precipitation deficit (dry condition or drought), while a positive one indicates surplus (humid condition or wet period).

Recent cooperation efforts among the Mexican Comisión Nacional del Agua, the U.S. National Oceanic and Atmospheric Administration, and the Meteorological Service of Canada have produced the North American Drought Monitor, which presents a monthly drought map of southern Canada, the United States, and Mexico. The drought monitor is the result of a process that synthesizes multiple indices, outlooks, and local impacts into an assessment that best represents current drought conditions. The final outcome of each drought monitor is a consensus of federal, state, and academic scientists (NCDC, 2003). An advantage of this small scale is that the climatic phenomena can be visualized in all its extension; a disadvantage is that the scale is not practical for important drought-affected areas.

The Mexican Drought Research Center (CEISS) applies the standardized precipitation index (SPI; chapter 9; McKee, 1993, 1995). Total monthly precipitation data recorded by the SMN in Chihuahua State is obtained by the CEISS during the first week of each month. The data are processed using specially designed software to compute the monthly SPI. The software stores the state historical precipitation database since 1970, which is automatically updated every month. SPI values are computed on an interval of three months (SPI-3) and 12 months (SPI-12). The SPI-3 is intended to help dry-land farming where production depends on seasonal rainfall, whereas SPI-12 can be used by federal and state agencies that manage state water resources. Finally, the SPI values are loaded on to a Geographic Information System (GIS) for showing spatial distribution of droughts and their severity levels (http://www.sequia.edu.mx/). Further GIS analyses are performed to estimate damages to agricultural areas, other natural resources, and society.

The advantage of using a large scale (state size) is that regional resources affected by drought can be better evaluated. A main disadvantage, however, is that the geographic limits of major drought events affecting large areas might not be included in the drought monitor. The current monitoring activities of the CEISS are confined to the state of Chihuahua. The center has plans to extend its monitoring activities to the northern part of Mexico. The SMN is also planning drought monitoring at the national level by applying the SPI methodology in addition to the precipitation deviation method described earlier.

Predicting Crop Yields using Standardized Precipitation Index

The SPI is a qualitative indicator of drought. To use it for quantitative assessment of agricultural drought, the quantitative relationship between SPI and yields of maize and bean crops was examined for the north-central states of Mexico. Table 10.2 shows the basic variation in main yields of maize and bean during the last two decades.

The rainfall data were collected from two databases (ERIC I, 1995; ERIC II, 1999) of the Instituto Mexicano de Tecnología del Agua (IMTA; i.e., Mexican Institute of Water Technology) to compute SPI-3. In addition, maize and bean yield data were collected from the Sistema Integral de Información Agroalimentaria y Pesquera (SIIAP; i.e., Agricultural and Food Information Integral System) created by the federal agency, Secretaría de Agricultura, Ganadería, Desarrollo Rural, Pesca y Alimentación (i.e., Agency of Agriculture, Cattle, Rural Development, Fishing and Food; SIIAP, 2003) and INEGI (1993, 1995, 1998, 1999a, 1999b, 2000b, 2000c).

Results of these correlations are presented in figure 10.4. The results show positive but low correlation coefficients (from 0.26 to 0.54) for bean and maize yields with SPI-3. In addition to SPI-3, the correlation analysis was attempted using SPI-4, SPI-5, and SPI-6, but the results showed even lower correlations. Aguascalientes, Durango, and Zacatecas are the states that indicate moderate correlations for both crops. Correlations for Chihuahua and Coahuila are the lowest (figure 10.4b,c). SPI-3 seasonal values for the state of Durango show the best correlation with the economic values of bean and maize crops.

A possible reason for low correlations is that the SPI values used in the correlation analysis were averaged for an entire state. One way to improve these results is by splitting the state into regional agricultural areas and

Table 10.2 Basic statistics showing the variation in bean and maize yields in the five north-central states of Mexico

State	Bean (tons/ha)				Maize (tons/ha)			
	Min	Max	Mean	Std. dev.	Min	Max	Mean	Std. dev.
Aguascalientes	0.06	0.36	0.17	0.07	0.17	0.93	0.36	0.18
Coahuila	0.20	0.60	0.40	0.09	0.33	0.91	0.62	0.14
Chihuahua	0.20	0.73	0.40	0.12	0.13	1.15	0.75	0.24
Durango	0.17	0.67	0.40	0.16	0.38	0.95	0.64	0.14
Zacatecas	0.14	0.57	0.39	0.12	0.50	1.02	0.68	0.13
National	0.32	0.57	0.47	0.07	1.39	2.05	1.68	0.16

carry out correlation analysis between regional yields and corresponding SPI values. Another reason for poor relationship is that SPI depends only on rainfall, which is just one of the several variables that affect a crop yield. Though a simple indicator, SPI cannot be used to satisfactorily predict crop yield and therefore cannot be used for quantitative assessment of agricultural droughts in Mexico. There is a need to develop an agricultural drought index that can be used to predict agricultural drought by predicting yields of major crops.

Drought Research Needs

Density of Weather Stations

During the last two decades, there has been a tendency to reduce the total number of weather stations. In 1980, there were about 2000 stations, whereas in 2002 only 800 remained. Also, the gauging network is unevenly distributed, with a higher density in the central states of Mexico and a lower density in the northern (or arid) part of the country. This uneven distribution is understandable considering a more mountainous terrain, more urban centers, and greater population in the south. However, the northern states need more weather stations for better monitoring of droughts. Recent efforts by the CNA and the SMN are focused on bridging the gap by installing automatic weather stations in areas that have difficult terrain.

Automation

Another issue for drought monitoring is transferring the database updating process, from the manual climatic stations to the SMN's main data collection and analysis center. Although there are about 800 weather stations in the nation, only about 160 stations send the daily climatic report to the SMN's analysis center. Therefore, there is an urgent need to expedite this process of data transfer from manual stations to the SMN, so that drought monitoring can be performed on real-time basis. The automation process is being accomplished using 74 automatic stations distributed all over Mexico since 1999. In the five states there are nine stations, six of them located at the mountainous southwestern part of Chihuahua, two in Durango, and one in Zacatecas. For the study of drought it would be better to install such stations at the desertic zones of the arid states (e.g., eastern part of Chihuahua).

Weather Variables

In Mexico, only a few weather variables are monitored daily at the manual stations. These variables include precipitation (24-h total), maximum and minimum temperature, evaporation, wind speed and direction. The new automatic stations linked via satellite (GOES-12) monitor the above variables plus relative humidity, solar radiation, and precipitation. Mean values of 10-min data are sent every 3 h to the SMN headquarters for rapid uploading on-line (http://smn.cna.gob.mx/productos/emas/emas.html).

Some of the common drought indices, such as the Palmer drought severity index (Palmer, 1965) require data on evapotranspiration, recharge, runoff, soil moisture characteristics, and so on, in addition to precipitation and temperature (Lohani and Loganathan, 1997). It would therefore be appropriate to monitor some additional variables, at least at limited number of stations.

Conclusions

Droughts have far reaching social, economic, and political implications. With limited hydrometeorological data available in Mexico, simple drought indicators such as the deviation from the mean rainfall are more suitable than a sophisticated indicator such as Palmer drought severity index. The standardized precipitation index, which depends only on precipitation, appears to be a qualitative tool for drought monitoring, but this index was not suitable for quantitative assessment of agricultural droughts at small scales. The correlation between maize and bean yields with SPI ranged from 0.26 to 0.54. However, the relationship was better in the case of bean than maize. Additional factors that influence crop yields were not analyzed. Low yields of maize and high probabilities of drought occurrence in the north-central states tend to discourage maize farming in these areas. However, bean is a less drought-resistant crop and therefore is more favorable for the region.

Drought research in Mexico is a relatively new but is catching up fast. New models and methodologies are being attempted to better understand the climatic processes responsible for drought occurrence. Currently, methodologies can estimate length, severity (intensity), and spatial distribution

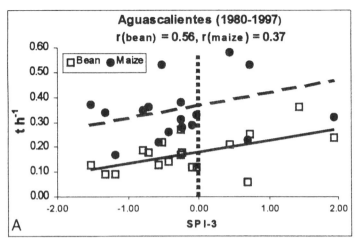
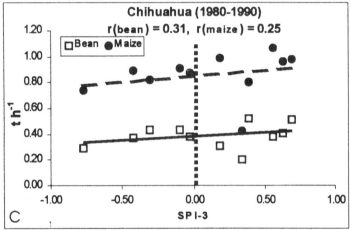
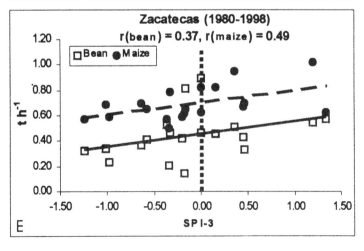

Figure 10.4 Correlations of dryland bean and maize yields (t/ha) with seasonal standardized precipitation index (SPI-3) for the five states.

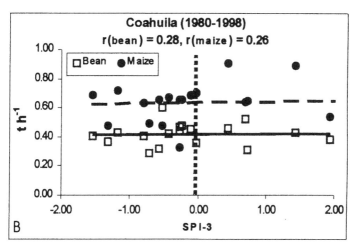

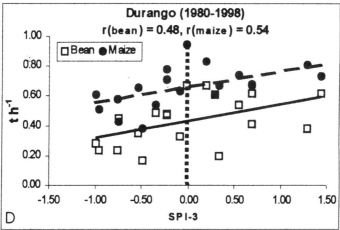

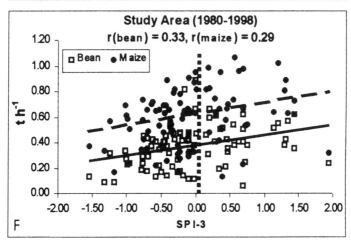

th^{-1} = ton/ha

of droughts on annual and seasonal bases. But there is a need for installing additional weather stations and expediting the process of data transfer from manual stations to the central data analysis center to improve drought monitoring. Installing automatic weather stations should also be considered for arid and desertic regions of Mexico. Also, there is a need to perform this analysis on a daily basis for better monitoring and predicting of droughts in Mexico.

References

El Heraldo de Chihuahua. 2002. Sequía: Daños por 265 Milliones. May 20, 2002, Chihuahua, Chihuahua, Mexico. Available http://www.online.com.mx/el_heraldo/locales/20mayo2002/1.html, accessed April 29, 2004.

ERIC I. 1995. Extractor rápido de información climatológica v. 1.0. CDROM. Instituto Mexicano de Tecnología del Agua, Jiutepec, Morelos, Mexico.

ERIC II. 1999. Extractor rápido de información climatológica v. 2.0. CDROM. Instituto Mexicano de Tecnología del Agua, Jiutepec, Morelos, Mexico.

García, E. 1981. Modificaciones al sistema de clasificación climática de Köppen Enriqueta Garcia de Miranda (ed.). Press offset, Larios, Mexico 18 DF.

INEGI. 1993. El Sector Alimentario en México. Instituto Nacional de Estadística, Geografía e Informática, Aguascalientes, Aguascalientes, Mexico.

INEGI. 1995. El Sector Alimentario en México. Instituto Nacional de Estadística, Geografía e Informática, Aguascalientes, Aguascalientes, Mexico.

INEGI. 1998. El Sector Alimentario en México. Instituto Nacional de Estadística, Geografía e Informática, Aguascalientes, Aguascalientes, Mexico.

INEGI. 1999a. El Sector Alimentario en México. Instituto Nacional de Estadística, Geografía e Informática, Aguascalientes, Aguascalientes, Mexico.

INEGI. 1999b. Estadísticas históricas de México, vol. I, 4th ed. Instituto Nacional de Estadística, Geografía e Informática, Aguascalientes, Aguascalientes, Mexico.

INEGI. 2000a. Indicadores de desarrollo sustentable en México. Agenda 21. Tierras afectadas por la desertificación. Instituto Nacional de Estadística, Geografía e Informática, Aguascalientes, Aguascalientes, Mexico.

INEGI. 2000b. Agenda Estadística de los Estados Unidos Mexicanos. Cuadro 1.2. Instituto Nacional de Estadística, Geografía e Informática, Aguascalientes, Aguascalientes, Mexico.

INEGI. 2000c. El Sector Alimentario en México. Instituto Nacional de Estadística, Geografía e Informática, Aguascalientes, Aguascalientes, Mexico.

Lohani, V.K., and G.V. Loganathan. 1997. An early warning system for drought management using the Palmer drought index. J. Am. Water Resources Assoc. 33:1375–1385.

Magaña, V.O., J. L. Vasquez, J.L. Pérez, and J.B. Pérez. 2003. Impact of El Niño on precipitation in México. Geofís. Int. 42:3:313–330.

McKee, T.B., N.J. Doesken, and J. Kleist. 1993. The relationship of drought frequency and duration to time scales (preprints). Eighth Conference on Applied Climatology, January 17–22, Anaheim, California, pp. 179–184.

McKee, T.B., N.J. Doesken, and J. Kleist. 1995. Drought monitoring with multiple time scales. In: Proceedings of the Ninth Conference on Applied Climatology. American Meteorological Society, Boston, pp. 233–236.

NCDC. 2003. North American Drought Monitor. National Climatic Data Center. Available http://www.ncdc.noaa.gov/oa/climate/monitoring/drought/nadm/#overview, accessed October 31, 2003.

Palmer, W.C. 1965. Meteorological drought research paper no. 45. Department of Commerce Weather Bureau, Washington, DC.

Schmidt, R.H. Jr. 1986. Chihuahuan Climate, Chihuahuan Desert—United States and Mexico. In: J.C. Barlow, A.M. Powell, and B.N. Timmermann (eds.), Second Symposium on Resources of the Chihuahuan Desert Region. Chihuahuan Desert Research Institute, Alpine, Mexico, pp. 40–63.

Schmidt, R.H. Jr. 1989. The arid zones of Mexico: Climatic extremes and conceptualization of the Sonoran Desert. J. Arid Environ. 16:241–256.

SIIAP. 2003. Sistema Integral de Información Agroalimentaria y Pesquera. Secretaría de Agricultura, Ganadería, Desarrollo Rural, Pesca y Alimentación. Available http://www.siea.sagarpa.gob.mx/ar_comagri.html, accessed October 16, 2003.

Tiscareño-López, M. 1999. Pronóstico climático y de cosechas para el ciclo Primavera-Verano 1999 en México. Inifap/Produce. SAGAR, Mexico, D. F. Marzo-Abril.

Tiscareño-López, M., C. Izaurralde, N.J. Rosenberg, A.D. Baéz-González, and J. Salinas-García. 2003. Modeling El Niño Southern Oscillation climate impact on Mexican agriculture. Geofís. Int. 42:3:331–339.

Toledo, V.M., J. Carabias, C. Mapes, and C. Toledo. 1985. Ecología y autosuficiencia alimentaria, 3rd ed. Siglo Veintiuno Editores, Mexico, DF.

Velasco, I., 1999. Severe droughts becoming recurrent, more persistent in Mexico. Drought Network News 11:1. Available http://www.drought.unl.edu/pubs/dnn/dnnarchive.htm, accessed October 5, 2003.

CHAPTER ELEVEN

Monitoring Agricultural Drought in the West Indies

ABRAHAM ANTHONY CHEN, TREVOR FALLOON,
AND MICHAEL TAYLOR

The core of the West Indies (figure 11.1) consists of the archipelago of islands that stretches southeast from the Yucatan and Florida peninsulas to Venezuela. Generally the term "West Indies" is synonymous with the "Antilles" and is therefore often used to refer to the islands that compose the Greater and Lesser Antilles. The islands of the Greater Antilles include Cuba, Hispaniola, Puerto Rico, and Jamaica—all located in the north Caribbean Sea—while the Lesser Antilles encompasses the smaller islands found to the south and east (figure 11.1). In total, the West Indies embraces about 25 island territories.

There are complex mountain ranges in the Greater Antilles, such as the Blue Mountains (2257 m) in central Jamaica and the Pico Duarte (3175 m) in the Dominican Republic, smaller volcanic peaks in the northeast island arc, and low-lying islands composing the remainder of the Lesser Antilles. The variation in local topography contributes significantly to the general rainfall pattern across the West Indian islands, as the windward sides of the larger and more mountainous islands are rainy and windswept, while the leeward sides are drier. In comparison, the low-lying eastern islands receive much less rainfall due to their lack of topographic relief and are much more dependent on seasonal rains.

It is, however, the location of the West Indian islands between the permanent high pressure zone of the subtropical north Atlantic (the Azores high) and the equatorial trough of low pressure that gives rise to the mean monthly West Indian rainfall depicted in figure 11.2.

Early in the year (December through March) and for a brief period in July, the Caribbean is dominated by subsidence from the inner zone of the Azores high and is at its driest. Rainfall during this period (barring July) is largely from the intrusion of fronts from North America. By the onset of the rainy season, however, the Azores high drifts farther north, resulting

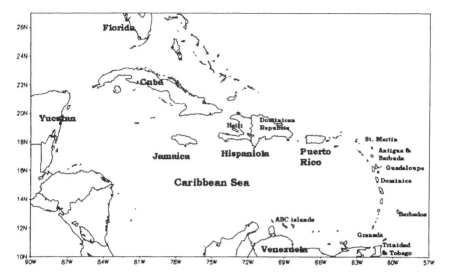

Figure 11.1 Map of the West Indies.

in weakened trade winds. At the same time, the Caribbean Sea warms up. These weakened trades and warmer sea-surface temperatures favor the convective development of easterly waves (weak troughs of low pressure). The waves begin their regular (every 3–5 days) trek across the tropical Atlantic and through the region during mid-June. Later, the easterly waves produce a summer with maximum rainfall and are largely responsible for the tropical storms and hurricanes that frequent the region and augment rainfall totals during this period.

Causative Factors of Drought

Drought, as defined in the West Indies, is most often related to disruptions of the seasonal rainfall cycle. The primary phenomenon associated with such seasonal disruptions is the El Niño/Southern Oscillation (ENSO). A detailed description of ENSO is provided in chapter 3. Other factors, such as the decadal fluctuations in Caribbean rainfall amounts, can also cause drought (Taylor et al., 2002). We do not consider these other factors due to the secondary role they play in producing interannual fluctuations in the West Indian rainfall.

The ENSO phenomenon impacts West Indian rainfall both directly and through a lag. When an El Niño event occurs, drier than normal conditions characterize the West Indies during the later months of the rainfall season, just before peak anomalies in the winter. Pacific sea-surface temperature (Giannini et al., 2000; Taylor et al., 2002). Analysis of National Meteorological Office data for Jamaica reveals that since 1960, El Niño events were the cause of many island-wide meteorological droughts in 1965, 1969,

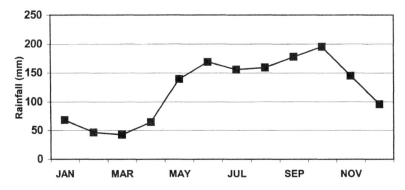

Figure 11.2 Rainfall climatology of the Caribbean in millimeters/month, showing the early (April–July) and late (August–November) rainfall season.

1972, 1976, 1982–83, 1991, and 1997. The worst drought conditions occurred in 1976 and 1991, when the island received 72% and 73%, respectively, of normal total annual rainfall with respect to a 30-year mean. Similar responses to El Niño events have also been noted elsewhere in the West Indies (e.g., rainfall totals of 150 mm less than average in Trinidad between August and October during the 1997–98 event; Pulwarty et al., 2001).

Because the sea-surface temperatures in the Atlantic and Caribbean lag those in the eastern equatorial Pacific by 4–6 months, La Niña events (cold equatorial Pacific waters), which peak in boreal winter, can also induce drought conditions in the early rainfall season of the following year. If the La Niña phenomenon persists beyond the winter months, then sea-surface temperatures in both the eastern equatorial Pacific and Caribbean Sea are simultaneously below normal during the early portion of the West Indian rainfall season. Chen and Taylor (2002) show that such conditions at that time of year are often coincident with droughts in the region. Island-wide meteorological droughts in Jamaica in 1971, 1974, 1975, 1985, 1989, and 2000 (J. Spooner, National Meteorological Office, Kingston, Jamaica, personal communication, 2002) are attributable to this scenario.

Some islands in the West Indies are large enough to generate their own weather due to orographic effects, and hence some territories are more prone to droughts than others. As an example, during the El Niño year of 1997, the regions of Jamaica that experienced drought conditions during the late rainfall season (August–October) were situated mostly on the southern and eastern half of the island in the parishes of St. Thomas, Kingston and St. Andrew, and St. Catherine (figure 11.3). Similarly, during the La Niña year of 2000, it was primarily St. Thomas and St. Catherine that were most severely affected in the early rainfall season (April–July), though the drought also extended to the west, on the southern plains of Clarendon and St. Elizabeth, and to the parish of Trelawny in the north of the island.

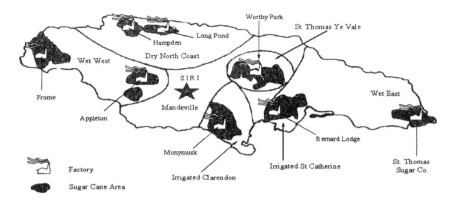

Figure 11.3 Ecological regions of the Jamaican sugar industry.

Agriculture in the West Indies

Agriculture in the West Indies is heavily dependent on the seasonal rainfall. This is true even of the larger territories of the Greater Antilles. The planting/harvesting cycles of both small farmers and large-scale agricultural producers revolve around the wet and dry seasons, especially since irrigation is hardly an option for the small farmers due to high implementation and maintenance costs. Estimates of agricultural production in Jamaica indicate that 95% of all domestic agricultural production is rain-fed (C. Thompson, personal communication, 2002).

Agriculture is important to the economies of the West Indian territories. Sugarcane, banana, and rice are the most important crops from the standpoints of export earnings, contributions to national economies, and the number of persons employed. Of the three, rice is largely produced in Guyana and Trinidad. None of the other smaller islands cultivate rice to the extent of even self-sufficiency.

Banana, the second most important agricultural crop of the West Indies, is the backbone of the economies of the Windward Islands of Dominica, Grenada, St. Vincent, St. Lucia, and the French island territories of Martinique and Guadeloupe. Banana cultivation accounts for half of the export earnings and generates employment ranging from 23 to 34% of the total agricultural employment in these islands (Dearden, 1996).

Sugarcane is the main agricultural crop in the West Indies. Several West Indian economies were founded on sugarcane, and though significant plantings have disappeared from islands such as Antigua and Grenada, sugarcane still remains the dominant crop in Barbados, Guyana, Trinidad and Tobago, St. Kitts, Belize, and Jamaica in the English-speaking Caribbean. Among Spanish-speaking territories, until the 1960s and early 1970s, Cuba was the world's largest sugar exporter, and the Dominican Republic still is a major producer. Sugarcane also dominates the agricultural sectors of the French islands of Guadeloupe and Martinique.

The West Indian territories composing the Sugar Association of the Caribbean (Barbados, Belize, Guyana, Jamaica, Trinidad and Tobago, and St. Kitts) estimate earnings from the 2000–01 sugar crop to be $261,370,000 from a total export of 638,017 tons. The 1999–2000 earnings were $310,000,000 from 685,877 tons exported (The Gleaner, 2001), and the major earners were Guyana, Jamaica, Belize, and Trinidad and Tobago. The importance of sugar is further illustrated by the fact that, for Jamaica, sugar supercedes the combined economic value of all other leading crops—banana, coffee, and citrus (Planning Institute, 2000). In Jamaica, where unemployment is about 15.5% (Planning Institute, 2000), sugar provides direct employment to some 36,500 persons. Approximately 46,000 ha are under cane cultivation, with an annual sugar output of just greater than 200,000 tons from nearly 2.3 million tons cane (SIRI, 1999). The estimated economic losses from poor crop yields, crop failure, and livestock fatality due to widespread drought in Jamaica, attributed to the 1997 El Niño event, was $1 million (Office for Disaster Preparedness and Emergency Management, 1999).

Drought Monitoring in the West Indies

Agricultural drought monitoring in the West Indies is largely restricted to determining the presence of ongoing agricultural drought. It is generally the responsibility of the Ministries of Agriculture in the individual territories or other appointed bodies. In all cases monitoring of agricultural drought by these bodies consists of monitoring two sets of indices: a set of simple rainfall indices to determine when meteorological drought has set in, and a set of equally simple agricultural production indices to determine when production levels are abnormal with reference to a base year.

The island of Jamaica is divided into 14 parishes, which are further subdivided into 4–5 agricultural extension areas each. It is the responsibility of an agricultural extension officer to collect both the rainfall indices and the agricultural production indices. The subset of rainfall indices are reported by an agricultural extension officer to the parish supervisor on a monthly basis. The indices include the number of rainfall days per month, the total rainfall received for the month, the highest rainfall received, and the days on which it was received. These are supplemented with similar data representing the means of the same indices for the same region and for the same time of year as calculated from data collected over a number of years. A comparison of the collected and mean values is used to determine if a meteorological drought occurred during the month.

Similarly, a subset of production indices is also collated by crop type for the region. These include hectares planted during the month and to date, hectares harvested during the month and to date, hectares currently growing, yields per acre, and total production in tons. As with the meteorological data, these are also supplemented with similar data for both the previous year and for a reference year which was deemed a good production year for the region on the basis of yields. The comparison between the collected and mean values determines anomalies in the agricultural production levels.

It is on the basis of both the reported monthly values of the rainfall and production indices that the ministry determines whether an agricultural drought is in the making. If the meteorological drought has continued (on the basis of previous month's reports) and there is at least a 5% reduction in production levels, then a drought alert is issued for the region. If the drought condition persists for consecutive months, it leads to a severe drought.

Agricultural drought is monitored in the other islands of the West Indies by monitoring both rainfall and production indices and by comparing these indices. This is true of both government-designated institutions that monitor domestic crop production and the privately funded institutes or large-scale producers of export crops. In some cases, however, the Jamaican subset of indices is augmented by one or more additional and easily measured indices as deemed relevant by the particular territory's monitoring institute.

It is generally true that for the West Indian region, the use of more complex drought indices such as the Palmer drought severity index, the standardized precipitation index (chapter 9), rainfall percentiles, or the crop moisture index, is not common in agricultural drought monitoring and management. Departures from the norm (calculated or intuitively deduced) are largely used to determine the presence of ongoing drought and to develop mitigation strategies.

Regional Drought Monitoring

In 1991, by an Agreement of Heads of Government of the Caribbean Community, the Caribbean Disaster Emergency Response Agency (CDERA; www.cdera.org) was established. The primary function of the CDERA is to make an immediate and coordinated response to any disastrous event affecting any participating state, once the state requests such an assistance.

Drought is one of the disasters to which CDERA reacts. Using the knowledge of the impact of El Niño and La Niña in the region, CDERA (in conjunction with other institutions) issues to the heads of government of its 16 participating territories information regarding the likelihood of drought conditions for specific periods of the El Niño event. The information also includes the potential impacts of the drought on various sectors including agriculture and ways to mitigate them.

Drought Monitoring for Sugarcane

Figure 11.3 shows the major sites and ecological regions in which sugar is produced in Jamaica. Roughly a third of the industry occupies the relatively arid and southern coastal lowlands, where irrigation is a necessity, and where average annual rainfall is just > 800 mm (table 11.1). The rest of the industry, however, is rain-fed and typically receives between 1300 and 1600 mm of rainfall per annum. Even in the irrigated areas there is a dependence on seasonal rainfall, which is estimated to be responsible for up to 50% of yields in good rainfall years and 30% in poor rainfall years (SIRI, 1973).

A boost in early growth of sugarcane usually occurs with the April–June (early season) rains. The July–September period is regarded as the "boom period of growth," because during this period the cane grows at its fastest rate (Shaw, 1963). Lack of rains during this period has a telling effect on sugarcane growth. In contrast, good rainfall during this period can dramatically compensate for severe growth retardation from drought in the earlier part of the year, as happened in 1998 at the Worthy Park estate, when the cane yield of 96.81 tons per hectare (tc/ha) was one of the highest ever recorded. The Frome cane-growing area, which lies in the wet west and receives nearly 2000 mm of rainfall per annum, shows a tendency toward bumper crops in years following below normal April–June rainfall (Felix, 1977).

The industry attempts to harvest as much of the crop as possible between December and April, taking advantage of the cool and dry period (figure 11.2), which favors sugarcane ripening (Biswas, 1988). The onset of rains in April–May usually triggers a sharp deterioration in sugar content. Often, when the dry period is too prolonged due to drought in the early rainfall season, cane yield begins to be adversely affected.

Four climate-related stresses lead to a poor harvest of sugarcane in Jamaica: (1) below-normal July–September rainfall, (2) below normal November–May rainfall, resulting in poor establishment of plantings and retardation of early growth, (3) above-normal temperatures and excessive rainfall during November–March (i.e., four to six weeks before harvest, and (4) excessive spring rains in poorly drained, flood-prone areas. These anomalies in seasonal rainfall and temperature are attributed to El Niño or La Niña events (Spence and Taylor, 2002; Stephenson and Chen, 2002). To illustrate this we briefly consider the production records from two sugar estates in Jamaica: Worthy Park, which is situated in the rain-fed area of St. Catherine in the Central Uplands of Jamaica (table 11.1 and figure 11.3).

An examination of the last 50 years of yield data for sugarcane at Worthy Park revealed that El Niño and La Niña years reduced yields in subsequent years. This happened in 10 different years, and the yield reduction was about 8% from the normal yield of 86 tons/ha (table 11.2). For example, the June–October rainfall in 1991 was only 46% of normal rainfall, and the December 1991 to February 1992 rainfall was only 61% of normal rainfall, and this significantly reduced yields in 1992. In addition, the minimum temperature was 2.8% above normal. All these weather anomalies were due to the 1991–92 El Niño. The 1965 El Niño caused

Table 11.1 Quarterly and annual rainfall (mm), 30-year means, in the major ecological sugarcane-growing areas of Jamaica

Ecological area	Jan.–March	Apr.–June	July–Aug.	Sept.–Dec.	Total	% Cane
Irrigated area	93	246	213	276	828	33.5
Wet west	210	514	540	385	1649	39.4
Dry north	198	335	315	359	1207	12.3
Wet east	261	409	484	637	1791	6.2
Central uplands	192	480	442	414	1528	8.6

268% of normal rainfall during the spring of 1996. In the following year, the sugarcane yield declined significantly.

In Trinidad, sugarcane is grown mainly on the drier western and central plains of the island, with a growing cycle similar to that of Jamaica. Pulwarty et al. (2001) found a correlation between ENSO values and sugarcane yield.

Monitoring Agricultural Drought in the Sugar Industry

It is evident that the effects of a drought on sugarcane may either be positive or negative, depending on the stage of the growth cycle at which it occurs. The Sugar Industry Research Institute (SIRI) in Jamaica collects and collates data on production and productivity factors, including rainfall, and other biophysical factors and management practices. A careful watch is kept on the development of drought, and attempts are made to predict its likely effect on sugar production.

SIRI (1976) reported that disastrous droughts occurred in 1976, when an average of 26.2 mm of rainfall fell on the irrigated area (and in some farms as little as 17.8 mm) against a long-term annual total of 115.9 mm for the region.

Drought Mitigation

At the national level, drought mitigation practices include drought preparedness, drought assistance, and drought rehabilitation. Usually the ministries of agriculture deliver educational materials to farmers to help them prepare for drought. A variety of mechanisms are adopted, such as area meetings with agricultural stakeholders, farm visitations, newsletters, and the use of radio and television advertisements. Education involves informing and preparing the agricultural stakeholders for the onset of dry conditions, highlighting the characteristics of an ongoing drought, and teaching or encouraging alternative farming practices for use in drought-prone areas. Four practices are generally encouraged in the West Indies: (1) mulching or covering the soil surface with dried plant matter to prevent evaporation of the surface water; (2) planting a "cover crop" such as peas or

Table 11.2 Worst harvests at Worthy Park estate in Jamaica with possible stress types and causative conditions

Year	% yield deviation from 1950–99 average yield	Stress 1 (% or normal rainfall)	Stress 2 (% of normal rainfall)	Stress 3 (% of normal min. temp)	Stress 4 (% of normal rainfall)	Prevailing conditions
1951	31	54%; 1950 (March–May, July)	27%; 1951 (Jan.–March)			1950–51 La Niña
1952	30	58%; 1951 (June, Sept, Oct.)	48%; Nov. '51–Feb. '52			1951–52 El Niño, Hurrican Charlie-Aug. 51
1977	23	64%; 1976 (Apr.–Oct.)	43%; Nov. '76–March '77	100.6%; Nov. '76–March '77		1976–77 El Niño
1987	21	62%; 1986 (May, July–Oct.)	59%; Nov. '86–March '87	102.8%; Jan. '87–March '87		1986–87 El Niño, Flood rains in June 86
1989	20	64%; 1988 (June, Aug, Oct.)	61%; Nov. '88–Jan. '89			1987–88 El Niño, Hurricane Gilbert-Sept. 88
1992	19	46%; 1991 (June–Oct.)	61%; Dec. '91–Feb. '92	102.8%; Nov. '91–March '92		1991–92 El Niño
1967	13				268%; 1966 (May, June)	1965–66 El Ninño
1983	11	49%; 1982 (June, Aug.–Dec.)		100.7%; Nov. '82–Dec. '82		1982–83 El Niño
1991	10	63%; 1990 (May, June, Aug. Sept.)	51%; Dec. '90–Apr. '91	104.4%; Nov. '90–March '91		1990–92 El Niño
1986	7	67%; 1985 (March–July, Sept. Oct.)				1984–85 La Niña

pumpkins in regions of porous (e.g., sandy-loam) soil to help retain surface water; (3) planting deep-rooted crops along the edges of hilly terrain to prevent soil erosion and loss of surface water; and (4) the use of intercropping, where tall crops are planted to shade a shorter crop.

In the advent of prolonged drought, drought-assistance strategies come into play. Often this involves trucking water to drought-stricken areas. A Rapid Response Unit under the Jamaican Ministry of Water and Housing is responsible for installing water tanks in households and on farms (at reduced cost) to mitigate drought impacts on humans, livestock, and crops. Finally, in the case of severe drought causing total loss of a season's crop, many West Indian governments implement drought rehabilitation programs, which include supply of fertilizers, seeds, land preparation assistance, planting materials, livestock feed, or the installation of mini-irrigation systems.

Between August 1997 and June 1998, Guyana experienced drier (75% of normal rainfall) than normal conditions attributable to El Niño. Sugarcane production declined by 35%, and rice production for the spring season fell by 37%. Upon the request of the Guyanese government, CDERA assessed the severity of the drought, its effects, and the resulting needs of the sector, and formulated response strategies on a local, regional and international scale. These responses included coordinating the purchase and installation of water storage facilities and pumps for farmers in the most severely affected areas; acquiring paddy seeds for replanting during the next season; and obtaining grants and food rations for affected families.

Following the devastating effects of the 1997 El Niño event, the government of Jamaica, through the Office of Disaster Preparedness and Emergency Management (ODPEM; www.odpem.org.jm), formulated a drought plan and activated a drought response team. The drought plan addresses domestic, agricultural, and industrial water needs throughout Jamaica, utilizing a multiple-agencies approach to drought management. It is divided into sections according to the disaster cycle, under the headings of preparedness, mitigation, emergency response, rehabilitation, and development. It calls for the activation of the response team during the period identified as a meteorological drought by the meteorological office.

For drought preparedness activities, the Climate Studies Group Mona was assigned the task of issuing advisories of impending drought to the ODPEM and the Meteorological Office of Jamaica when appropriate. The meteorological office, which maintains the water and climatological station network in Jamaica and monitors meteorological drought indices, was assigned the task of issuing warnings and alerts to the ODPEM and other agencies, including the Ministry of Agriculture and the Rural Agricultural Development Agencies (www.radajamaica.com.jm).

Conclusions and Recommendations

With agriculture being a major contributor to the economies of the individual territories of the West Indies, monitoring the onset and evolution of agricultural drought plays a significant role. Agricultural drought monitoring in the region is largely carried out by agricultural ministries in each territory and relies on collating and monitoring both rainfall and agricultural production indices. Drought mitigation strategies generally encompass sensitizing agricultural producers to ongoing drought and teaching water retention strategies, providing drought assistance in the form of water, or providing agricultural necessities to rehabilitate the farmer after a devastating drought. The mitigation strategies are largely responsive rather than proactive.

Only a few regional entities such as the Caribbean Institute for Meteorology and Hydrology (http://inaccs.com.bb/carimet/top.htm) and The Climate Studies Group Mona of the University of the West Indies (http://www.mona.uwi.edu/physics/Research/csg/index.htm) are actively involved in such endeavors, and their research is in the early stages. Encouragement of such research efforts in addition to other research investigating links between climate and crop yields would enable the development of truly proactive drought monitoring and mitigation system for the West Indies. Though few West Indian territories possess such a plan, the Jamaican government provides a model for the structure and potential use of such a plan. Finally, we note that the potential exists for the use of crop models and Geographical Information Systems (GIS) to assist in drought monitoring and management. Neither of these tools are currently employed on a significant scale in the territories of the West Indies.

ACKNOWLEDGMENTS Worthy Park data were obtained from the Sugar Research Institute, Mandeville, Jamaica, through A. Amarakoon of the physics department, University of the West Indies. We thank C. Thompson of the Ministry of Agriculture (Jamaica; http://www.moa.gov.jm) for many useful discussions.

References

Biswas, B. 1988. Agroclimatology of the sugar cane crop. WMO Technical Note 193. World Meteorological Organization, Geneva.

Chen, A.A., and M. Taylor. 2002. Investigating the link between early season Caribbean rainfall and the El Niño+1 year. Intl. J. Climatol. 22:87–106.

Dearden, S. 1996. The EU Banana Regime and the Caribbean Island Economies. DSA European Development Policy Study Group, Discussion Paper no. 1. Manchester Metropolitan University, Manchester, UK.

Felix, F. 1977. Effect of rainfall on cane yield at Frome. J. Jamaica Assoc. Sugar Technol. 28:24–31.

Giannini, A., Y. Kushnir, and M.A. Cane. 2000. Interannual variability of Caribbean rainfall, ENSO and the Atlantic Ocean. J. Climate 13:297–311.

Office for Disaster Preparedness and Emergency Management. 1999. Drought Review 1997. Mitigation Planning and Research Division. Camp Road, Kingston, Jamaica.

Planning Institute. 2000. Value of agricultural exports, 1995–2000. Economic and Social Survey. Grenada Way, Jamaica, sec. 8.2. Kingston, Jamaica.

Pulwarty, R.S., J. Eischeid, and H. Pulwarty. 2001. The impact of El Niño on rainfall and sugar production in Trinidad. In: Proceedings of the West Indies

Sugar Technologists Conference, April 23–27, Trinidad and Tobaga. Available http://wistonline.org.

Shaw, M.E.A. 1963. The growth pattern and yield of annual cane planted at different seasons, and the effects of nitrogen and irrigation treatments. In: Proceedings of the Annual Conference of the Jamaica Association of Sugar Technologists, pp. 48–60.

SIRI. 1973. Factors affecting the 1973 crop. Annual Report of the Sugar Industry Research Institute, Kendal Road, Mendeville, Jamaica, pp. 2–3.

SIRI. 1976. Weather. Annual Report of the Sugar Industry Research Institute, Kendal Road, Mendeville, Jamaica, pp. 3–7.

SIRI. 1999. Agricultural Production and Extension Services. Annual Report of the Sugar Industry Research Institute, Kendal Road, Mendeville, Jamaica, p. 34.

Spence, J., and M. Taylor. 2002. Examining the effects of concurrent SST anomalies on Caribbean rainfall. In: Proceedings of 13th Symposium on Global Change and Climate Variations, January 13–17, 2002, Orlando, Florida. American Meteorological Society, Boston, pp. 149–151.

Stephenson, T., and A. Chen. 2002. Analyzing and understanding climate variability in the Caribbean Islands. In: Proceedings of 13th Symposium on Global Change and Climate Variations, January 13–17, 2002, Orlando, Florida. American Meteorological Society, Boston, pp. 152–154.

Taylor, M.A., D.B. Enfield, and A.A. Chen. 2002. The influence of the tropical Atlantic vs. the tropical Pacific on Caribbean rainfall. J. Geophys. Res. 107 (C9)3127, doi:10.1029/2001JC001097.

The Gleaner. 2001. Caribbean sugar earning decline. October 22, p. C6. Gleaner Company Ltd., Kingston, Jamaica.

CHAPTER TWELVE

Agricultural Drought Phenomenon in Latin America with Focus on Brazil

ORIVALDO BRUNINI, PEDRO LEITE DA SILVA DIAS,

ALICE M. GRIMM, EDUARDO DELGADO ASSAD,

AND VIJENDRA K. BOKEN

Latin America encompasses a vast territory between 12°30'N and 55°30'S latitude and between 29°W and 82°W longitude. This subcontinent has 13 countries with complex climatic conditions. Extremely humid weather is typical closer to the equator, while semiarid, arid, and desertic conditions prevail in the Bolivian and Chilean high plains (figure 12.1).

The wide variation in climatic conditions leads to distinct agricultural conditions across Latin America. For example, forests, equatorial fruits, and perennial vegetation exist throughout the Amazonian region. Farther from the equator, toward the Andes and at higher latitudes, there is a noticeable change in agricultural systems. There is a greater emphasis on growing cereal/grain crops in Argentina and Brazil.

The countries that compose the Amazon River basin experience a higher amount of annual precipitation, and drought is not a characteristic phenomenon there, except during high-intensity El Niño years (Marengo et al., 2001). In contrast, drought is a regular event commonly observed in parts of Peru, Chile, Paraguay, Argentina (Scian and Donnari, 1996), Uruguay, and Brazil. The Atacama Desert in Chile is one of the most arid regions on the earth, where the average annual precipitation is as low as 0.8 mm in Arika or even 0.5 mm in other regions of this desert.

Figure 12.2 provides a more detailed description on climatic conditions of Brazil. Although the average annual precipitation in the northeastern region is less than 300 mm, it exceeds 2500 mm in some other regions of Brazil (Grimm, 2003). Agricultural operations take place during the rainy season (March–October). The northeast region is drought prone, but the central, west, and southeast regions are traditionally grain-producing regions. In the northeast and central-west regions, water deficiency is higher, which seriously affects food production. Table 12.1 shows production losses in Brazil due to climate anomalies including droughts that occurred

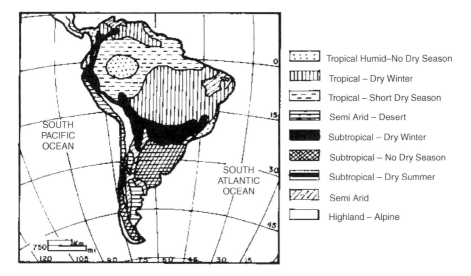

Figure 12.1 Spatial distribution of climatic regions of South America (http://www.fao.org).

during 1978–1986 (Mota, 1979) and 1991–1994 (Rossetti, 2001). About 33% (about 50% in the northeast region) of these losses were attributed to droughts. Maize production also significantly declined due to drought that occurred during 1990–91, 1993–94, 1996–97, and 1997–98 (figure 12.3).

Causative Factors of Agricultural Drought

Precipitation and Land Management

The causes for agricultural drought in Latin America can be analyzed from two points of view. In the first case, drought is intrinsically related to precipitation deficit during a given period of time, which negatively affects crops and water resources (Brunini et al., 2002). Second, the poor management of agricultural lands and exploitation of natural resources have turned highly productive areas into degraded lands. The decaying of organic matter increases the risk of drought due to the decrease in water retention capacity of the soil. Also, soil compaction contributes to drought occurrence.

El Niño/Southern Oscillation

The most important meteorological phenomenon that causes large-scale droughts (both spatially and temporally) is El Niño/Southern Oscillation (ENSO) and La Niña (Alves and Repelli, 1992; Cunha et al., 2001; Marengo et al., 2001; chapter 3). In Brazil, the northern part of the northeastern

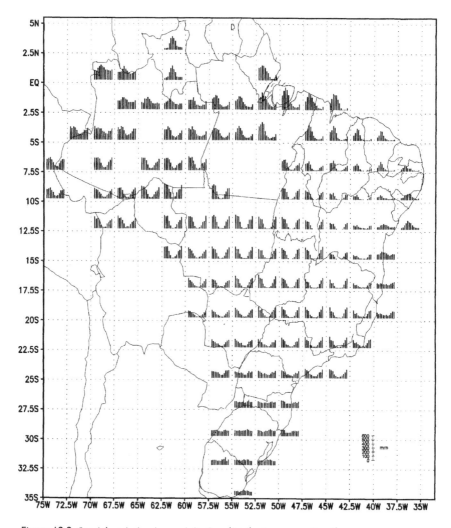

Figure 12.2 Spatial variation in precipitation distribution across Brazil.

region, the eastern part of the Amazon (tropical zone), and the southern region (extra-tropical zone) are influenced most by the ENSO phenomenon. Normally, during the positive phase of the ENSO, the northeastern region of Brazil experiences a decrease in seasonal rainfall and an increase in the intensity of the droughts, but the south of Brazil experiences an increase in precipitation (Cunha, 1999; Martelo, 2000).

Monitoring Agricultural Drought

Although droughts reduce agricultural production significantly in Latin America, only limited attempts have been made to monitor or mitigate them (Brunini et al., 2000). One of the institutions that is directly involved

Table 12.1 Production losses due to climate anomalies including droughts that occurred in Brazil during 1978–85 (Mota, 1987) and 1991–94 (Rossetti, 2001)

	Percentage of loss		
	1978–85	1991–94	
Crop	Entire Brazil	Entire Brazil	Northeastern Brazil
Corn	23	41	71
Soybeans	29	23	—
Rice		37	47
Beans		32	41
Average	26	33	53

in drought monitoring is FUNCEME (Fundação Cearense de Meteorologia, Ceará State Meteorological Foundation), in northeastern Brazil. The FUNCEME carries out research on monitoring and conducting a hydrometeorological surveillance in the northeast Brazil; however, no specific indices are used to monitor droughts. Recently, the Agronomic Institute in São Paulo state has begun a weekly drought monitoring evaluation by considering both agricultural and hydrological aspects. Such an evaluation will contribute to the development of public policy for drought mitigation and natural disaster preparedness.

Precipitation Anomaly from Potential Evapotranspiration

The anomaly of annual precipitation (P) from potential evapotranspiration (PET), is shown for some locations in Brazil, Argentina, Chile,

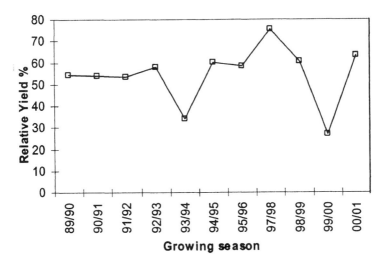

Figure 12.3 The variation in maize production from 1989 to 2000 for the central southern region of Brazil. Data for 1996–97 was not available.

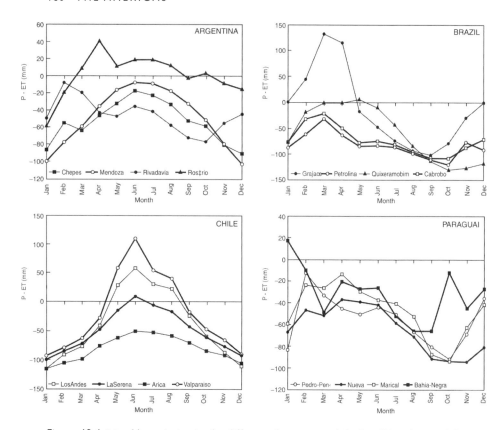

Figure 12.4 Monthly variation in the difference between precipitation (P) and potential evapotranspiration (PET) for different locations in Argentina, Brazil, Chile, and Paraguay.

and Paraguay in figure 12.4. This anomaly can be used as a drought indicator. The anomaly is positive for some hot and humid regions, negative during certain times of the year for other regions, and negative throughout the year for semiarid and arid regions where drought is a common phenomenon.

Aridity Index

The aridity index (AI) is the ratio of annual water deficiency (DEF) to PET which is determined by the methodology proposed by Thornthwaite and Mather (1955):

$$\mathrm{AI} = \left(\frac{\mathrm{DEF}}{\mathrm{PET}}\right) 100 \qquad [12.1]$$

Based on the above index, various locations in Argentina, Brazil, Chile, and Paraguay have been categorized into different climate types (table 12.2).

Table 12.2 Climatic classification and aridity index based on methodology by Thornthwaite and Mather (1955) for distinctive locations in Argentina, Brazil, Chile, and Paraguay

Country	Location	Aridity index	Climate
Argentina	Rivadavia	0.48	Semiarid
	Mendoza	0.75	Arid
	Chepes	0.66	Arid
Brazil	Quixeramobim (NE)	0.48	Semiarid
	Petrolina (NE)	0.75	Arid
	Jaguaribe (NE)	0.58	Semiarid
	Fortaleza (NE)	0.33	Subhumid, humid
	Cuiabá (W)	0.13	Subhumid dry
	Ribeirão Preto (SE)	0.09	Humid
	Barretos (SE)	0.13	Subhumid, humid
	Campinas (SE)	0.03	Humid
Chile	Los Andes	0.63	Semiarid
	Patrerilios	0.89	Arid
	La Serena	0.79	Arid
	Antofogasta	0.99	Arid
Paraguay	Pedro Peña	0.54	Semiarid
	Nueva	0.54	Semiarid
	Mariscal	0.45	Semiarid

Distribution of Dry Spells

The occurrence of dry spells lasting more than 10 days in the month of January seriously affects the production of field and grain crops, particularly field and grain crops in the southeastern and central parts of the country (Alfonsi et al., 1979), and the knowledge of the distribution of the dry spells is helpful in identifying, quantifying, and mapping droughts in the region (Arruda and Pinto, 1980).

Drought monitoring and mitigation in Brazil is in its initial stage. Only the state of São Paulo conducts weekly drought monitoring—through its Integrated Agrometeorological Information Center (CIIAGRO) and the Instituto Agronômico de Campinas (IAC-APTA) of the Agriculture and Supply Secretariat (Government of the State of São Paulo)—using various indices (Brunini et al., 1998) that are described below.

Soil Moisture and Potential Evapotranspiration

The ratio of the actual water availability in soil (W) to the maximum water availability in soil (Wx) and PET can be used to monitor drought conditions, as shown in table 12.3.

Currently a new index, the agricultural drought index (ADI), as shown in table 12.4, is being implemented to monitor crop development stages and drought conditions. The ADI takes into account not only the actual soil water availability, but also the crop phenological stages and rainfall distribution.

Table 12.3 Categorization of drought conditions based on soil water and potential evapotranspiration for the state of São Paulo, Brazil

Criterion		
P and ET_p	W/W_x	Categorization of drought condition
$P \leq ET_p$	≥ 0.60	Nil
$P \leq ET_p$	$0.4 \leq W/W_x < 0.6$	Moderate
$P < ET_p$	$0 < W/W_x < 0.4$	Severe
$P = 0$	0	Extremely severe

Note: P is accumulated rainfall, ET_p is potential evapotranspiration, W is the actual water availability in the soil, and W_x is maximum water availability in the soil.

Table 12.4 Agricultural drought index for São Paulo State based on crop maximum evapotranspiration (CME) and rainfall (P)

Agricultural drought index	Crop development conditions
$0.80 < \text{ADI} \leq 1.0$	Good
$0.60 < \text{ADI} \leq 0.80$	Favorable
$0.40 < \text{ADI} \leq 0.60$	Reasonable
$0.20 < \text{ADI} \leq 0.40$	Unfavorable
$\text{ADI} \leq 0.20$	Critical

Precipitation Anomaly

The monthly precipitation anomaly for a location is the difference between monthly precipitation and a historic average (normal) of precipitation for that month. Such an anomaly is the simplest way to monitor drought conditions and rainfall variability from year to year.

Water Deficit Anomaly

Water deficit anomaly is the difference between water deficit for a month and the average water deficit for the month. Water deficit is computed using a water balance methodology (Thornthwaite and Mather, 1955) through a software developed by Brunini and Caputi (2000).

Palmer Drought Severity Index

Currently the CIIAGRO determines the Palmer drought severity index (PDSI; Palmer, 1965) for the São Paulo State and makes it available online (http://ciiagro.iac.br). The PDSI is determined as

$$\text{PDS}_i = 0.897\,\text{PDSI}_{i-1} + (z/3)$$
$$z = (P_i - \overline{P}_i)K_i \qquad [12.2]$$

where z is the monthly value of the precipitation anomaly, and k is climatic characterization of a location.

The normalized value of K_i was determined experimentally by Palmer

(1965) using nine locations in the United States. For the São Paulo State (Brunini et al., 2002), K_i was obtained using data collected at 93 locations across São Paulo State. K_i is defined as:

$$K_i = \frac{21.14 K_i''}{\sum_{1}^{12} D_i K_i''} \quad [12.3]$$

where K'' is a climatic characterization factor related to moisture departure, and D is the difference between observed precipitation and expected precipitation for a month.

The Palmer index, modified for Brazil, has been used throughout the state of São Paulo for monitoring drought conditions both on a monthly and a dekadal basis. Figure 12.5 shows the relationship between PDSI and maize/corn yield.

Standardized Precipitation Index

The standardized precipitation index (SPI) was developed for monitoring precipitation anomalies in the United States (McKee et al., 1993; chapter 9). The SPI is determined based on probabilistic density functions that describe historic precipitation series for different durations (1–72 months). The SPI is simply a z-score of the normal distribution variable, Z_i:

$$\text{SPI} = Zi = (Pi - \overline{P}i)/\delta i \quad [12.4]$$

Currently the SPI maps are produced for the state of São Paulo on a monthly basis, with recurrent periods of 1, 3, 9, 12 and 24 months (SPI-1, SPI-3, SPI-9, SPI-12, and SPI-24) to monitor droughts in the state.

Crop Drought Index

The crop drought index (CDI) is defined as

$$\text{CDI} = 1 - (\text{AET}/\text{PET}) \quad [12.5]$$

where AET is actual evapotranspiration. The CDI is computed on a 10-day basis considering maximum water holding capacity of the soil as 125 mm. This index, together with the Crop Moisture Index (CMI), the Standardized Precipitation Index (SPI), and the Palmer Drought Severity Index (PDSI) are used on a routine basis to forecast and to monitor drought and dry spells in São Paulo state.

The analysis of the results obtained for the state of São Paulo show that the above indices (SPI, PDSI, precipitation and deficit anomalies, and the CDI) are helpful in monitoring drought conditions in the state. Yields of various crops were also satisfactorily monitored by PDSI, CDI, and SPI (Mota, 1979; Brunini et al, 2002).

Use of Satellite Data

The use of the satellite data for drought monitoring in Brazil is at its preliminary stage. Nevertheless, studies are being conducted by the Center of

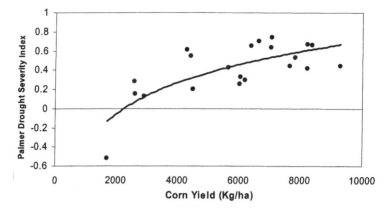

Figure 12.5 The relationship between Palmer drought severity index and corn yield.

Study and Research on Agriculture (CEPAGRI—Unicamp; www.cpa.uni camp.br) and mainly by the National Institute of Space Research (Instituto Nacional de Pesquisas Espaciais, www.inpe.br, which are trying to develop technologies for monitoring agricultural droughts.

Agrometeorological Warning System

The Instituto Agronômico de Campinas (IAC) of the Agriculture and Supply Department began an Operational Advisory System in 1988 to aid agricultural activities. The counseling and agrometeorological analyses are conducted by the CIIAGRO (Brunini et al., 1996). This system, a network of 128 weather stations across the state of São Paulo, is an operational framework that provides agrometeorological information to farmers and extension services in regard to soil type, crop development, agricultural practices, pest management, irrigation requirements, climatic risks (frost, drought, dry spell), water balance, crop yield, and weather forecast. In short, the information disseminated by the system can be used by the agribusiness community for decision making about agricultural practices, food production and civil defense, and drought preparedness and mitigation (Brunini et al., 1998). Using daily temperature and precipitation data twice a week from the weather stations, the system executes a water-balancing approach (Thornthwaite and Mather, 1955), as well as the relationships between the ratio AE/PET and available soil water. Weekly maps describing drought conditions, dry spells, crop development and climate anomalies are issued. More details can be obtained from the CIIAGRO's Web site (http://ciiagro.iac.br). Figure 12.6 is a sample of the product used for monitoring droughts using some of the drought indices for the state of São Paulo; figure 12.6a shows drought conditions for one month period using SPI (SPI-1), whereas figure 12.6b indicates drought conditions based on the weekly CDI.

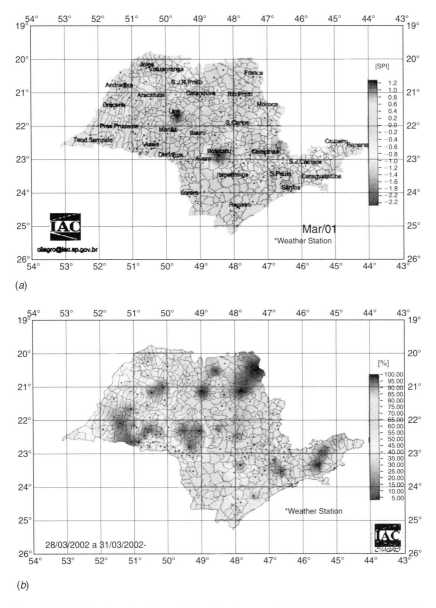

Figure 12.6 Monitoring drought conditions for the state of São Paulo (a) for March 2001 using a monthly standardized precipitation index (SPI-1) and (b) for March 2002 using the crop drought index.

Drought Mitigation and Drought Research Needs

No large-scale measures have been adopted to mitigate droughts in Latin America. The majority of the governmental agencies act on a post-facto circumstances seeking to minimize the social effects of the drought phenomenon. Nevertheless, the following measures are taken to mitigate droughts.

Climatic Risk Zoning

The climatic risk zoning defines the best sowing period for each crop to reduce the risk of drought. This approach is being developed for all of Brazil and is coordinated by the agricultural ministry (Assad, 2001). Before the implementation of this zoning, the agricultural losses due to droughts were as high as 30%, but currently the loss is reduced to under 2% in the areas where the approach was implemented. There is also a need to develop better irrigation techniques to further improve water-use efficiency and drought mitigation.

Better Planting Techniques

The use of appropriate agronomic techniques, ranging from soil management to crop selection, can help ensure survival of plant species leading to reasonable production. In the state of São Paulo, direct planting of maize and wheat is better than the conventional planting to withstand drought conditions, because the direct planting allows for better establishment of plants under the conditions of up to 45 days without rainfall (Cunha, 1999). It was also observed that the use of a drought-tolerant variety of maize increased the yield and resistance to drought conditions. Further research needs to be conducted to evaluate planting systems that allow the root system of plants to grow deeper and exploit a larger soil volume to mitigate impacts of droughts.

Conclusions

There is a wide variation in climatic conditions in South American countries. ENSO activities cause droughts in some regions. There is a need to strengthen relationship between ENSO activities and drought occurrence. Better models need to be developed by combining ENSO data, drought indices, and satellite data for monitoring and predicting agricultural droughts more effectively across South America.

References

Alfonsi, R.R., H.S. Pinto, and H.V. Arruda. 1979. Freqüência de verânicos (dry spell) em regiões rizícolas do Estado de São Paulo. Proceedings Reunião de

Técnicos em Rizicultura do Estado de São Paulo—Coordenadoria de Assistência Técnica e Integral Campinas, Estado de São Paulo, Março 1979, pp. 147–151.

Alves, J.M.B., and C.A.A. Repelli. 1992. A variabilidade pluviométrica no setor nordeste e os eventos El Niño—oscilação sul (ENOS). Rev. Brasil. Meteorol. 7:583–592.

Arruda, H.V. de, and H.S. Pinto. 1980. An alternative model for dry spell probability analysis. Month. Weath. Rev. 108:823–825.

Assad, E.D. 2001. Climatic risks zoning in Brazil. Available http://marsv54.agricultura.gov.br/ma, accessed 2002.

Brunini, O., G.C. Blain, A.P.C. Brunini, R.L. dos Santos, R.S. Brigante, E.L. de Almeida. 2002. Avaliação do índice de severidade de seca de Palmer para quantificação da seca agrícola no Estado de São Paulo. In: XII Congresso Brasileiro de Meteorologia, Foz do Iguaçu. Proceedings-CD, vol. 1. Sociedade Brasileira de Meteorologia, Foz do Iguaçu, pp. 1140–1147.

Brunini, O., and E. Caputi. 2000. Software para cálculo do Balanço Hídrico. Software for Estimation of Water Balance. Fundação de Apoio à Pesquisa Agrícola, São Paulo, Brazil.

Brunini, O., H.S. Pinto, M. Fujiwara, R.C.M. Pires, E. Sakai, and F. Arruda. 1998. Sistema de Informações Agrometeorológicas. In: Anais-II Congresso Brasileiro de Biometeorologia, Goiânia, vol. 2. Sociedade Brasileira de Biometeorologia, São Paulo, pp. 15–37.

Brunini, O., H.S. Pinto, J. Zullo Jr., M.T. Barbano, M.B.P. Camargo, R.R. Alfonsi, G.C. Blain, M.J. Pedro Jr., and C.Q. Pellegrino. 2000. Drought quantification and preparadness in Brazil—the example of São Paulo State. Expert Group Meeting on Early Warning Systems for Drought Preparedness and Drought Management, Part II, Lisbon, Portugal, 2000.

Brunini, O., M.A. Santos, R.V. Calheiros, E. Caputi, J.M. Santos, A.G. Piccini, and H.S. Pinto. 1996. Integrated Center of Agrometeorological Information. In: Proceedings—Buenos Aires—AR. Latin American and Iberian Federation of Meteorological Societies, Buenos Aires, pp. 133–135.

Cunha, G.R. 1999. El-Niño Southern Oscillation and climate outlooks applied to agricultural management in Southern Brazil. Rev. Bras. Agrometeorol. Santa Maria. 7:277–284.

Cunha, G.R., G.A. Dalmago, V. Estefanel, A. Pasinato, and M.B. Moreira. 2001. El Niño–Southern oscillation and its impacts on the barley crop in Brazil. In: Anais XII Congresso Brasileiro de Agrometeorologia e III Reunião Latino Americana de Agrometeorologia, July 3–6, 2001. Sociedade Brasileira de Agrometeorologia, Fortaleza, pp. 17–18.

Grimm, A.M. 2003. The El Niño impact on the summer monsoon in Brazil: Regional processes versus remote influences. Journal of Climate 16:263–280.

Marengo, J.A., I.F.A. Cavalcanti, H. Camargo, M. Sanches, G. Sampaio, and D. Mendes. 2001. Climate simulation and assessment of predictability of rainfall in the Northeast Brazil Region Using the CPTEC/COLATMOSPHERE Model. In: Anais XII Congresso Brasileiro de Agrometeorologia e III Reunião Latino Americana de Agrometeorologia, July 3–6, 2001. Sociedade Brasileira de Agrometeorologia, Fortaleza, pp. 25–26.

Martelo, M.T. 2000. Study of the possible influence of the El-Niño Southern Oscillation Phenomena (ENOS) on the Los Llanos in Venezuela. In: Reunión de expertos de las Asociaciones Regionales III y IV sobre fenômenos adversos,

Caracas, Venezuela, July 12–14, 1999. Organización Meteorologica Mundial, Geneva, pp. 111–118.

McKee, T.B., N.J. Doesken, and J. Kleist. 1993. The relationship of drought frequency and duration to time scale. In: Proceedings of the 8th Conference on Applied Climatology, January 17–22, 1993. American Meteorological Society, Boston, pp. 179–184.

Mota, F.S. 1979. Metodologia para caracterização da seca agronômica no Brasil. Interciência 4:344–350.

Palmer, W.C. 1965. Meteorological Drought Research Paper no. 45. Weather Bureau, Washington, DC.

Rosseti, L.A. 2001. Zoneamento agrícola em aplicações de crédito e securidade rural no Brasil: aspectos atuais e de política agrícola. Rev. Brasil. Agrometeorol. 9:386–399.

Scian, B., and M. Donnari. 1996. Historic analysis of the droughts in the semi arid Pampa region, Argentina. In: Anais VII Congreso Argentino de Metorologia y VII Congreso Latinoamericano e Ibérico de Meteorologia, Buenos Aires, September 2–6, 1996. Centro Argentino de Meteorológos, Buenos Aires, pp. 333–334.

Thornthwaite, C.W., and J.R. Mather. 1955. The Water Balance. Drexel Institute of Technology, Centerton, NJ.

PART IV

EUROPE, RUSSIA, AND THE NEAR EAST

CHAPTER THIRTEEN

Monitoring Agricultural Drought in Poland

ZBIGNIEW BOCHENEK, KATARZYNA
DABROWSKA-ZIELINSKA, ANDRZEJ CIOLKOSZ,
STANISLAW DRUPKA, AND VIJENDRA K. BOKEN

Poland is situated in the Great European Plain between the Baltic Sea and the Carpathian and Sudety mountains. Its territory includes lowlands (91.3%), highlands (7.7%), and mountains (1%). Most of Poland's soils are light soils of podsolic origin, which are usually of poor quality. It is for this reason that only 25% of the agricultural land, which accounts for 60% of the total territory and engages about 12% of population, is used for producing wheat, barley, sugar beets, rape seed, and vegetables. Average yields of main crops in Poland are lower than in the majority of West European countries. But the higher harvest areas put Poland sixth in Europe in the production of wheat, second in the production of rye and potatoes, and fourth in the production of sugar beet. The variation in the production of these crops during 1990–2000 is shown in figure 13.1.

Private farms cover about 84% of the total agricultural land. About 55% of the farms have an individual area < 2 ha. Liquidation of state farms and substantial reduction in the number of cooperative and collective farms have impacted the size of individual farms and increased their importance in agricultural production and Polish export. Since 1980, the average area of individual farms increased from 6.5 to 7.8 ha.

Poland is located in the region where precipitation exceeds transpiration. But since the 1960s, annual rainfall has gradually decreased by about 70 mm (Slota et al., 1992). Due to the shortage of precipitation, high temperature fluctuations in the spring, and cool weather during summertime, yields of the main crops have decreased and drought frequency has increased, particularly during the last decade (figure 13.2).

Spatial Distribution of Droughts and Their Causes

Drought usually begins in western Poland, moves through the central part, and eventually reaches eastern side (between 51°N and 54°N), which is

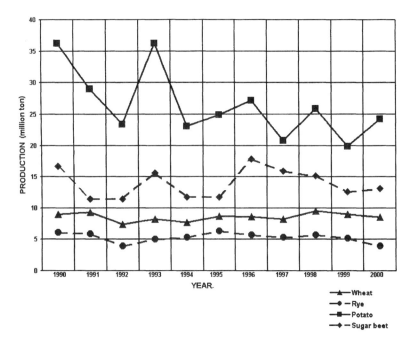

Figure 13.1 Production of the main crops in Poland.

highly susceptible to droughts. Regions located above 54°N are in the zone of Baltic Sea climate characterized by higher rainfall (600–700 mm) and hence are less prone to drought than the rest of Poland. The distribution of drought-prone areas is shown in figure 13.3.

Droughts occurred 19 times in Poland during 20th century (i.e., once every 4–5 years; Slota et al., 1992). These droughts can be linked to the processes of air circulation over Europe and the adjacent ocean. Thermal-humidity anomalies caused droughts in 1977, 1983, 1989, and 1992 (the year of the great European drought). In 1992, drought affected more than 90% of the land, and the total precipitation was 40–50% below normal. This severe drought was caused by translocation of very warm, dry tropical masses of air in the beginning of summer, which increased evapotranspiration and reduced precipitation, forming long dry spells. In addition, poor drainage, accelerated land development, and acid rains also contributed to occurrence of poor yields and agricultural droughts in Poland.

Drought Monitoring Methods

To study agricultural droughts in Poland, a special index was constructed which was found to be a better indicator of drought than the precipitation alone. The index is defined as EP/P_{veg}, where EP is potential evapotranspiration and P_{veg} is the average sum of rainfall during vegetation period (April–

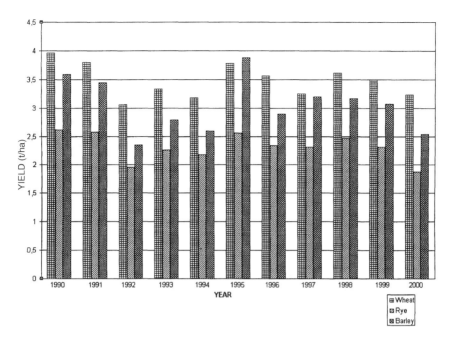

Figure 13.2 Average yield of the main cereals in Poland.

September). In Poland, the index ranges from 1.6 for areas heavily prone to drought to 1.0 for areas where sum of summer precipitation exceeds potential evapotranspiration. Such areas exist in mountainous regions in southern Poland and are characterized by high rainfall in summer months. In addition, field-level assessment of potential yield of crops is done several times throughout the growing season by experts in the Central Statistical Office. A wealth of information about actual crop conditions is available from agricultural correspondents posted in different regions of the country (www.stat.gov.pl). The results of these predictions are published in official bulletins and delivered to the Ministry of Agriculture and other governmental agencies throughout Poland.

Until about 1990, drought monitoring was limited to assessing drought conditions as described above as well as by studying agrometeorological data collected from 60 meteorological stations across Poland. Drought is considered to have occurred if some threshold conditions are met. Agrometeorological variables and the threshold conditions (in parentheses) are (1) precipitation deficit from April 1 to September 1 (<50% of multiyear mean); (2) precipitation deficit during the 3 dekads (1 dekad = 10-day period) preceding the drought (<25% of multiyear mean); (3) the number of successive nonrainy days (>17); (4) the difference between total precipitation and evapotranspiration from June 1 to September 1 (< 75 mm); (5) number of days for which the mean daily soil temperature at 5-cm depth

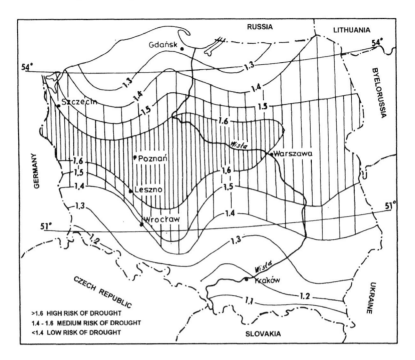

Figure 13.3 Map of drought-prone areas in Poland.

in each dekad after June 1 was higher than 25° C (>2); (6) if soil moisture was extremely low; and (7) if shortage of water was reported by farmers for at least 25% of farm wells.

Every indicator described above is assigned 1 point if the threshold condition is met and 0 point if the condition is not met. The total sum of the assigned points is denoted as the drought index (DI), which is used to assess drought conditions. A drought is classified as no drought if DI = 0; drought if DI = 1 or 2; heavy drought if DI = 3, 4, or 5; and very heavy drought if DI = 6 or 7.

The Institute of Meteorology and Water Management (www.imgw.pl) developed the above criterion of drought assessment based on point measurement (Slota et al., 1992). Figure 13.4 shows classification of drought during 1992 using the above criterion. But this criterion could not be used because it was too time consuming and labor intensive. Therefore, another approach based on satellite data was developed during the 1990s at the Remote Sensing Center of the Institute of Geodesy and Cartography, Warsaw (www.igik.edu.pl) for monitoring crop conditions and drought.

Remote Sensing-Based Crop Condition Assessment System

Using the National Atmospheric Oceanic Administration (NOAA) satellite's Advanced Very High Resolution Radiometer (AVHRR) data, three

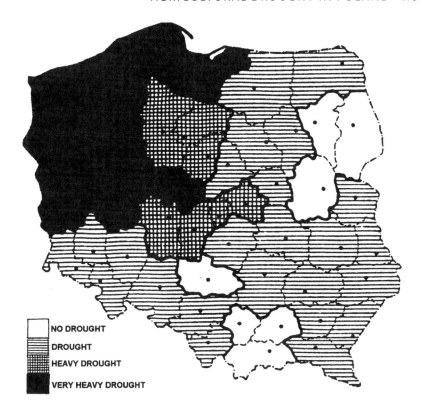

Figure 13.4 Map of distribution of drought conditions at the end of June 1992.

indices, vegetation condition index (VCI), normalized difference vegetation index (NDVI), and temperature condition index (TCI), were used to monitor crop conditions and droughts in Poland. Chapter 6 provides a detailed description of derivation of these indices. Indices were computed for every week for 1985–98. During this period of 14 years, crop yields varied significantly and so did the indices. The following yield models were developed by regressing the deviations in yields and indices.

Analysis of TCI and VCI in different years revealed that the most important for crop yield assessment are the TCI values in weeks 16 and 22 and the VCI values in week 25. Based on these findings the model was developed using 1985–97 data, which relates yield deviation from mean (Y/Y_{mean}) with TCI and VCI values as described below:

$$Y/Y_{mean} = 103.492 - 0.273(TCI_{16})$$
$$+ 0.214(TCI_{22}) + 0.004(VCI_{25}) \qquad [13.1]$$

The above model was used to make yield predictions for 1998 for Poland. The yield predictions were issued four weeks before harvest. The results were compared with yield estimates produced with the use of conventional methods by the Central Statistical Office. The mean error of cereal yield assessment for 49 administrative units was about 4%. The results of cereal yield estimation were accepted by agricultural experts and statisticians, creating a basis for operational crop condition assessment in Poland with the use of remotely sensed data. In addition, using TCI, the 1992 drought was also detected effectively, as shown in figure 13.5 (Dabrowska-Zielinska et al., 1998, 2002). It can be seen from figure 13.5 that western part of Poland was affected by drought already in May.

Two important phases of crop development had the highest correlation with crop yield. The increase of water demand by plants responds to the increase of sensitivity of TCI during these periods. The period of significant correlation during early summer (Julian week 22–25) is critical because cereals pass through reproductive phases, when cooler weather is favorable for crop development and yield formation. The second important period was spring (week 14–16); the negative correlation for this period indicates the low temperature during spring yields of cereals in Poland. It was also found that TCI could be used to interpret soil moisture conditions over large areas.

The data used in the above system were based on using 4-km GAC (global area coverage) NOAA/AVHRR data (www.saa.noaa.gov/cocoon/nsaa/products/) for global monitoring of crop and drought conditions. However, the diversified cropping and small land holdings that characterize Polish agriculture make 1-km LAC (local area coverage) NOAA-AVHRR data more useful for monitoring drought conditions. The research for this purpose began in cooperation with the Canada Centre of Remote Sensing, Ottawa (www.ccrs.nrcan.gc.ca), in 1996. The global land data (1-km resolution) collected under International Geosphere Biosphere Program for 1992–95 were used for this research. A Geographic Information System database was developed and used for making regression analyses and for determining relationships between indices derived from remotely sensed data and parameters characterizing agricultural production (Walker, 1988; Wood, 1993; Yang et al., 1997; Boken and Shaykewich, 2002).

Figure 13.5 Development of drought conditions in Poland in 1992, characterized by temperature condition index.

Crop conditions were monitored during 1997–2001 vegetation periods by comparing dekadal NDVI values with the previous year's values and the long-term mean values for each administrative unit in Poland. Administrative units were grouped into regions to study regional differences in NDVI levels. All of the above mentioned materials were used for analyzing crop development in Poland. For example, figure 13.6 illustrates crop growth conditions in Poland in 2001 based on analysis of NDVI data.

These NDVI-based outputs were delivered to the Central Statistical Office for comparative analysis. In addition, research conducted at the Institute of Geodesy and Cartography found the following indices useful for monitoring vegetation and drought conditions in Poland:

1. Crop growth index (CGI), the ratio of NDVI to radiation temperature (NDVI/T_s). This index more precisely reflects the deficits in soil water content than the NDVI or temperature alone. The lower the CGI, the worse the vegetation conditions.
2. Soil-adjusted vegetation index, a function of NDVI, taking into account reflection from soil background (Huete, 1988; chapter 5).
3. Ratio of sensible heat to latent heat (H/LE); the higher the index, the worse the crop conditions. Figure 13.7 shows spatial variation of H/LE index for 1998 and 2000.
4. Water deficit index, the ratio of actual to potential latent heat, LE/LE_p (Moran et al., 1994).

Drought-Mitigating Measures

Efforts to mitigate drought in Poland began in the beginning of 20th century in the central part of Poland (Wielkopolska and Kujawy regions),

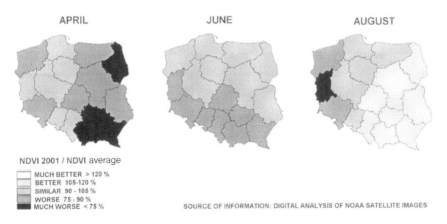

Figure 13.6 Changes of crop growth conditions in 2001 as compared to normal year expressed by NDVI ratios.

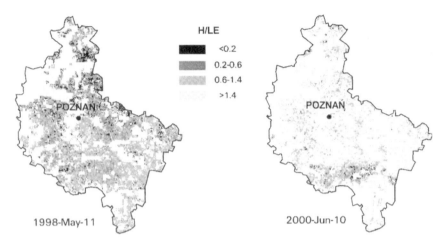

Figure 13.7 Distribution of sensible heat to latent heat index illustrating soil moisture conditions for western Poland (Wielkopolska region). A darker tone indicates higher soil moisture condition.

which is characterized by the lowest rainfall and the highest potential evapotranspiration in the country. Measures for drought mitigation can be grouped into direct and indirect methods. Direct methods are applied while drought is in effect and include various forms of irrigation (e.g., sprinkler, microirrigation) suitable for soils with 25–35 cm of humus layer. These methods are used for vegetable and potato crops and apple orchards. However, subsoil irrigation is usually applied in the case of grasslands that grow mainly in river valleys on organic and peat soils.

Indirect methods are preventive measures such as (1) the use of drought-resistant varieties of crops; (2) rotation of crops with less number of crop species in annual rotation (only one main crop instead of three crops per year); (3) preference for planting cereals on farms susceptible to drought; (4) application of cultivation methods and fertilizers so as to accelerate crop development and maturity in order to shift the critical phase of plant development to the period endowed with higher availability of soil moisture; (5) harrowing and shallow plowing after harvest to minimize moisture loss; (6) deep tillage before winter to increase soil roughness; (7) raising hedgerows and tree stripes at field boundaries to help preserve winter snowfalls and limit evapotranspiration; and (8) using sluice gates to prevent or reduce water outflow from fields.

Potential Drought Research Needs

Current state of drought monitoring in Poland requires further research to improve both traditional and remote sensing-based methods. More profitable drought-resistant crop varieties need to be developed for drought-prone regions. Local and regional irrigation needs for agriculture should be identified following a detailed assessment of water resources in Poland. Determining the probability and frequency of droughts using long-term climatic data will be an important part of a drought early-warning system. Vegetation and soil moisture indices derived from remotely sensed data must be refined for early detection of drought conditions and to improve yield models and early warning systems.

Conclusions

Drought has become a severe problem in Polish agriculture. It is the main reason for decline in crop yields. Drought early warning can help minimize losses in agriculture. Drought information is available from different institutions in Poland, but it is directed more to research than to applications. There is a good network of meteorological stations in Poland, however, drought monitoring and prediction are developed using point measurements. Combining remotely sensed and meteorological data will enhance spatial information about drought and its progress. In particular, data provided by NOAA meteorological satellites (1-km resolution) offer rapid and frequent information about drought development. Extensive research needs to be carried out in Poland to further develop an operational system for assessing crop conditions and monitoring and predicting droughts.

References

Boken, V.K., and C.F. Shaykewich. 2002. Improving an operational wheat yield model for the Canadian Prairies using phenological-stage-based normalized difference vegetation index. Int. J. Remote Sens. 23:4157–4170.

Dabrowska-Zielinska, K., F.N. Kogan, A. Ciolkosz, M. Gruszczynska, U. Raczka, W. Kowalik, and R. Jankowski. 1998. New method of drought detection based on NOAA satellites and its impact on Polish agriculture. In: Proceedings of the ASPRS-RTI 1998 Annual Conference, Tampa, Florida. American Society for Photogrammetry and Remote Sensing, pp. 1501–1504.

Dabrowska-Zielinska, K., F.N. Kogan, A. Ciolkosz, M. Gruszczynska, and W. Kowalik. 2002. Modelling of crop growth conditions and crop yield in Poland using AVHRR based indices. Intl. J. Remote Sens. 23:1109–1123.

Huete, A.R. 1988. A soil-adjusted vegetation index (SAVI). Remote Sens. Environ. 25:295–309.

Moran, M.S., T. Clarke, Y. Inoue, and A. Vidal. 1994. Estimating crop water deficit using the relation between surface-air temperature and spectral vegetation index. Remote Sens. Environ. 49:246–263.

Slota, H., E. Bobinski, A. Dobrowolski, B. Fal, S. Galka, R. Korol, H. Lorenc, M. Mierkiewicz, Z. Rutkowski, T. Tomaszewska, and J. Zelazinski. 1992. Drought 1992: Range, intensity, reasons and effects, conclusions. Materialy Badawcze, Seria: Hydrologia i Oceanologia, Instytut Meteorologii i Gospodarki Wodnej, Warsaw, Poland.

Walker, G.K. 1988. Model for operational forecasting of Western Canada wheat yield. Agric. Forest Meteorol. 44:339–351.

Wood, D. 1993. The use of multitemporal NDVI measurements from AVHRR data for crop yield estimation and prediction. Intl. J. Remote Sens. 14:190–210.

Yang, W., L. Yang, and J.W. Merchant. 1997. An assessment of AVHRR/NDVI ecological relations in Nebraska, USA, Intl. J. Remote Sens. 18:2161–2180.

CHAPTER FOURTEEN

Monitoring Agricultural Drought in Mainland Portugal

FÁTIMA ESPÍRITO SANTO, RITA GUERREIRO,
VANDA CABRINHA PIRES, LUÍS E. V. PESSANHA,
AND ISABEL M. GOMES

Mainland Portugal (37°–42° N latitude) is located in the transitional region between the subtropical anticyclone and the subpolar depression zones. In addition to latitude, its orography and the effect of the Atlantic Ocean are the major factors affecting the climate of the mainland Portugal. The highest altitudes vary from 1000 m to 1500 m, with the exception of the Serra da Estrela range (figure 14.1), whose peak is just below 2000 m. The regions farthest from the Atlantic Ocean are around 220 km away.

Portugal has a total area of 9.2 million ha, 41% of which is devoted to cropland. Although agricultural practices in some regions are still traditional and not competitive, they are slowly becoming more and more industrialized and employ about 20% of the active population, which includes employment in agricultural industries as well. The inclusion of Portugal to the European Community in 1986, on the one hand, provided financial resources, but, on the other hand, caused agricultural policy to become more dependent on community policies. Farm size has doubled since the 1970s with increased mechanization.

Wine production is the most important agricultural activity, which contributes significantly to Portugal's economy (GPPAA, 1999). Among fruit trees, apples and citrus fruit have the highest productions, about 250,000 tons and 230,000 tons per year, respectively. Olive trees produce about 40,000 tons of olive oil per year (GPPAA, 1999). Food grain production is relatively low and not enough for domestic use. This is mainly due to the prevailing climate.

Influence of Climate Agriculture

The Portuguese climate is classified as Mediterranean, with some variations depending on orography and latitude. The annual cycles of monthly mean

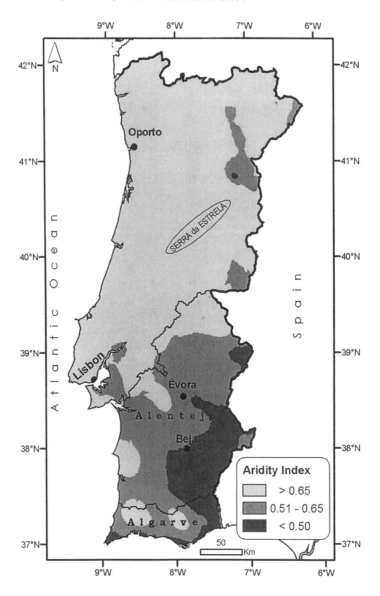

Figure 14.1 Climate index identifying drought-prone areas in Portugal during the 1961–90 period.

precipitation and temperature (minimum and maximum) reveal that warm and dry summers are more pronounced in the southern regions. This type of climate presents several drawbacks to agriculture, the major one being insufficient rainfall during the summer or spring seasons (Pinto and Brandão, 2002). Most of the rainfall occurs during winter season, from November until March. The majority of soils in Portugal are badly drained and suffer from water logging during the rainy season. A significant decrease in spring precipitation has been observed for last two decades.

There is a lag between the rainy season and the growing season. The critical period for water requirement by crops is in the spring. The lack of the coincidence between radiation and heat on the one hand, and soil water availability on the other, limits the agricultural productivity and crop yields in drier regions that do not have irrigation facilities (Pinto and Brandão, 2002). If in some years the cold season is dry, it leads to water stress and damages dry land crops, mainly wheat, rye, oats, and barley.

Vulnerability to Drought

An aridity index (the ratio of annual precipitation to potential evapotranspiration) was computed for the 1961–90 period (figure 14.1). Evapotranspiration was computed using the Penman method (Penman, 1948), as proposed by the United Nations Environment Programme (UNEP). In accordance with this criterion, dry subhumid regions represent 23%, and semiarid regions represent 11% of the territory.[1] This means that measures should be taken in about 34% of the territory to prevent further degradation of the soil in the most drought prone areas subject to desertification. These problem areas are located in the southeastern zone (Alentejo) and the northeastern zone (figure 14.1).

Alentejo represents almost one-half of the total cropland of the mainland. However, 60% of irrigated land is in the northern and central regions (GPPAA, 1999). In Alentejo, few areas are irrigated, although there has been an increase in the last few years. This region of Portugal is particularly vulnerable to droughts, which have frequently occurred here during the last few years. Due to the seasonal distribution and high variability in precipitation, agricultural practices are highly dependent on irrigation during spring and summer seasons (IHERA, 1999). A strong Mediterranean influence and declining spring precipitation have contributed to the desertification of some rural areas, mainly in the southern and northeastern regions. In addition, conventional farming practices have contributed, to some extent, to land degradation and subsequently to desertification.

Main Causes of Drought

General Meteorological Conditions

The geography of the mainland Portugal favors the occurrence of droughts. The subtropical anticyclone in the North Atlantic is situated in a blocking position that prevents disturbances on the polar front from reaching the Iberian Peninsula. Although the overall climate for mainland Portugal varies little, significant variations, temporal as well as spatial, exist in temperature and precipitation, causing droughts.

Precipitation Mean annual precipitation in mainland Portugal is around 900 mm. The northwest region of Portugal is one of the wettest spots in Europe, with mean annual precipitation exceeding 3000 mm, but the average rainfall in the interior of Alentejo and in the northeastern part of the territory is about 500 mm and is characterized by high variability.

On average, about 42% of the annual precipitation falls during the 3-month winter season (December–February). The lowest precipitation (only 6% of the annual precipitation) occurs during the summer season (June–August). During the transition months (March–May, October–November), precipitation is highly variable.

A statistical analysis of long-term records (1931–2002) of annual precipitation data over mainland Portugal shows that, since 1982, only 6 years had precipitation values above the 1961–90 mean. Precipitation significantly declined during the spring season, slightly declined during the winter season, and slightly increased during other seasons. The decrease in annual precipitation in Portugal during 1976–2002 accompanied the same percentage decrease in the number of wet days. The number of consecutive dry days for several stations in Portugal showed an increasing trend from the 1970s onward. This trend was more evident in the southern region. In general, the number of wet days and dry days show a weak tendency for more extreme events during 1976–2002, especially in the south of Portugal.

Temperature The mean annual air temperature varies between 7°C in the inner highlands of central Portugal and 18°C in the southern coastal area. The annual number of days with minimum temperature below 0°C (frost days) reaches a peak in the highlands of northern and central inland, with more than 100 days/year, and is nil in the western coastal and southern zones. The number of days with minimum temperature above 20°C (tropical nights) and maximum temperature above 25°C (summer days) and above 35°C (hot days) is higher in the inner center of the country, the eastern part of Alentejo and the seaside Algarve. This indicates cold and warm spells in the Portuguese climate, which have significant impacts on agriculture.

An analysis of temperature data (1931–2002) in mainland Portugal revealed a general trend toward an increase in the mean annual temperature since 1972. The year 1997 was the hottest of the last 70 years, and it was a year with a severe drought. The 6 warmest years occurred in the last 12 years, and 2002 was the 16th consecutive year with a minimum temperature above normal (i.e., above the 1961–90 mean). There has been a reduction in the frequency of extreme low temperatures without an equivalent increase in the frequency of extreme high temperatures.

Droughts in Mainland Portugal

Droughts are common occurrences in mainland Portugal. During the 1976–2002 periods, severe droughts occurred during 1980–81, 1982–83, 1991–92, 1994–95, and 1997.

Between September 1991 and May 1992, the absence of rainfall caused a long drought that directly affected winter cereals and pastures. This agricultural year was classified as extremely dry in the whole territory (Report on Climate Conditions, 1992).

Between September 1994 and October 1995, a severe drought occurred; this period was classified as extremely dry, especially in the southern region of Portugal. These conditions caused serious damage to winter cereals (Report on Climate Conditions, 1995).

The meteorological drought event of 1997 was so intense that, within three months, it gave rise to a severe agricultural drought. In these three months (February, March, and April) it did not rain at all and the temperature was well above the average, which caused dryness of the soil that affected winter cereals. In Alentejo, 70% of the cereal production was lost, and a state of crop calamity was declared (Report on Climate Conditions, 1997).

Yield Losses of Main Crops

Figure 14.2 shows the variation in wheat yields for Portugal for the 1970–99 period. A higher average yield with higher variability can be observed for the second half of the period. From 1977 to 1979, too much rainfall caused a very low yield for wheat, as well as for other crops. Yield losses from 1981 to 1983 were caused by a prolonged and severe drought. In 1992 and 1995 also, drought caused significant yield losses. The low yields in 1997 and 1998 were caused by irregular rainfall distribution (i.e., surplus in autumn and in the beginning of winter, but deficit in February and March) and untimely heat during critical growing phases (GPPAA, 1999).

Drought Monitoring

Drought is monitored using different indices. To characterize drought spells in mainland Portugal, some simple indices such as the percentage of normal have been used. To monitor drought situations in Portugal, the Palmer drought severity index (PDSI) (Palmer, 1965) is now being used. The next step will be the implementation of the standardized precipitation index (McKee et al., 1995) and its comparison with PDSI.

Precipitation Deviation and Deciles

The percentage of the normal precipitation is one of the simplest methods of quantifying rainfall for a given location, and it is very effective when used for a single region or a single season. However this criterion can be misleading, because it is not standardized for varying environments.

The decile was developed by Gibbs and Maher (1967). The distribution of occurrences over a long-term precipitation record is divided into tenths of the distribution. Each of these categories is called a decile. The first decile is the rainfall amount not exceeded by the lowest 10% of the precipitation, and so on, until the rainfall amount identified by the tenth decile is the

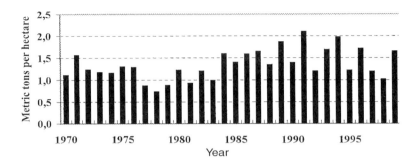

Figure 14.2 Wheat yield variation in mainland Portugal (from GPPAA, 1999).

largest precipitation amount within the long-term record. By definition, the fifth decile is the median, and it is the precipitation amount not exceeded by 50% of the total precipitation over the period of record (Hayes, 2002).

Palmer Drought Severity Index

The Palmer drought severity index (PDSI) was developed by Palmer (1965). This index is based on the supply-and-demand concept of the water-balancing equation. The objective of the PDSI was to quantify moisture conditions that were standardized, so that both spatial and temporal comparisons in drought conditions could be made (Palmer, 1965). The PDSI is calculated on the basis of precipitation and temperature data, as well as on the locally available water content (AWC) of the soil. The PDSI varies roughly between −6.0 and +6.0. A PDSI value between −2.0 and −3.0 refers to a moderate drought; between −3.0 and −4.0 refers to a severe drought; and below −4.0 refers to an extreme drought.

The PDSI is applied within the United States but has little acceptance elsewhere (Kogan, 1995). One explanation for this is provided by Smith et al. (1993), who suggested that it did not perform well in regions with extreme variability in rainfall or runoff. To overcome this situation, the PDSI was adapted and calibrated to the specific climatic conditions of mainland Portugal. The calculations pertaining to runoff, the procedure for water balancing, and the identification of the beginning and the end of a drought or a wet spells were modified. The climatic coefficient (K) was prepared using drought periods from the time series and from different regions of Portugal (Pires, 2003).

To begin with, the PDSI was studied to provide spatial and temporal representations of historical droughts. The PDSI was calculated on a monthly basis for three southern stations with a long time series (Lisbon, Évora, and Beja) for the period 1901–2000. On evaluating the trends of moderate, severe, and extreme droughts, the results indicated that, generally, some categories of droughts occurred more frequently just near the end of the time series, especially in the last 20 years. In Beja, 50% of the extreme droughts occurred after 1980. In fact, during the 1901–2000 period, this station recorded the greatest frequency of extreme droughts (4.0%), with Lisbon and Évora showing similar frequencies (2.0%; Pires, 2003).

The PDSI time series for the three meteorological stations is shown in figure 14.3. They reveal a high-frequency oscillation of the PDSI between negative and positive values, superimposed by periods of consecutive months with negative or positive values, which are almost coincident for the three stations presented. With respect to the change in variability of the PDSI, the negative values seem to dominate the last 20 years of the 20th century, especially in the south inland stations of Évora and Beja. The 1980s begin with a sudden and large decrease in the PDSI, maintaining a trend for negative values through several years. According to figure 14.3, the values of the PDSI in the cooling period 1946–75 are less negative than in the warming period 1976–2000, suggesting an increased frequency of droughts in the south of Portugal (Pires, 2003).

A Geographical Information System (GIS) is used to map the PDSI and monitor the historical evolution of the index in the southern regions of Portugal that are the most affected by droughts (Pires, 2003). The percentage of time with mild, moderate, and severe drought was mapped for the southern region from 1961 to 2000 (Pires, 2003). During this period of 40 years, the percentage of time with mild drought (PDSI values below − 1.0) in a great part of the region occurs between 30 and 40% of time. The percentage of time in moderate drought is lower; however, a large area with 15–20% of time in moderate drought is still observed. The percentage of time with severe drought is low, although in a small region it reaches nearly 10%.

A statistical analysis of long climatological series of the PDSI was made for the southern region of mainland Portugal. The PDSI average was calculated for the last four decades since 1961. An increase in severity is observed in most of the months in the first three decades, while in the last decade the drought intensity, although it has increased in some regions, does not increase significantly as compared with the previous one. In the period from February to April the increase is more significant, changing from normal conditions (PDSI with small positive values) to conditions of mild and moderate drought, especially in the months of February and March (figure 14.4). As these are decade-average values, this change is very significant. No relevant change is noticed during the summer period (June to August) over the four decades because this period is normally very dry (Pires, 2003).

As a means of validation of the PDSI as a tool to monitor agricultural droughts in Portugal, three situations, which occurred in the 1990s (1991–92, 1994–95, and 1997), were analyzed. The example of the meteorological station of Beja is presented in table 14.1 (Pires, 2003). In these three periods of drought, the PDSI values show the occurrence of severe and

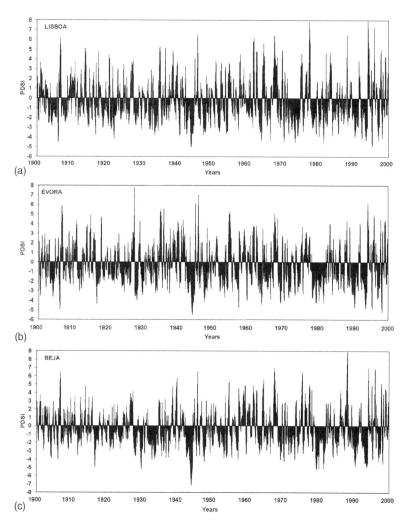

Figure 14.3 Time series of Palmer drought severity index for Lisbon, Évora, and Beja stations in southern Portugal (from Pires, 2003).

extreme droughts, as expected due to the consequences on agricultural production.

The next step to improve agricultural drought monitoring is to implement the crop moisture index (CMI), an index also developed by Palmer (1968) and based on PDSI procedures. Whereas the PDSI monitors long-term meteorological wet and dry spells, the CMI was designed to evaluate short-term moisture, as it is based on the mean temperature and total precipitation for each week and also for the CMI value from the previous week. The CMI responds rapidly to changing conditions and, like PDSI, it is weighted by location and time, allowing the preparation of maps covering moisture conditions at different locations (Hayes, 2002).

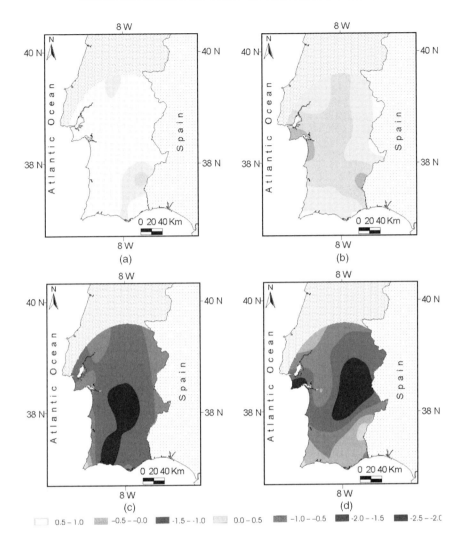

Figure 14.4 Average March Palmer drought severity index for (a) 1961–70, (b) 1971–80, (c) 1981–90, and (d) 1991–2000 (from Pires, 2003).

Drought Monitoring Using Satellite Data

The Portuguese Meteorological Institute (IM) routinely calculates normalized difference vegetation indices (NDVI) based on the National Oceanic and Atmospheric Administration's Advanced Very High Resolution Radiometer (AVHRR) satellite data (Tucker and Sellers, 1986; Gomes et al., 1989; Duchemin and Maisongrande, 2002; chapters 5 and 6) to quantify vegetation stress and monitor droughts. Since 1987 drought effects have been monitored on the main region of cereal winter crops in Alentejo. Usually, only one parameter (NDVI) is used to monitor land, which fails to fully characterize the surface. Thermal infrared channels can provide

Table 14.1 Palmer drought severity index values for Beja, Portugal, during three drought spells (Pires, 2003)

Year	Month	PDSI
1991	Sept.	−2.17
1991	Oct.	−1.98
1991	Nov.	−2.64
1991	Dec.	−3.56
1992	Jan.	−4.11
1992	Feb.	−4.28
1992	March	−4.80
1992	Apr.	−4.13
1992	May	−3.62
1994	Sept.	−2.20
1994	Oct.	−2.50
1994	Nov.	−3.22
1994	Dec.	−4.31
1995	Jan.	−4.51
1995	Feb.	−4.02
1995	March	−4.37
1995	Apr.	−4.29
1995	May	−4.65
1995	June	−4.91
1995	July	−4.76
1995	Aug.	−4.15
1995	Sept.	−3.40
1995	Oct.	−4.59
1997	Feb.	−1.67
1997	March	−3.97
1997	Apr.	−3.75

additional information on surface conditions. These channels are used to estimate mass and energy fluxes between the surface and the atmosphere (Cooper et al., 1989). The difference between surface temperature (T_s) and atmospheric temperature (T_a) is a decreasing function of the plant's real evapotranspiration (Jackson et al., 1977).

Taking into account this relationship, another index, the forest fire index (FFRI), is computed at the IM on an operational basis:

$$\text{FFRI} = k(T_s - T_a)/[R_g(\text{NDVI}_{max})] \qquad [14.1]$$

where k is a constant, R_g is the global radiation, and NDVI_{max} is 10-day maximum NDVI value. The above index is calculated on a daily basis; T_s is calculated using AVHRR data, and T_a and R_g are collected from the IM's meteorological stations. The surface temperature, T_s, is calculated using a split window method (Melia et al., 1991) as expressed by equation 14.2:

$$T_s = T_4 + [1.31 + 0.27(T_4 - T_5)](T_4 - T_5) + 1.16 \qquad [14.2]$$

where T_4 and T_5 are the brightness temperatures of respectively channel

4 and channel 5 of the AVHRR. The temporal variation in the FFRI is analyzed for different test sites and compared with the temporal variation at each respective site for a representative year to detect drought conditions.

Satellite Application Facilities (SAFs) are specialized centers within the European Organization for Exploitation of Meteorological Satellites Applications Ground Segment and are hosted by the European National Meteorological Services in member states. The Portuguese Institute of Meteorology (www.meteo.pt) is responsible for the development of the SAF for land surface analysis in Portugal (www.meteo.pt/landsaf). The main purpose of this SAF is to enhance the benefits of the EUMETSAT satellite systems, Spinning Enhanced Visible and InfraRed Imager/Meteosat Second Generation and EUMETSAT Polar System related to land, land–atmosphere interactions, and biophysical applications by developing the techniques, products, and algorithms for more effective use of the satellite data. The land SAF involves the near real-time generation, archiving, and distribution of a coherent set of products that characterize the land surface by surface temperature, albedo, evapotranspiration, snow/ice cover, soil moisture, and vegetation parameters that are especially relevant to drought management. Some of these products include the leaf area index (LAI), the fractional vegetation cover (FVC), and the fraction of absorbed photosynthetic active radiation (fAPAR) or the fraction of green vegetation (FGV). The quality of these products depends to a large extent on the sensor characteristics (spectral, radiometric, and geometric), cloud detection, atmospheric correction, and angular distribution of the observations.

The algorithms used to estimate biophysical parameters in the land SAF depend on empirical relationships for vegetation indices (Asrar et al., 1985), inversion models (Roujean et al., 1992; Knyazikhin et al., 1998; Bicheron and Leroy, 1999), and physical and empirical models (Qin and Goel, 1995; Weiss and Baret, 1999; Lacaze and Roujean, 2001). Two complementary inversion approaches for the retrieval of FVC and LAI include kernel-driven reflectance models (Roujean et al., 1992) that obtain nadir-zenith reflectance as an input, before applying a more robust technique—namely, variable multiple endmember spectral mixture analysis (VMESMA; García-Haro et al., 2002a). With VMESMA it will be possible to estimate the subpixel abundance of vegetation, soils, and other spectrally distinct materials that fundamentally contribute to the spectral signal of the mixed pixels. Although the primary output is FVC, some empirical relationships can be then used to derive LAI (Lacaze and Roujean, 2001). The application of a directional strategy relates the Bidirectional Reflectance Distribution Function of the surface with the directional signatures of vegetation and soil and meaningful biophysical parameters (García-Haro et al., 2002b). Reflectance of an individual pixel is assumed to consist of an area-weighted linear combination of the soil and vegetation radiances. Canopy geometrical effects are considered in the first-order scattering. The model makes a simple treatment of the multiple scattering effects using an approximate analytical solution to the radiative transfer equation.

The volume scattering formulation is similar to the G-function and Hot Spot model. However, the probability between the crown and gap is formulated in terms of the FVC and a geometric variable (η) associated with the shape of plants. The canopy model can be coupled with a leaf-level radiative transfer model, which provides a stronger connection between leaf optical properties and biochemical and structural parameters. The model can be inverted to estimate both FVC and LAI.

Early Warning Systems

In Portugal an early warning system is being developed involving data acquisition, analysis leading to drought monitoring, and prediction using a software application (Pires, 2003). Such a system will help planners devise the best strategy for drought management. In the near future this product will be available on the Web for direct access by users.

A project is also being developed to identify, characterize, and predict local and regional droughts (INTERREG IIIB, 2002) using general circulation models, soil water balancing, and drought indices (e.g., PDSI, SPI) and probabilistic and stochastic modeling of regional droughts.

Crop Growth Models

In Portugal, crop growth simulation models have been used to assess the impacts of climate variability on agriculture. The models used are CERES models included in the Decision Support System for Agrotechnology Transfer series (Pinto and Brandão, 2002; Jones et al., 2003) and, more recently, the World Food Studies model (Supit and van der Goot, 2002). These models will be used for yield estimation, together with seasonal and annual forecasting, to complete the early warning system.

Drought-Mitigating Measures

Portugal is reducing risks associated with the occurrence of droughts using mitigation measures and early warning systems. Portugal is in the process of building dams and artificial lakes to store water and develop irrigation techniques to cope with water shortage, especially during drought periods. Alqueva Dam in Alentejo will create a large artificial lake. Other methods for drought mitigation include carefully choosing appropriate crops for a given region and local climate, developing crop varieties that are better adapted to higher temperatures and water stress (in the southern regions of Portugal, introduction of subtropical and tropical plants is an important strategy), improving carbon sink capacity of Portuguese forests through aforestation/reforestation, avoiding land degradation, and improving irrigation and drainage systems.

Conclusions

Drought is a significant problem for Portugal. Droughts have become more frequent and severe since the 1980s, seriously affecting agricultural production. A drought early warning system is being developed. In this regard, various drought indices and satellite data are being studied. Such a system will help better manage droughts and alleviate economic and social impacts. Cooperation between governmental and associate institutions is a must for such a system to succeed and help concerned farmers and industries.

ACKNOWLEDGMENTS We are grateful to the Ministry for Agriculture, Rural Development and Fisheries, Institute of Hydraulics, Rural Engineering and Environment (IHERA), Office of Planning and Agro-Food Policy (GPPAA), Directorate General for Forests (DGF), and Institute of Financing and Support to Agriculture and Fisheries Development (IFADAP) for supplying relevant information incorporated in this chapter. We also thank Álvaro Silva and Sofia Moita of the Climate and Environment Department of Portuguese Meteorological Institute for providing GIS maps and to Mario Pereira for his suggestions in reviewing the text.

Note

1. "Semiarid" and "dry subhumid" mean areas, other than polar and subpolar regions, where the ratio of annual precipitation to potential evapotranspiration falls within the range from 0.05 to 0.65 (United Nations Convention to Combat Desertification, 1997).

References

Asrar, G., E.R. Kanemasu, and M. Yoshida. 1985. Estimates of leaf area index from spectral reflectance of wheat under different cultural practices and solar angle. Remote Sens. Environ. 17:1–11.

Bicheron, P., and M. Leroy. 1999. A method of biophysical parameter retrieval at global scale by inversion of a vegetation reflectance model. Remote Sens. Environ. 67:251–266.

Cooper, D.I., C. Asrar, and T.R. Harris. 1989. Radiative surface temperatures of the burned and unburned areas in tallgrass prairie. University of Arizona, Tucson.

Duchemin, B., and P. Maisongrande. 2002. Normalisation of directional effects in 10-day global synthesis derived from VEGETATION/SPOT: I. Investigation of concepts based on simulation. Remote Sens. Environ. 81:90–100.

García-Haro, F.J, S. Sommer, and T. Kemper. 2002a. Variable multiple endmember spectral mixture analysis (VMESMA): A high performance computing and environment analysis tool. Remote Sens. Environ. (Under Review)

García-Haro, F.J., F. Camacho-de Coca, and J. Meliá. 2002b. Retrieval of biophysical parameters using directional spectral mixture analysis. In: Recent Advances in Quantitative Remote Sensing, J. Sobrino (Ed.), Universitat de Valencia, Torrente, Valencia, September 16–20, 2002, pp. 963–971.

Gibbs, W.J., and J.V. Maher. 1967. Rainfall deciles as drought indicators. Bureau of Meteorology Bulletin. 48. Commonwealth of Australia, Melbourne.

Gomes, I., T. Abrantes, L. Sousa, and C. Tavares. 1989. Classification des Donnés NOAA-AVHRR Selon l'Occupation du Sol. Joint Research Center, Ispra, Italy.

GPPAA. 1999. Panorama Agricultura. Ministério da Agricultura, do Desenvolvimento Rural e das Pescas, Gabinete de Planeamento e Política Agro-Alimentar, Lisboa.

Hayes, M. 2002. National Drought Mitigation Center, Drought Indices. Available http://www.drought.unl.edu, accessed 2002.

Heim, R. 2002. A review of twentieth-century drought indices used in the United States. Bulletin of Am. Meteorol. Soc. 83:1149–1165.

IHERA. 1999. Novos Regadios para o Período 2000–2006. Ministério da Agricultura, do Desenvolvimento Rural e das Pescas, Instituto de Hidráulica, Engenharia Rural e Ambiente, Lisboa.

INTERREG IIIB. 2002. Drought and desertification: Drought monitoring using meteorological and soil water balance indices. Project submitted to Technology and Science Foundation, Portugal, Lisbon.

Jackson, R.D., R.J. Reginato, and S.B. Idso. 1977. Wheat canopy temperature: A pratical tool for evaluating water requirements. Water Resour. Res., 13:651–656.

Jones, J.W., G. Hoogenboom, C.H. Porter and K.J. Boote. 2003. The DSSAT cropping system model. European Journal of Agronomy 18:235–265.

Knyazikhin, Y., J.V. Martonchik, R.B. Myneni, D.J. Diner, and S.W. Running. 1998. Synergistic algorithm for estimating vegetation canopy leaf area index and fraction of absorbed photosynthetically active radiation from MODIS and MISR data. J. Geophys. Res. 103:32.257–32.275.

Kogan, F.N. 1995. Droughts of the late1980s in the United States as derived from NOAA polar-orbiting satellite data. Bull. Am. Meteorol. Soc. 76:655–668.

Lacaze, R., and J.L. Roujean. 2001. Retrieval of biophysical parameters over land based on POLDER directional and hot spot measurements. Remote Sens. Environ. 76:67–80.

McKee, T.B., N.J. Doesken, and J. Kleist. 1995. Drought monitoring with multiple time scales. In: 9th Conference on Applied Climatology, January 15–20, Dallas, TX, pp. 233–236.

Melia, J., E. Lopez-Baeza, V. Casalles, D. Segara, J.A. Sobrino, M.A. Gilabert, J. Morino, C. Coll. 1991. EFEDA Annual report, EPOC-CT90-0030 (LNBE). University of Valencia, Valencia, Spain.

Palmer, W.C. 1965. Meteorological drought. Research Paper no. 45, U.S. Department of Commerce Weather Bureau, Washington, DC.

Palmer, W.C. 1968. Keeping track of crop moisture conditions nationwide: The new crop moisture index. Weatherwise 21:156–161.

Penman, H.L. 1948. Natural evaporation from open water, bare soil and grass. Proc. R. Soc. Lond. A 193:120–145.

Pinto, P.A., and A.P. Brandão. 2002. Agriculture. In: F.D. Santos, K. Forbes, and R. Moita (eds.), Climate Change in Portugal. Scenarios, Impacts and Adaptation Measures. SIAM Project, Lisbon, pp. 223–239.

Pires, V.C. 2003. Intensity and frequency of extreme meteorological phenomena associated to precipitation. Developing a monitoring system for drought in Mainland Portugal [in Portuguese]. Master's dissertation, University of Lisbon.

Qin, W., and N.S. Goel. 1995. An evaluation of hotspot models for vegetation canopies. Remote Sens. Rev. 13:121–159.

Report on Climate Conditions. 1992. Meteorological information of the 1991/1992 drought. Instituto Nacional de Meteorologia e Geofísica, Lisboa, Portugal.

Report on Climate Conditions. 1995. Meteorological information of the 1994/1995 drought. Instituto Nacional de Meteorologia e Geofísica, Lisboa, Portugal.

Report on Climate Conditions. 1997. Meteorological information of the 1997 drought. Instituto Nacional de Meteorologia e Geofísica, Lisbon.

Roujean, J.L., M. Leroy, and P.Y. Dechamps. 1992. A bidirectional re-flectance model of the earth's surface for the correction of remote sensing data. J. Geophys. Res. 97:20455–20468.

Smith, D.I., M.F. Hutchinson, and R.J. McArthur. 1993. Australian climatic and agricultural drought: payments and policy. Drought Network News 5(3):11–12.

Supit, I., and E. van der Goot. 2002. Updated System Description of the WOFOST Crop Growth Simulation Model as Implemented in the Crop Growth Monitoring System Applied by the European Commission. Available http://www.iwan-supit.cistron.nl/~iwan-supit/contents/, accessed January 2002.

Tucker, C.J., and P.J. Sellers. 1986. Satellite remote sensing of primary production. Intl. J. Remote Sens. 7:1395–1416.

United Nations Convention to Combat Desertification. 1997. United Nations convention to combat desertification in countries experiencing serious drought and/or desertification, particularly in Africa. The Interim Secretariat for the convention to combat desertification (CCD), GE.97-00389-February 1997. DPCSD/CCD/97/1.

Weiss, M., and F. Baret. 1999. Evaluation of canopy biophysical variable retrieval performances from the accumulation of large swath satellite data. Remote Sens. Environ. 70:293–306.

CHAPTER FIFTEEN

Monitoring Agricultural Drought in Russia
ALEXANDER D. KLESCHENKO, ERODI K. ZOIDZE,
AND VIJENDRA K. BOKEN

Drought has been posing serious problems for agricultural production in Russia. A well-known Russian scientist, Vavilov (1931), noted that droughts characterize Russian farming. Recently, in some Russian Federation regions, there has been a high probability of severe or extremely severe droughts (Pasechnyuk et al., 1977; ARRIAM, 2000; Kleschenko, 2000; Ulanova and Strashnaya, 2000; Zoidze and Khomyakova, 2000; table 15.1).

Numerous definitions of drought are available in the Russian literature (Bova, 1946; Alpatiev and Ivanova, 1958; David, 1965; Kalinin, 1981; Polevoy, 1992; Khomyakova and Zoidze, 2001). However, Kleschenko (2000) noted that all definitions are similar. Droughts are most frequently observed in Russia (Povolzhie, North Caucasus, Central-Chernozem regions, Ural, West and East Siberia) as well as in other Commonwealth of Independent States (CIS) countries: Ukraine, Moldova, Kazakhstan, Uzbekistan, Turkmenistan, Georgia, and Armenia.

The Povolzhie, North-Caucasus, and Central-Chernozem regions contribute significantly to the Russian economy because these regions have fertile chernozems soils and produce most (about two-thirds) of the food grains—wheat and rye during the winter season and wheat, maize, and barley during the spring season. In recent moisture-favorable or nondrought years (1978, 1990 and 2001), the total grain production was 130 million tons, while during drought years (1975, 1981, 1995 and 1998), the production declined by half (Ulanova and Strashnaya, 2000).

Decline in food grain yields was observed from 1917 to 1990 in the former USSR, and since 1990 in the post-Soviet Russia. Rudenko (1958) reported that Ukraine experienced severe droughts during 1875, 1889, 1918, and 1921, when the spring wheat yield was 70% of the mean yield. A sudden depression in the winter rye yield was observed in Povolzhie region

Table 15.1 Recurrence (%) of extremely severe and severe droughts in different regions of Russia during 1946–95 period

Territory	May	June	July
Central-Chernozem Economic region (averaged)	23	18	16
Lipetzk region	30	26	10
Tambov region	40	30	14
Kursk region	26	16	16
Belgorod region	26	24	24
Voronezh region	30	34	16
Povolzhsky economic region (averaged)	30	30	40
Volgograd region	50	56	60
Saratov region	38	52	50
Samara region	36	30	30
Penza region	38	22	12
Ulianovsk region	34	20	14
Tatarstan	38	24	14
Astrakhan region	78	82	96
Kalmykia	56	54	72
North-Caucasus economic region (averaged)	16	18	26
Rostov region	26	26	48
Krasnodar territory	16	14	22
Stavropol territory	14	20	32

Source: Ulanova and Strashnaya (2000).

during severe droughts of 1890, 1898, and 1911, when the yield was less than 60%, and during 1906, when the yield was only 25% of the mean yield. During severe droughts in Russia during 1972, 1975, 1979, 1984, and 1995, the crop yield deviated by an average of 17–42% in Russia as a whole, up to 19–91% in the Central-Chernozem regions, up to 45–100% in Povolzhie region, 27–36% in the North-Caucasus region, and 21–100% in the Ural region (table 15.2).

Figure 15.1 shows the variation in winter and spring wheat yields, depending on the drought intensity during various developmental phases of

Table 15.2 Grain yield deviation (%) as compared to the averaged yield in drought years in Russia and some of its regions

Territory	Drought year (yield deviation from the average, %)					
Russian Federation	1975	1979	1981	1984	1995	
	(32)	(19)	(37)	(17)	(42)	
Central-Chernozem region	1975	1979	1981	1984	1995	
	(19)	(88)	(89)	(77)	(91)	
Povolzhsky region	1972	1975	1979	1981	1984	1995
	(53)	(100)	(45)	(68)	(72)	(79)
North-Caucasus region	1972	1975	1979	1995		
	(27)	(30)	(36)	(33)		
Ural region	1975	1981	1984	1987	1995	
	(100)	(21)	(45)	(41)	(31)	

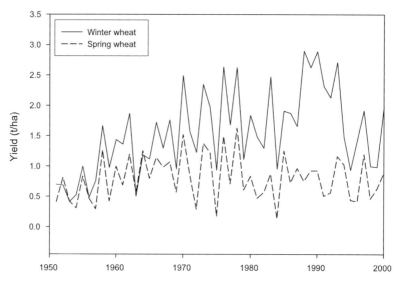

Figure 15.1 The variation in winter and spring wheat yields in Volgograd region, Russia, for the 1950–2000 period.

the crop. For the areas under study, the winter wheat yield was twice as high as the spring wheat yield. For example, in Volgograd region as a whole, the mean winter wheat yield was 1.6 tons/ha, whereas the spring wheat yield was 0.8 tons/ha, during the 1951–2000 period. The same ratio (1.2 tons/ha in winter, and 0.6 tons/ha in spring season) was observed in Novoakhtubinsk area of Volgograd region during the 1972–2000 period (figure 15.2).

Both winter and spring wheat yields react to arid conditions. In Novoakhtubinsk region, for example, during severe droughts of 1972, 1975, 1979, 1984, and 1998, the spring wheat yield ranged from 0.01 to 0.1 tons/ha and the winter wheat yield had the same range during 1984, 1998, and 1999. The drought problems were also severe for Moldova, Kazakhstan, and Uzbekistan.

Main Reasons for Drought

The main reason for drought formation in arid regions of Russia and other CIS regions is the penetration of anticyclonic air masses from the Arctic (Selyaninov, 1928, 1958, 1966). These air masses have low temperature and low water content and are mostly transparent for solar radiation under clear weather conditions. When these air masses move south, their temperature rises quickly, which increases the water vapor deficit in the atmosphere without changing its absolute water content. As a result, the relative humidity falls to about 10% or lower, causing droughts. When a system of arctic anticyclones moves to the west, it acts in parallel with the

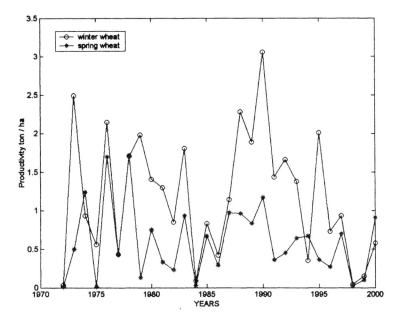

Figure 15.2 The variation in winter and spring wheat yields in Novoakhtubinsk area (Volgograd region, Russia) for the 1970–2000 period.

ring of subtropical anticyclones at the southern (40–50°) and high (≈75°) latitudes. Normal air circulation in the Northern Hemisphere is disturbed, and dry weather conditions in the former USSR European territory, West Siberia, and Kazakhstan are formed. As the Atlantic air masses in subtropical anticyclones move away from the ocean, they are heated more slowly as compared to polar intrusions. However, having passed a longer way over the continent (from the Bay of Biscay to Dnieper), water vapor in the air is exhausted, and drought phenomena become more pronounced to the west of Dnieper.

Most often, the arctic and the Azores intrusions are combined in the Lower Volga and the Lower Yuzhy Ural, and pronounced drought conditions occur in the former USSR regions. At times, the geographical position of subtropical and the Azores anticyclones vary from year to year and even from month to month. As a result, the geographical distribution of droughts varies on a temporal scale. The Azores anticyclone may move to 55° latitude, bringing drought conditions to the whole southern part of the former USSR. In the case of the Meridional polar intrusions, not only does air dry up quickly and intensely, but also the movement of moist Atlantic air masses to the east is retarded, and droughts occur during the whole vegetation period.

Khomyakova and Zoidze (2001) determined the probabilities of drought occurrence of various regions. The degree of drought intensity (DDI) was obtained using nine indices (Zoidze and Ovcharenko, 2000).

The study of severe and extremely severe droughts in the primary arid Russian Federation regions has shown that there is no certain periodicity in droughts (Khomyakova and Zoidze, 2001). Droughts may occur during some years in succession as well as with a 1-, 2-, 3-, and even 16-year interval. Therefore, the prediction of this phenomenon becomes much more complicated, and drought monitoring at regular basis is more important.

Procedures for Drought Monitoring

At early stages of studying droughts in Russia, much effort was directed to the development of argometeorological indices to assess drought conditions. Russian scientists have developed more than 50 indices (Ivanov, 1949; Koloskov, 1958; Konstantinov et al., 1976; Loginov et al., 1976; Drosdov, 1980; Ulanova, 1988; Khomyakova and Zoidze, 2001). However, in this chapter we discuss only those indices that are widely used for drought monitoring in Russia.

Hydrothermal Coefficient

A hydrothermal coefficient (HTC) developed by Selyaninov (1928) is widely used to monitor droughts in Russia. The HTC is defined as

$$\text{HTC}_i = \frac{\sum p_{[i+(i-1)+(i-2)]}}{0.1 \sum T_{\geq 10°C[i+(i-1)+(i-2)]}} \quad [15.1]$$

where p is monthly precipitation in millimeters and T is temperature in degrees Celsius, i is the ordinary number of a dekad (10-day period) to be estimated, and $\sum T_{\geq 10°C}$ is the sum of temperatures above or equal to 10°C.

Moisture Index

Moisture index (Md) was developed by Shashko (1985) and is also widely used. It is the ratio between the total precipitation during a month and the diurnal mean values of water vapor deficit (the diurnal mean deficit sum of air humidity), as expressed by the following:

$$\text{Md}_i = \frac{\sum R_{[i+(i-1)+(i-2)]}}{\sum d_{[i+(i-1)+(i-2)]}} \quad [15.2]$$

where $\sum R$ is the precipitation sum in millimeters, and $\sum d$ is the sum of water vapor pressure in air (the sum of air humidity deficits) in hecto Pascal (hPa).

Water Supply Index

The water supply index (V_i) is calculated as

$$V_i = \frac{|W_{0-100(i-1)} - W_{0-100(i)} + R_i|}{0.375 \sum d_i} (100) \quad [15.3]$$

where W means water storage in the corresponding soil layer available to plants, 0–100 (i − 1) is the index showing thickness of relevant soil layer, and *i* is the number of the dekad.

The above indices are widely used in all CIS countries. In Ukraine the drought conditions are also estimated from the index developed by Dmitrienko (1978, 1992):

$$\xi = \eta(W)\beta + \eta(T)\eta(R)\left[1 - \gamma(d)\right](1 - \beta) \qquad [15.4]$$

where ξ is the complex dryness index in terms of relative crop yield; $\eta(W)$ is the field crop productivity coefficient derived from soil water reserves, W; β is the weight coefficient; $\eta(T), \eta(R)$ are the temperature, T, and precipitation, R, coefficients of field crop productivity; and γ is the index of crop depression from the hot wind impact, d.

Drought Index

Specialists and scientists working in the field of agrometeorology accept the drought index developed by Ped (1975):

$$S_{i(\tau)} = \frac{\Delta T}{\sigma_T} - \frac{\Delta R}{\sigma_R} - \frac{\Delta E}{\sigma_E} \qquad [15.5]$$

where S_i is the drought index which is the sum of anomalous weather conditions, i is the site; τ is the time; ΔT is the mean air temperature deviation from the normal (T_N) during a period T; ΔR is the deviation in precipitation from the normal (R_N) during a period (R); ΔE is the deviation of moisture reserves in a 1-m soil layer (E) from the normal (E_N); and σ_T, σ_R, and σ_E are the root-mean-square deviation in temperature, precipitation, and moisture reserve, respectively.

Zoidze and Khomyakova (2000) analyzed the above indices and also the following indices to describe drought conditions: (1) maximum air temperatures above 30, 35, 40, and 45°C in dekads, (2) rainless periods of different durations, (3) relative air humidity of 30% or less, (4) productive moisture reserves in the 0–20 and 0–100 cm soil layers during winter, early and later spring cereals at the beginning of sowing and mass development phase as well as for individual vegetation months of these crops, (5) water vapor saturation deficit in air at 1500 h at various wind velocities, and (6) crop yield deviations depending on arid conditions.

Until 1917, the Science Committee of Meteorological Agency of the Central Land Organization and Farming Administration monitored droughts in Russia and published map showing probability of dry dekads in the European Russia (Brounov, 1913); a dekad during which the precipitation did not exceed 5 mm was considered a dry dekad. In Soviet and post-Soviet periods, the Hydrometeorological Service monitored drought. This agency collects information about droughts in the Russian Federation regions and submits it to the Ministry of Agriculture and other federal authorities. For the past 20–30 years, the agency has published and updated various agroclimatic reference manuals. For example, two guide books, *Agroclimatic Guide Book* and *Agroclimatic Resources*, were published for all regions, territories, and republics of the former USSR. These books show intensity of drought (e.g., weak, moderate, severe, extremely severe) for different regions based on the average number of drought days in a month during vegetation or warm periods and probabilities of drought occurrence.

Yield Risk Estimation

According to Zhukov et al. (1989), the following equation can be used to estimate the yield risk (i.e., the deviation of expected crop yield from the maximum yield):

$$P_{ij} = 1 - \prod_{k=1}^{M} \left[1 - g_{ij}(k)\right] \qquad [15.6]$$

where P_{ij} is the pure risk standard of the system climate-yield transition from the initial ith state into the jth state in a given time interval; $g_{ij}(k)$ is the transitional probability of the system weather-yield, and k is the dekad number.

Agricultural drought monitoring in Russia has become more systematic after two publications based on ground and satellite data (Zoidze and Khomyakova, 2000). In the first publication, degree of aridity was determined with the help of a mathematical model based on an image recognition procedure (Danielov and Zhukov, 1984; Zhukov et al., 1989; Zhukov and Svyatkina, 2000). Zoidze and Ovcharenko (2000) prepared a map showing the distribution of the degree of aridity in the Russian Federation regions based on an aridity index. The index value was more than 60 (highest) for Kalmykia region and between 51 and 60 for Astrakhan region. About 14% of the RF regions had index value greater than 30.

The operative system of drought assessment is a computer-based system that uses agrometeorological data and various drought indices to regularly monitor the onset and development of droughts of different intensity during the entire vegetation period. The following drought indices are used: (1) Selyaninov's HTC, (2) Shashko's index of moisture (Md), (3) Protserov's water supply index (V; Protserov, 1949), (4) the number of days with relative air humidity <30% (N_0), (5) the number of days with maximum air temperature > 30°C (N_T), (6) productive moisture reserves in soil layers of 0–20 cm (W_{0-20}), 0–50 cm (W_{0-50}), and 0–100 cm (W_{0-100}) under winter, early spring, and late spring grain crops. All of these indices can be determined using standard hydrometeorological data: precipitation (R), air temperature ($T°C$), water vapor pressure in air (d), productive water reserves in soil (W), the number of days with the relative air humidity <30% (N_0), and the number of days with the maximum air temperature >30°C (N_T).

Based on the range of the above values, a climatic event can be categorized as extremely severe, severe, moderate, weak, or no drought (table 15.3). Drought is referred to that category of intensity where its average measure of closeness (P) is maximum:

Table 15.3 Drought categorization based on various indices used in Russia

Drought assessment index	Category of drought intensity				
	No drought (class 5)	Weak (class 4)	Moderate (class 3)	Severe (class 2)	Extremely severe (class 1)
1. Selyaninov's hydrothermal index (HTC)	≥ 0.76	0.61–0.75	0.40–0.60	0.20–0.39	≤ 0.19
2. Shashko's index of moisture (Md)	≥ 0.41	0.31–0.40	0.20–0.30	0.10–0.19	≤ 0.09
3. Protserov's index of water supply (V, %)	≥ 71	61–70	51–60	41–50	≤ 40
4. Number of days with relative air humidity $\leq 30\%$ (N_O)	0	1–2	3–5	6–7	8–11
5. Number of days with maximum air temperature $>30°C$ (N_T)	0	1–2	3–4	6–7	8–11
6. Productive moisture reserves (mm) in the 0–20 cm soil layer (W_{0-20})	≥ 21	16–20	11–15	6–10	≤ 5
7. Productive moisture reserves (mm) in the 0–50 cm soil layer (W_{0-50})	≥ 46	36–45	26–35	16–25	≤ 15
8. Productive moisture reserves (mm) in the 0–100 cm soil layer (W_{0-100})	≥ 81	61–80	41–60	26–40	≤ 25

$$\overline{P_i} = \frac{1}{n}\sum_{j=1}^{n}\delta_j\alpha_{ji}, \qquad [15.7]$$

where $\overline{P_i}$ is the mean value of measures of closeness to the category for the ith dekad; n is the number of indices; δ_j is the information weight of indices given in advance; α_{ji} is the metrics characterizing the measure of closeness of the actual value of the jth index for the ith dekad relative to a chosen criterion (boundary value) in the category; and j is the ordinary number of an index.

The metrix α_{ji} is calculated from the following formula:

$$\alpha_{ij} = 1 - \frac{|x_{ij} - \varphi'_{ij}|}{\varphi''_{ij\alpha} - \varphi'_{ij\alpha}}, \quad \text{if } x_{ij} < \varphi'_{ij} \qquad [15.8]$$

$$\alpha_{ij} = 1 - \frac{|x_{ij} - \varphi''_{ij}|}{\varphi''_{ij\alpha} - \varphi'_{ij\alpha}}, \quad \text{if } x_{ij} > \varphi''_{ij} \qquad [15.9]$$

$$\alpha_{ij} = 1, \quad \text{if } \varphi'_{ij} \leq x_{ij} \leq \varphi''_{ij} \qquad [15.10]$$

where x_{ji} is the actual value of the jth index of drought assessment in the ith dekad; φ_j' and φ_j'' are the range (min, max) of the jth index within the drought category; and $\varphi'_{ij\alpha}$, $\varphi''_{ij\alpha}$ are the absolute ranges (minimum and maximum values) of the jth index of drought assessment. The minimum value is zero for every index listed in table 15.3, while the maximum values are 5, 3, 100, 11, 11, 70, 140, and 280 for each index listed in table 15.3, respectively.

Recently, a Russian meteorological satellite (Meteor) has been used for drought monitoring, in addition to satellites of the U.S. National Oceanic and Atmospheric Administration (NOAA) (chapters 5 and 6). Beginning in 1995, RF Roshydromet has carried out regular drought monitoring from ground-based and satellite information (Kleschenko, 1998). Subsequent studies in this direction will be aimed at obtaining a coordinated map of aridity from satellite and ground data. These activities will be performed by a Drought Monitoring Centre (DMC) that will be located in the territory of the All-Russian Research Institute of Agricultural Meteorology (ARRIAM, Obninsk, Russia).

Drought Early Warning System

In Russia and most of the CIS countries a drought warning system is in its infancy. The Interstate Council for Hydrometeorology has taken a decision to establish the DMC at the ARRIAM. One of the primary objectives of this center is to develop the procedures for monitoring agricultural droughts (intensity, area, duration), their early diagnosis, drought impact on the condition of crops, pastures, meadows, and crop yields. A dekadal bulletin containing drought information will be issued for the period from April to October. The following steps are involved in the development of this system: (1) agrometeorological information in a form of agrotelegrams for the ith dekad is received from observation site of RF Hydrometeorological Service by ARRIAM (DMC), (2) ARRIAM estimates drought intensity during i^{th} dekad, (3) the hydrometeorological center makes a monthly weather forecast in the territory of Russia, which is presented to ARRIAM in early month, and (4) ARRIAM forecasts the expected drought development using the estimated drought category for the ith dekad and informs possible users.

Measures for Drought Mitigation

Various biological, agrotechnical, and agrometeorological methods are used to mitigate the adverse impact of arid conditions (Gringof, 2000). These include (1) selecting drought-resistant crop varieties; (2) following complex water-saving practices (crop rotation, leaving fallows, snow retention, establishing field windbreaks, fertilizing, weed control, and effective irrigation including drip irrigation); (3) shifting the sowing dates; (4) optimizing the winter and spring crop ratio; (5) presowing drying of germinated seeds (i.e., inducing artificial drought before sowing that provokes the deep physiological–biochemical plant reconstruction resulting in increased drought resistance); (6) salt hardening, which increases chloride, sulfate, and carbonate resistance of plants (presowing seed treatment with sodium chloride, magnesium sulfate, and sodium carbonate solutions decreases cytoplasm membrane permeability and increases the toxic salt effect threshold); (7) presowing heating (up to 35–38°C) of potato bulbs; and (8) presowing hardening of fodder grass seeds.

Some of the above measures, however, are very expensive and cannot be adopted in the whole territory affected by droughts. In addition, because of the long-term impacts of irrigation and fertilizers, there is soil degradation, salting, environmental pollution, and so on. In this case, farming adaptations to the specific arid conditions seems to be most effective. Therefore, it is necessary to determine where, when, in what form, with what intensity, and with what probability the arid conditions exist, and to determine their durations and geographical distributions. By knowing these characteristics, it is possible to develop a specific farming type that allows the adverse drought effects under these specific conditions to be mitigated.

Conclusions

Highly intensive agricultural droughts are experienced in some regions of Russia, Ukraine, Moldova, Kazakhstan, and other CIS countries. Povozhie, North Caucasus, and Central-Chernozem are arid regions that produce two-thirds of the total grain production in Russia. In severe drought years, crop yields are reduced to 40–60% as compared to nondrought years. The Russian system of early drought monitoring is in its developmental stage. However, its isolated components have been in operation for some years.

References

Alpatiev, A.M., and V.N. Ivanova. 1958. Characteristic and geographical distribution of droughts. In: A.I. Rudenko (ed.), Droughts in the USSR. Their origin, recurrence and crop impact. Hydrometeoizdat, Leningrad, pp. 31–45.

AARIAM. 2000. Problems of drought monitoring. Proceedings of ARRIAM, vol. 33.

Bova, N.V. 1946. On climatic studies of droughts in the south-east of the USSR. Izv. AN SSSR. Ser. Geogr. Geogr. 5:150–156.

Brounov, P.I. 1913. Schematic charts of the probability of drought dekads in the European Russia. Atlas on Agricultural Meteorology. Meteorological Bureau, St. Petersburg.

Danielov, S.A., and V.A. Zhukov. 1984. Typization of weather conditions in a vegetation period by image recognition methods. Proc. ARRIAM, Vyp. 12:111–119.

David, R.E. 1965. Studies of drought and ways to control it. In: Isolated Papers on Agricultural Meteorology. Hydrometeoizdat, Leningrad, pp. 154–215.

Dmitrenko, V.P. 1978. On a complex agrometeorological index of dryness. In:

Proceedings of the Ukranian Research Hydrometeorological Institute, Leningrad, vol. 169, pp. 3–24.

Dmitrenko, V.P. 1992. Methodical instructions on complex assessment of drought impact on the cereals and sugar beet yields. Moscow Branch of Hydrometeoizdat.

Drosdov, O.A. 1980. Droughts and moistening dynamics. Hydrometeoizdat, St. Petersburg.

Gringof, I.G. 2000. Droughts and desertification—ecological problems of today. In: Proceedings of All-Russian Research Institute of Agricultural Meteorology, vol. 33, pp. 14–40.

Ivanov, N.N. 1949. Landscape-climate global zones. In: Notes of VGO AN SSSR. V. I (new version). Hydrometeoizdat, St. Petersburg.

Kalinin, N.I. 1981. A principal scheme of agrometeorological assessment of droughts, territory dryness, and crop drought-resistance. Hydrometeoizdat, Leningrad.

Khomyakova, T.V., and E.K. Zoidze. 2001. Drought phenomena in Russia and experience in their assessment. Literature review. Manuscript in All-Russian Research Hydrometeorological Institute-World Data Center, Obninsk.

Kleschenko, A.D. 1998. Drought monitoring on the basis of remote sensing data. In: International Conference on Early Warning Systems for the Reduction of Natural Disasters. Potsdam, Germany, September 1998.

Kleschenko, A.D. 2000. Modern problems of drought monitoring. In: Proceedings of ARRIAM, vol. 33, pp. 3–13. Hydrometeoizdat, St. Petersburg.

Koloskov, P.I. 1958. Problems of agroclimatic zonation in the USSR. Proc. Res. Inst. Aeroclimatol. 6:5–51. Hydrometeoizdat, Leningrad.

Konstantinov, A.R., V.V. Svirina, and N.K. Chernova. 1976. Drought assessment in the European USSR territory as applied to winter wheat. Proc. IEM. 6(57):136–146. Hydrometeoizdat, Leningrad.

Loginov, V.F., A.I. Neushkin, and E.V. Rogova. 1976. Droughts, their possible causes, prerequisites, and prediction: review. VNIIGMI-MCD, Obninsk. Hydrometeoizdat, Leningrad.

Pasechnyuk, L.E., V.A. Zhukov, E.K. Zoidze, and L.G. Mamaeva. 1977. Drought characteristics and propagation in the USSR territory. Proc. IEM 11(79):3–18.

Ped, D.A. 1975. On the drought and supersaturation index. Proc. USSR Hydrometcenter. 156:19–38. Hydrometoizdat, Leningrad.

Polevoy, A.N. 1992. Agricultural Meteorology. Hydrometeoizdat, St. Petersburg.

Protserov, A.V. 1949. Common drought characteristics and main peculiarities of the drought of 1946. In: Proc. TsIP 13(40):19–28. Hydrometeoizdat, Leningrad.

Rudenko, A.I. (ed.). 1958. Droughts in the USSR. Their Origin, Recurrence and Crop Impact. Hydrometeoizdat, Leningrad.

Selyaninov, G.T. 1928. On agricultural assesment of climate. In: Proceedings of Agricultural Meteorology, vol. 20, pp. 165–177.

Selyaninov, G.T. 1958. Drought origin and dynamics. In: Droughts in the USSR. Their Origin, Recurrence and Crop Impact. Hydrometeoizdat, Leningrad, pp. 5–29.

Selyaninov, G.T. 1966. Agroclimatic Map of the World. Hydrometeoizdat, Leningrad.

Shashko, D.I. 1985. Agroclimatic Resources of the USSR. Hydrometeoizdat, Leningrad.

Ulanova, E.S. 1988. Methods for Assessing Agroclimatic Conditions and Cereals Yield Forecasts. Hydrometeoizdat, Leningrad.
Ulanova, E.S., and A.I. Strashnaya. 2000. Droughts in Russia and their impact on the grain crop yields. In: Proceedings of ARRIAM, vol. 33, pp. 64–83.
Vavilov, N.I. 1931. World resources of dry-resistant varieties. In: Reports of the All-Union Conference on Drought Control. Bulletin no. 2, pp. 18–29. Hydrometeoizdat, Leningrad.
Zhukov, V.A., A.N. Polevoy, A.N. Vitchenko, and S.A. Danielov. 1989. Mathematical methods for assessing agroclimatic resources. Hydrometeoizdat, Leningrad.
Zhukov, V.A., and O.A. Svyatkina. 2000. Agroclimatic characteristic of droughts, assessment of their impact on crop yields and consideration in planning and organizing the agricultural production. In: Proceedings of ARRIAM, vol. 33, pp. 101–117.
Zoidze, E.K., and T.V. Khomyakova. 2000. Assessment of drought phenomena in Russian Federation. In: Proceedings of ARRIAM, vol. 33, pp. 118–133.
Zoidze, E.K., and L.I. Ovcharenko. 2000. Comparative Evaluation of Agricultural Climate Potential of the RF Territory and the Degree of Usage of Its Agroclimatic Resources by Crops. Hydrometeoizdat, St. Petersburg.

CHAPTER SIXTEEN

Monitoring Agricultural Drought in the Near East

EDDY DE PAUW

The countries of North Africa and West Asia, hereafter referred to as the "Near East," cover a large part of the world (more than 7,200,000 km^2). This region is characterized by diverse but generally dry climates, in which evaporation exceeds precipitation. The level of aridity is indicated by the aridity index, the ratio of annual precipitation to annual potential evapotranspiration, calculated by the Penman method (UNESCO, 1979). The degree of aridity is shown spatially in figure 16.1 and summarized per country in table 16.1. These data show that the region is characterized by humid, subhumid, semiarid, and arid to hyperarid moisture regimes. In addition, temperature regimes vary considerably, particularly due to the differences in altitudes and, to a lesser extent, due to the oceanic/continental influences. For most of the region, the precipitation generally occurs during the October–April period and thus is concentrated over the winter season.

Table 16.1 shows that, with more than 90% of the land area in hyperarid, arid, or semiarid moisture regimes, aridity is very significant in the Near East. Turkey is better endowed with surface and groundwater resources due to the orographic capture of Atlantic cyclonal precipitation, but much of the interior is semiarid. If one excludes the hyperarid zones, which cover the driest deserts and have no potential for agricultural use, nearly 34% of the region, or about 2,460,000 km^2, is dryland (i.e., the area with arid or semiarid moisture regime). These are the areas with some potential for either dryland farming (in semiarid zones) or for extensive rangeland (in arid zones).

In the Near East countries, agriculture contributes about 10–20% to the gross domestic product and is therefore a major pillar of their economies. However, the indirect importance of agriculture is larger because it provides the primary goods that constitute the majority of merchandise exports and because of the relatively high number of people employed in agriculture (table 16.2).

AGRICULTURAL DROUGHT IN THE NEAR EAST 209

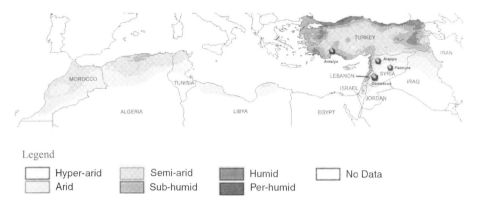

Figure 16.1 Aridity in North Africa and West Asia.

Because of the high degree of aridity in large parts of the region, agriculture in the Near East is particularly vulnerable to drought. Most of the agricultural systems depend on rainfall. Irrigation water is scarce, and although the area under irrigation is expanding, supply constraints are likely to increase. The reasons are limitations on the total size of the extractable water resources, consideration of environmental and socioeconomic impacts of large dam-building programs, continued population growth coupled with increasing urbanization, and competition among communities, industrial and service sectors, and agriculture for increasingly limited water resources. Irrigated agriculture currently consumes an average of about

Table 16.1 Climatic moisture regimes in North Africa and West Asia

	Moisture regime						
Country/region	% Hyper-arid	% Arid	% Semi-arid	% Sub-humid	% Humid	% Per-humid	Area (km²)
Algeria	71.7	15.9	8.6	3.8	0.0	0.0	2,381,741
Egypt	91.5	8.5	0.0	0.0	0.0	0.0	997,739
Gaza Strip	0.0	16.9	83.1	0.0	0.0	0.0	363
Iraq	11.6	69.9	17.3	1.2	0.0	0.0	435,052
Israel	2.6	55.3	34.7	7.4	0.0	0.0	20,700
Jordan	23.5	69.8	6.7	0.0	0.0	0.0	88,946
Lebanon	0.0	0.0	19.5	55.3	15.1	10.1	10,230
Libya	80.9	17.6	1.5	0.0	0.0	0.0	1,757,000
Morocco	0.0	41.8	55.5	2.7	0.0	0.0	458,730
Syria	0.0	71.3	23.1	3.3	1.4	0.9	185,180
Tunisia	14.4	52.3	30.3	3.0	0.0	0.0	164,150
Turkey	0.0	0.0	29.6	48.7	18.5	3.2	779,452
West Bank	0.0	20.4	74.4	5.2	0.0	0.0	5,900
% of total area	56.8	21.4	12.4	6.9	2.0	0.4	7,285,183

Source: Computed from the GIS data archived at the International Centre for Agricultural Research in Dry Areas (ICARDA).

Table 16.2 Agriculture in the Near East countries

Country	Importance of agriculture		Main crops (in order of importance, most important first)[a]						
	% GDP[b]	% Agricultural workers[c]	1	2	3	4	5	6	7
Algeria	11	26.0	WHE	POT	BAR	WAT	CIT	ONI	OLI
Egypt	18	40.0	SUC	WHE	MAI	CIT	SUB	POT	WAT
Iraq	n.a.	16.0	WHE	BAR	WAT	CIT	POT	GRA	MAI
Israel	n.a.	4.1	CIT	WAT	POT	WHE	COT	GRA	ONI
Jordan	3	15.3	CIT	POT	OLI	WAT	WHE	ONI	BAR
Lebanon	12	7.3	CIT	SUB	POT	GRA	WAT	OLI	ONI
Libya	n.a.	10.9	WAT	POT	OLI	ONI	WHE	CIT	BAR
Morocco	25	44.7	WHE	SUB	BAR	CIT	SUC	POT	OLI
Syria	n.a.	33.2	WHE	SUB	COT	BAR	OLI	GRA	POT
Tunisia	13	28.1	WHE	OLI	BAR	WAT	POT	SUB	ONI
Turkey	15	53.1	WHE	SUB	BAR	POT	WAT	GRA	ONI

Source: Food and Agriculture Organization (FAO) on-line FAOSTAT database (http://www.fao.org) and Word Resources Institute, Washington, DC (http://earthtrends.wri.org).

[a]BAR: barley; CIT: citrus; COT: cotton; GRA: grape; MAI: maize; OLI: olive; ONI: onion; POT: potato; SUB: sugar beet; SUC: sugarcane; WAT: watermelon; WHE: wheat. Ranking based on production figures.

[b]Percentage contribution to GDP from agriculture.

[c]Agricultural workers as a percentage of the total labor force.

80% of all the water used in the Near East (Margat and Vallée, 1999). In summary, the livelihoods and food security of large population segments in the Near East depend directly or indirectly on weather conditions.

Drought in the Near East

Regardless of the degree of aridity, precipitation variability is considerable in the Near East. Figure 16.2 shows the annual rainfall variations for three stations in different moisture regimes across a considerable annual precipitation gradient (150–1000 mm). It is evident from figure 16.2 that high rainfall variability is not confined to the low rainfall areas of the region. The large amplitude of the variations is typical for the region and predisposes it to drought. The patterns of drought in the region are extremely variable in their spatial and temporal dimensions. Some droughts are severe enough to affect the entire region from Morocco to Iran, and well beyond, into Afghanistan and Tajikistan.

Although droughts can have a regional scale, their effects can be very local. In Syria the drought of 1999 initiated a severe decline in the productivity of the rangelands and barley areas at the steppe margins, which continued for several years. However, the drought had comparatively little effect on the production of wheat and tree crops in the higher rainfall areas, which recovered from 2000 onward. This variability of drought patterns at

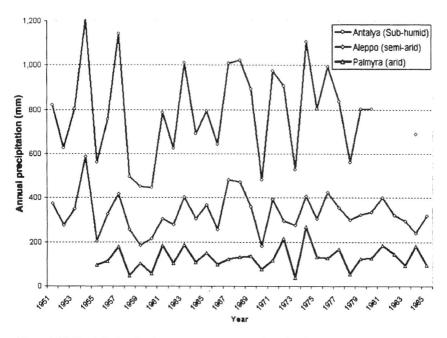

Figure 16.2 Precipitation for three stations in Syria and Turkey located in different moisture zones.

the local level is illustrated in figure 16.3. It shows the cumulative rainfall deviations for Aleppo, Damascus, and Palmyra in Syria, three locations less than 200 km apart but with different patterns of longer term positive or negative anomalies.

Throughout the region, drought can strike at any time of the growing season. Early-season, mid-season, and late-season droughts are all possible. Table 16.3 shows the occurrences of severe and mild droughts within the growing season at Tel Hadya, Syria, for the 1978–2001 period. For this particular example, a severe drought is defined as a precipitation total of less than 50% of the long-term average, and a mild drought as 50–70% of the long-term average. The periods considered are November–December (the early season), January–February (the mid-season), and March–April (the late season).

Causes of Drought Occurrence

The causes of drought in the Near East are complex, attributed to the geographical extent and exposure to different oceanic/continental influences and the wind systems in the western versus eastern parts of the region.

In North Africa, climate is still poorly understood. According to Ward et al. (1999), most of the precipitation is generated as a result of depressions, which are steered southward from the North Atlantic during

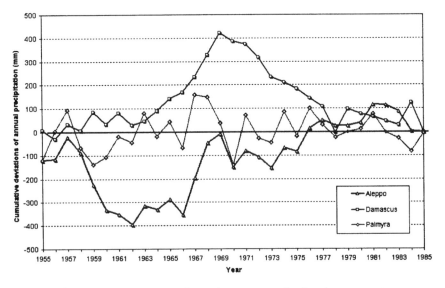

Figure 16.3 Cumulative deviations of annual precipitation for three locations in Syria.

blocking episodes by mid-latitude high-pressure cells. The strength of the blocking is, to a large extent, linked with the North Atlantic Oscillation (NAO; Lamb and Peppler, 1987), a large-scale mode of climate variability in the Northern Hemisphere at monthly, seasonal, interannual, and decadal timescales. According to Hurrell (1995), the NAO is the dominant mode of interannual–decadal climate variability for the Atlantic sector, accounting for 20–60% of the variance over the last 150 years. A simple index of the NAO is the sea-level air pressure difference between the Azores, an island group in the Atlantic Ocean at about 38° northern latitude, and Iceland. A positive NAO phase is characterized by the strengthened westerly winds across the mid-latitude North Atlantic, leading to mild and relatively wet winters in northern Europe (van Loon and Rogers, 1978) but to anomalously dry conditions in the Iberian Peninsula (Zorita et al., 1992) and the Maghreb countries (Lamb and Peppler, 1987). A negative NAO tends to be linked with moist air in the Mediterranean and cold air in northern Europe. The NAO index varies from year to year but tends to remain in one phase for intervals lasting several years. Since the late 1970s, the winter NAO index tends to be positive (Hurrell, 1995), while precipitation in subtropical northwestern Africa and the Iberian Peninsula is on a declining trend (Ward et al., 1999).

In the eastern Mediterranean, both the Atlantic Ocean and the Mediterranean Sea are the primary source regions for the formation of winter precipitation in the form of mid-latitude cyclones (Turkes, 1996). These migratory low-pressure systems have four primary cyclonic centers near Crete, Cyprus, southern Italy, and the Gulf of Genoa. Cullen and de-Menocal (2000) found a physical link between precipitation in the eastern

Table 16.3 Intraseasonal droughts at Tel Hadya, Syria[a]

	Early season	Mid-season	Late season
1978–79	1		1
1979–80			
1980–81			
1981–82			
1982–83			
1983–84		1	
1984–85			2
1985–86	1		1
1986–87			
1987–88			
1988–89		2	2
1989–90		1	2
1990–91	2		
1991–92			
1992–93			
1993–94	1		2
1994–95		2	
1995–96			
1996–97		1	
1997–98			
1998–99			
1999–00	2		
2000–01			1

Note: 1: mild drought; 2: severe drought.

Mediterranean and the NAO and its impact on the stream flow of the Tigris and Euphrates. Their analysis indicates that during positive-NAO years, Turkey (and to a lesser extent northern Syria) become cooler and drier, whereas during negative-NAO years anomalous warmer and wetter conditions prevail. Both in the Maghreb countries, particularly Morocco, and in the eastern Mediterranean a link thus appears to exist with North Atlantic sources of climatic variability. The predictability of this phenomenon and its application in early warning systems will be explored further on.

Impact of Drought on Agriculture

The major crops of the Near East include cereals (wheat, barley, maize), potatoes, olives, sugarcane, sugar beets, fruits (especially citrus, grapes, and watermelons), cotton, and vegetables (especially onions). The relative importance of each of these crops is shown on a country basis in table 16.2. Of these crops, barley is one of the most important in the Near East because of its link with livestock production and its value for the economic exploitation of agriculturally marginal lands. It covered about 11 Mha (during 1995–2002), accounting for approximately 13% of the region's total cropland as calculated from the FAOSTAT on-line database (FAOSTAT, 2002; http://www.fao.org).

Agricultural production in the region is strongly influenced by fluctuations in precipitation. To show the impact of weather on production in the region, barley is a suitable indicator crop because its production nearly always relies on rainfall. If rainfall is inadequate for grain formation, barley is not even harvested. Instead, its biomass is used for sheep grazing, thereby accentuating the swings in national production statistics.

An example of the absolute production fluctuations for barley in Morocco during the 1961–2000 period is shown in figure 16.4. The straight line represents the production trend. The relative scale of barley production fluctuations in the major producing countries of North Africa and West Asia is shown as a percentage deviation from the trend production in figures 16.5 and 16.6. Severe production fluctuations closely follow the precipitation fluctuations. An exception is Turkey, where the production fluctuations are more attenuated because the higher precipitation levels ensure that the failure of barley crop is unlikely even in drought years.

Reports of severe droughts are common in the Near East. Cullen and deMenocal (2000) reported droughts in Turkey during 1973, 1984, 1989, and 1990. Prolonged drought periods were also reported in Morocco during 1979–84 (Cullen and deMenocal, 2000) and 1994–95 (Zakaria, 2001). In Tunisia, droughts occurred during 1988–89 and 1994–95 (Louati et al., 1999). During the 1998–2000 period, both North Africa and West Asia experienced the worst regional drought in decades. In West Asia drought severely reduced food production in Jordan, Iran, Iraq, and Syria. In 1999, aggregate cereal production in the subregion was 16% lower than in the previous year and 12% lower than the average over the last five years. In Turkey, which normally contributes approximately 50% of subregional grain production, output fell by 6% as compared to the five-year average (GIEWS, 1999a). As Turkey is also the subregion's main cereal exporter, the volume of exports from the subregion, normally around 5 million tons, declined by about 50% (GIEWS, 1999a). In 2000, the drought forced Iran to import 7 million tons of wheat, making it the world's largest importer (United Nations Interagency Assessment Mission, 2001).

The situation in the Maghreb countries (Morocco, Algeria, Tunisia) was equally detrimental. As indicated by figure 16.5, the subregion's crop production (excluding Egypt, where much of the crop is irrigated) has widely fluctuated in recent years due to recurring droughts. From 1990 to 2000, aggregate cereal production ranged from 4 to 8 million tons in 5 drought years, and from 10 to 18 million tons in 5 good years (GIEWS, 2000a). The 1999 cereal crop was also affected by drought, with the production estimated at 8 million tons, which was 31% below the previous year's harvest. The year 2000 was the second consecutive year of reduced harvests in the subregion, particularly in Morocco and Algeria, with increased cereal imports putting more pressure on national budgets (GIEWS, 2000a).

Although droughts appear prominently in national-level agricultural statistics, their effects at subnational and local levels can be devastating. The drought episode of 1979–84 in Morocco reduced the small ruminant

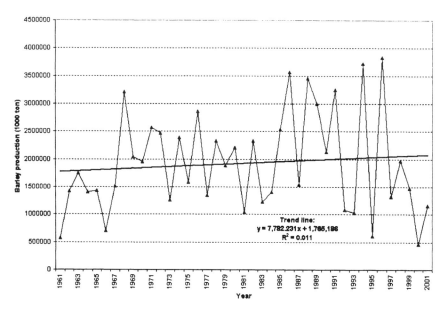

Figure 16.4 Fluctuation in barley production in Morocco during 1961–2000 (from FAOSTAT, 2002).

population by 40–50% (Berkat, 2001). The drought in Iran, which entered its third consecutive year in 2001, completely destroyed rain-fed agriculture in most areas visited by the UN Assessment Team (United Nations Interagency Assessment Mission, 2001). More than 200,000 livestock owners lost their only source of livelihood. More than 500 villages in Kerman Province had no drinking water. Many of these villages were abandoned, and the population moved to the edges of bigger cities, adding pressure on the urban water supplies (Siadat and Shariati, 2001; United Nations Interagency Assessment Mission, 2001). The same drought reduced the flow in the Tigris and Euphrates rivers in Iraq to about 20% of their average flow, seriously constraining irrigated production, which constitutes more than 70% of cultivated area (GIEWS, 2000b). In Syria, approximately 47,000 nomadic households (329,000 people) had to liquidate their livestock assets and became vulnerable to food shortages, requiring urgent food assistance (GIEWS, 1999b).

These cases amply illustrate the extent of the social problems induced by drought in the region. For this region, there is an urgent need to develop permanent and institutionalized drought early warning and monitoring systems.

Status of Drought Monitoring Systems

Comprehensive drought planning requires the integration of the following functions: monitoring and early warning, risk assessment, and mitigation

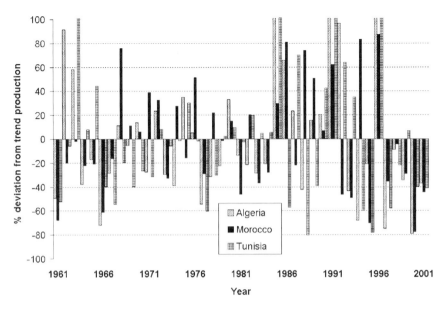

Figure 16.5 Relative production fluctuations for barley in the major producing countries of North Africa (from FAOSTAT, 2002).

and response (Wilhite and Svoboda, 2000). These functions can be exercised by different ministries, authorities, communities, or nongovernmental organizations (NGOs) but require excellent coordination to make use of all available human and financial resources. The common approach to drought in the Near East has been one of crisis management rather than long-term drought planning. When a drought becomes apparent, a government drought mitigation program is set up, usually steered by an intergovernmental committee, headed by a lead ministry. When the drought subsides, the program is terminated and the committee disbands. Given this ad-hoc approach to drought management, it is not surprising that drought monitoring systems that are integrated at national level are not operational. Although drought affects major segments of society in the region, in general there has been limited coordination of information from sources, such as water supply or irrigation authorities, agricultural extension services, meteorological departments, and NGOs about the extent and impact of drought (De Pauw, 2000).

Generally speaking, most countries in the region do not have the well-functioning drought monitoring systems that would allow them to take timely action to mitigate the effects of droughts. For this reason, they often appeal to the United Nations for assistance in the form of rapid assessment missions to accurately assess the severity and extent of droughts and to estimate food imports and emergency assistance requirements.

Particularly missing in this region are integrated spatial frameworks, such as maps of agroecological and production system zones. Given the

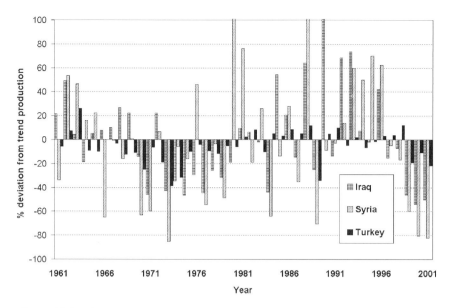

Figure 16.6 Relative production fluctuations of barley in the major producing countries of West Asia (from FAOSTAT, 2002).

tremendous diversity in agroecological conditions and livelihood systems, such information is vital for assessing the vulnerability to and potential impact of drought. In any drought monitoring system, meteorological services play a critical role. Without meteorological data and analytical tools to transform these data into relevant drought indicators, droughts cannot be adequately monitored. In most countries of the Near East, the meteorological networks are adequate and well equipped and represent major agroecological and agricultural production areas. Nevertheless, improvements are always possible, especially in highly arid zones (De Pauw et al., 2000). At the moment, meteorological departments of the region are poorly prepared to function effectively for drought early warning systems due to inadequate analytical tools required for drought monitoring, unsuitable information products, and insufficient data sharing.

Most services still define drought as a negative anomaly from normal precipitation, in terms of absolute or percentile deviations. This information is not used to monitor drought, but rather to characterize the climate during the ongoing year, month, and part of the year or agricultural season. Well-established drought indicators such as the Palmer drought severity index (Palmer, 1965) the standardized precipitation index (Guttman, 1998) or deciles (Gibbs and Maher, 1967) are not used for operational monitoring. Work on developing suitable drought indicators has just begun recently in Morocco (e.g., Yacoubi et al., 1998) and Syria but is not yet incorporated into operational drought monitoring systems. For the agricultural user community or other stakeholders, there are no regular bulletins that communicate early warnings or impacts of seasonal droughts based on the interpretation of the available raw data. A major problem is that the interinstitutional partnerships required to produce specialized drought information bulletins are hampered by the common practice of meteorological services of a region charging a fee for meteorological data, even to other government departments. With some exceptions (e.g., Turkey), the charges are often prohibitive and make no economic sense. As a result, the meteorological databases, which are indispensable for basic analyses such as drought risk assessment cannot be accessed by the agricultural user and research community, which has the highest data requirements but the lowest financial resources of all potential users of meteorological data.

From Crisis Control to Drought Management

In most countries of the region, governments have come to realize the shortcomings of their current responses to drought and have begun developing institutional arrangements for a shift from a crisis management approach to long-term drought management. The government of Morocco has recently established a National Drought Observatory (NDO), with the goal of collecting, analyzing, and delivering drought-related information in a timely manner (Ameziane, 2001). As part of its mandate, the NDO has to characterize drought, conduct vulnerability assessments, establish criteria for declaring drought and triggering mitigation and response activities, and establish procedures for evaluating the effectiveness of drought programs. The NDO is thus clearly a technical coordinating unit at the apex of a virtual structure composed of technical experts from different government administrations, dealing with different aspects of drought management through technical committees and working groups.

Syria has requested assistance from the Food and Agriculture Organization of the United Nations (FAO) to develop a national drought policy and strategy. The goal of the policy is to reduce vulnerability to drought, minimize drought impact, and facilitate postdrought recovery (Sweet, 2001). The FAO assistance will also help Syria specify how the policy will be implemented and translated into drought management plans for different ecological zones and livelihood systems.

In Tunisia the Ministry of Agriculture has developed detailed guidelines for drought management. The guidelines list a range of drought mitigation activities, which are specified according to drought stage (drought planning, management, and post-drought recovery), drought scenario (dry autumn, dry winter, dry spring, and dry year), and economic sector (Louati et al., 1999).

Regardless of the diversity of approaches, the governments of the Near East countries will need to consider the general experience gained elsewhere, which is that the institutional arrangements may determine the effectiveness of any drought monitoring system. The first requirement is to establish a multidisciplinary central drought management unit (CDMU), which has a legal status, a mandate, and its own core staff. Given the multi-institutional and multidisciplinary nature of drought management, the free flow of information is of critical importance. For this reason the CDMU would be most effective if housed in a coordinating ministry, such as a prime minister's or president's office, or a planning ministry, rather than a line ministry. This way it would have access to the multidisciplinary manpower and information sources in other ministries.

The international experience also indicates that countries can have excellent policies that are not implemented because there is no strategy or government mechanism to translate them into concrete action plans. Major responsibilities of the CDMU would be to develop or update policy to facilitate drought mitigation and ensure a linkage between drought policy and drought management strategies.

Under the general supervision of the CDMU, but not necessarily in the same ministry, a drought early warning unit needs to be established with a more technical character. This unit would compile and interpret all data required to monitor drought extent and impact and report through regular or special bulletins to the CDMU. The experience of international food security information systems, such as the Global Information and Early Warning System (GIEWS) or the Famine Early Warning System (FEWS; chapter 19), although not fully transferable in a drought and different economic context, may be useful to develop the specifications for national drought early warning systems in North Africa and West Asia.

Drought Research Needs

Feasibility of Drought Prediction

Weather patterns in many parts of the world appear to be related to different phases of the El Niño/Southern Oscillation (ENSO; chapter 3) cycle. The existence of such linkages is now being used in operational early warning systems, such as FEWS, to forecast rainfall patterns for the coming cropping season. The basis for forecasting is that a particularly strong linkage exists between the warm-ocean phase of the ENSO and drought in southern Africa. Such correlations have proved useful in tropical areas, and it may soon be possible to predict, for Southern and Eastern Africa, certain climatic conditions associated with ENSO events more than a year in advance.

In the North Africa and West Asia, no significant relationships have been confirmed between droughts and ENSO events. This is probably because the effect of this global ocean–weather linkage is substantially modified by more localized weather phenomena, such as the NAO, but also by very site-specific factors, in particular the topography and the nearness of large desert landmasses in the Sahara and the Arabian Peninsula. Nevertheless, according to Ward et al. (1999), sufficient evidence exists to suggest an

ENSO influence on sea-surface temperatures in the northern Atlantic and on late-season (March–April) precipitation levels in northwestern Africa. Particularly in Morocco, west and north of the Atlas Mountains, late-season precipitation tends to be below normal during warm ENSO years.

A major research need is to investigate whether the NAO, like ENSO, is a coupled ocean–atmosphere phenomenon as opposed to a random atmospheric phenomenon. If the NAO were a coupled phenomenon, it would enhance the use of an NAO index in drought early warning systems (Cullen and deMenocal, 2000). However, as Iglesias (2001) points out, the large variability of climate in the region, spanning time scales from the intraseasonal to the decadal, poses particular challenges for the management of agriculture. It is therefore expected that, even if this research leads to a successful outcome, the contribution of seasonal forecasts to the stabilization of agricultural production in the Near East will be relatively modest as compared to other management strategies adapted to highly variable climates.

Spatialization of Drought in Data-Insufficient Areas

The Near East is a region with relatively sparse and heterogeneous climatic data coverage. With the exception of Turkey, which has a good and homogenous coverage throughout the country, most climatic stations in the region are concentrated in coastal and agriculturally important areas. Rangeland (arid) areas and deserts are very poorly covered. Due to the pressure of the population increase, these areas are becoming increasingly important from an ecosystem function perspective. As the region is also diverse in terms of landscapes and topography, temperature regimes are not uniform, which needs to be taken into consideration while using drought indicators based on the water balance.

In such environments characterized by high spatial variations in moisture and temperature regimes, the delineation of drought can be considerably improved by advanced methods of spatial interpolation. Several statistical techniques are now available that make use of digital elevation models (DEM) to improve the spatialization of climatic parameters (De Pauw et al., 2000). In view of the strong linkages between climatic variables (especially temperature, but also rainfall, humidity, and sunshine) and topography, the most promising techniques for spatialization in climatology are multivariate approaches because the latter permit the use of terrain variables as auxiliary variables in the interpolation process. In contrast to the climatic target variables themselves, which are only known for a limited number of sample points, terrain variables have the advantage that they can be known for all locations in between, which increases the precision of the interpolated climatic variables significantly. Co-kriging (e.g., Bogaert et al., 1995) and co-splining (e.g., Hutchinson, 1995) are methods that in most cases lead to excellent interpolations. ICARDA has successfully combined the co-splining approach of Hutchinson with the GTOPO30 digital elevation model (Gesch and Larson, 1996) for regional-level mapping of various basic and derived climatic variables at 1-km resolution (De Pauw, 2002). Another important tool for spatialization is remote sensing. Remote sensing has become a standard tool in most food security early warning systems, such as FEWS, GIEWS, and MARS (Monitoring Agriculture with Remote Sensing, a project of the Joint Research Center of the European Commission; http://mars.aris.sai.jrc.it). This development has been promoted by the decreasing costs of satellite data products and image analysis tools, the difficulty of obtaining timely climatic data, and significant correlations between soil moisture status or biomass productivity and some parameters derived from spectral analysis (e.g., normalized difference vegetation index). Further details on remote-sensing techniques for drought monitoring are provided in chapters 5–8.

Drought Vulnerability Mapping

A first task for an early warning system is to understand the spatial variations of drought risk. If good time series data exist for spatially well-distributed climatic stations, drought risk can be spatialized from the probability surfaces of selected drought indicators. However, for drought planning it is also essential to go beyond the symptoms of drought, as they appear from the meteorological or hydrological records. As experienced in the region, access to and the stability of the natural resource capital (particularly natural vegetation, climate, soil, and irrigation water) are major determinants of the resilience of rural livelihood systems against drought. Understanding the underlying causes of vulnerability and anticipating the impact of drought thus requires an integrated approach, which considers both the differences in agroecological and socioeconomic characteristics between different areas.

The basis for mapping vulnerability to drought could be a spatial framework of combined agroecological zones and production systems zones. The agroecological zones can be established by integrating available climatic, soil, terrain, and land cover digital data sets. The production system zones can be derived from remote sensing in combination with farming systems information. By integrating the spatial agroecological and socioeconomic data, "agroecozones" can be established, which have unique characteristics in terms of climate, soil, and water resources, population characteristics, and livelihood systems. The agroecozones offer a useful framework for selecting sampling areas as part of a regular drought-monitoring program.

Conclusions

The Near East is a region with a high degree of aridity, and it experiences frequent droughts. Agriculture is a major and sensitive sector of the region's economy, consuming most of the available water resources. Agricultural production of major grain crops is strongly affected by fluctuations in precipitation. Crop and livestock losses attributed to drought can have severe repercussions on both the countries' balance of payments and on the livelihoods of individual producers. The response of the region's governments to drought has so far aimed at mitigating the worst effects of drought rather than treating it as a structural problem that can be incorporated into government policies and long-term management plans. While the required institutional changes are slowly taking place, the governments of the region are fully aware that drought has to be tackled within the context of a comprehensive dryland management strategy. The latter, in addition to its traditional drought mitigation aspects, has dimensions of agricultural stability and productivity and environmental sustainability.

Drought monitoring systems are important tools to implement such strategy. Strategic issues to be addressed are the institutional arrangements to ensure free information flow between monitoring and rapid response action teams at different decision-making levels. Particular drought research needs include the feasibility of drought prediction, the spatialization of drought, and the assessment of drought vulnerability from agroecological and livelihood perspectives.

References

Ameziane, T. 2001. How to cope with drought in Morocco: Integrating monitoring, assessment and management. Lecture module for the Advanced Course on Management Strategies to Mitigate Drought in the Mediterranean: Monitoring, Risk Analysis and Contingency Planning, held in Rabat, Morocco, May 21–26, 2001. CIHEAM-IAMZ and IAV Hassan II, Rabat, Morocco.

Berkat, O. 2001. Evaluation de l'impact de la sécheresse sur les parcours et systèmes d'élevage. Lecture module for the Advanced Course on Management Strategies to Mitigate Drought in the Mediterranean: Monitoring, Risk Analysis and Contingency Planning, held in Rabat, Morocco, May 21–26, 2001. CIHEAM-IAMZ and IAV Hassan II, Rabat, Morocco.

Bogaert, P., P. Mahau, and F. Beckers. 1995. The spatial interpolation of agroclimatic data. Cokriging software and source data. User's manual. FAO Agrometeorology Series 12. Food and Agriculture Organization, Rome.

Cullen, H.M., and P.B. deMenocal. 2000. North Atlantic influence on Tigris-Euphrates streamflow. Intl. J. Climatol. 20:853–863.

De Pauw, E. 2000. Drought early warning systems in North Africa and West Asia. In: D.A. Wilhite, M.V.K. Sivakumar, and D.A. Wood (eds.), Early Warning Systems for Drought Preparedness and Drought Management, WMO/TD no. 1037. World Meteorological Organization, Geneva, pp. 65–85.

De Pauw, E., W. Goebel, and H. Adam. 2000. Agrometeorological aspects of agriculture and forestry in the arid zones. Agric. Forest Meteorol. 2793(2000):1–16.

De Pauw, E. 2002. An agroecological exploration of the Arabian Peninsula. International Center for Agricultural Research in Dry Areas. Aleppo, Syria.

FAOSTAT. 2002. FAO statistical databases. Available http://apps.fao.org, accessed April 22, 2004.

Gesch, D.B., and K.S. Larson. 1996. Techniques for development of global 1-

kilometer digital elevation models. In: Pecora 13, Human interactions with the environment. Available http://edcdaac.usgs.gov/gtopo30/README.html, accessed April 22, 2004.

Gibbs, W.J., and J.V. Maher. 1967. Rainfall deciles as drought indicators. Bureau of Meteorology Bulletin no. 48. Bureau of Meteorology, Melbourne, Australia.

GIEWS. 1999a. Special report: Drought causes extensive crop damage in the Near East raising concerns for food supply difficulties in some parts. 16 July 1999. FAO Global Information and Early Warning System on Food and Agriculture, Rome. Available http://www.fao.org/giews/english/alertes/1999/SRNEA997.htm.

GIEWS. 1999b. Special Report: FAO/WFP Crop and Food Supply Assessment Mission to the Syrian Arab Republic, August 23, 1999. FAO Global Information and Early Warning System on Food and Agriculture, Rome. Available http://www.fao.org/giews/english/alertes/1999/SRSYR99.htm.

GIEWS. 2000a. Special Alert 304, April 5, 2000. FAO Global Information and Early Warning System on Food and Agriculture, Rome. Available http://www.fao.org/giews/english/alertes/2000/SA304NAF.htm.

GIEWS. 2000b. Special Alert 308, May 11, 2000. FAO Global Information and Early Warning System on Food and Agriculture, Rome. Available http://www.fao.org/giews/english/alertes/2000/SSSA308m.htm.

Guttman, N.B. 1998. Comparing the Palmer drought index and the standardized precipitation index. J. Am. Water Resources Assoc. 34:113–121.

Hurrell, J. 1995. Decadal trends in the North Atlantic Oscillation: Regional temperatures and precipitation. Science 269:676–679.

Hutchinson, M.F. 1995. Interpolating mean rainfall using thin plate smoothing splines. Intl. J. Geogr. Informat. Syst. 9:385–403.

Iglesias, A. 2001. Drought predictability, seasonal forecast and climate change. Lecture module for the Advanced Course on Management Strategies to Mitigate Drought in the Mediterranean: Monitoring, Risk Analysis and Contingency Planning, held in Rabat, Morocco, May 21–26, 2001. CIHEAM-IAMZ and IAV Hassan II, Rabat, Morocco.

Lamb, P.J., and R.A. Peppler. 1987. North Atlantic Oscillation: Concept and application. Bull. Am. Metereol. Soc. 68:1218–1225.

Louati, M. el Hedi, R. Khanfir, A. Al-Ouini, M.L. El Echi, L. Erigui, and A. Marzouk. 1999. Guide pratique de gestion de la sécheresse en Tunisie. Approche méthodologique. Ministère de l'Agriculture, Tunis, République Tunisienne.

Margat J., and D. Vallée. 1999. Mediterranean vision on water, population and the environment for the XXIst century. PNUE, PAM, Plan Bleu, Valbonne, France. Available http://www.planbleu.org/indexa.htm.

Palmer, W.C. 1965. Meteorological Drought. Research Paper no.45. U.S. Weather Bureau, Washington, DC.

Siadat, H., and M.R. Shariati. 2001. Drought in Iran: A Country Report. Internal document. Ministry of Agriculture, Tehran, Islamic Republic of Iran.

Sweet, J. 2001. Planning a National Drought Strategy for the Syrian Arab Republic. Internal document. Badia Directorate, Ministry of Agriculture and Agrarian Reform, and Food and Agriculture Organization of the United Nations, Rome.

Turkes, M. 1996. Spatial and temporal analysis of annual rainfall variations in Turkey. Intl. J. Climatol. 16:1057–1076.

UNESCO. 1979. Map of the world distribution of arid regions. Map at scale 1:25,000,000 with explanatory note. UNESCO, Paris.

United Nations Interagency Assessment Mission. 2001. United Nations Interagency Assessment Report on the extreme drought in the Islamic Republic of Iran. Tehran, Islamic Republic of Iran.

van Loon, H., and J.C. Rogers. 1978. The seasaw in winter temperatures between Greenland and northern Europe. Part I: Winter. Month. Weath. Rev. 104:365–380.

Ward, M.N., P.J. Lamb, D.H. Portis, M. El Hamly, and R. Sebbari. 1999. Climate variability in Northern Africa: understanding droughts in the Sahel and the Maghreb. In: A. Navarra (ed.), Beyond El Niño—Decadal variability in the Climate System. Springer Verlag, Berlin, pp. 119–140.

Wilhite, D.A. and M.D. Svoboda. 2000. Drought early warning systems in the context of drought preparedness and mitigation. In: D.A. Wilhite, M.V.K. Sivakumar, and D.A. Wood (eds.), Early Warning Systems for Drought Preparedness and Drought Management. WMO/TD No. 1037. World Meteorological Organization, Geneva.

Yacoubi, M., M. El Mourid, N. Chbouki, and C.O. Stöckle. 1998. Typologie de la sécheresse et recherché d'indicateurs d'alerte en climat semi-aride marocain. Sécheresse 9:269–276.

Zakaria, A. 2001. Gestion des effets de la sécheresse au Maroc. Paper presented at the Expert Consultation and Workshop on Drought Mitigation for the Near East and the Mediterranean, May 27–31, 2001. ICARDA, Aleppo, Syria.

Zorita, E., V. Kharin, and H. von Storch. 1992. The atmospheric circulation and sea surface temperature in the North Atlantic area in winter: The interaction and relevance for Iberian precipitation. J. Climate 5:1097–1108.

PART V

AFRICA

CHAPTER SEVENTEEN

Agricultural Drought in Ethiopia
ENGIDA MERSHA AND VIJENDRA K. BOKEN

In Ethiopia, 85% of the population is engaged in agriculture (CSA, 1999). Agriculture supplies a significant proportion of the raw materials for the agro-industries, and accounts for 52% of the gross product and 90% of the export earnings. A wide range of climatic, ecological, and socioeconomic diversities influence Ethiopian agriculture. The dependency of most of the population on rain-fed agriculture has made the country's economy extremely vulnerable to the effects of weather and climate, which are highly variable both temporally and spatially. If rains fail in one season, the farmer is unable to satisfy his needs and pay his obligations (tax, credit, etc). Farmers remain in the bottom line of poverty and lead a risky life. Moreover, due to climatic change and other human-induced factors, areas affected by drought and desertification are expanding in Ethiopia (NMSA, 1996a; WMO, 1986).

There are three major food supply systems in Ethiopia (IGADD, 1988; Teshome, 1996): crop, livestock, and market-dependent systems. Crop-based systems are practiced principally over the highlands of the country and comprise a very diverse range of production, depending on altitude, rainfall, soil type, and topography. Any surplus above the farmer's need is largely dependent on, for example, good weather conditions, absence of pests and diseases, availability of adequate human and animal power. Failure of rains during any cropping season means shortage of food supply that affects farmers and others. The livestock system constitutes about 10% of the total population, which is largely based in arid and semiarid zones of the country. This system is well adapted to highly variable climatic conditions and mainly depends on animals for milk and meat and is usually supplemented by grains during nondrought years. Approximately 15% of the Ethiopian population is market dependent and is affected by the preceding two food supply systems. Its food supply (grain, pulses, and oil seeds) has been facing serious shortages due to recurring droughts. People's purchasing power determines access to food in the market-dependent food supply system.

Drought Frequency

In Ethiopia, an agricultural drought is assessed using the concept of the length of growing period (LGP). The growing period is defined as the period (in days) during a year when rainfall exceeds half the potential evapotranspiration (PET), plus a period required to evapotranspire water from excess rainfall stored in the soil profile (FAO, 1978). One-half of the PET has been considered sufficient to meet the crop water requirement (FAO, 1978; Doorenbos and Kassam, 1979). A year with an LGP less than 90 days is considered to be a drought year. Because most crops in Ethiopia, with the exception of some pulses and very low-yield varieties of tef (*Eragrostis tef*) and wheat, require a growing period of at least 90 days (Henricksen and Durkin, 1985). The number of this below-90 day-LGP-years divided by the total number of years gives the severity level of drought. Based on this ratio, Reddy (1990) classified drought severity as ranging from low to very high. There are two rainy seasons in Ethiopia: *belg* (the first rainy season, from February to May) and *Kiremt* (the second rainy season, from June to September), which do not apply equally for the whole country.

The results of assessing agricultural drought (figure 17.1) based on the above explanation indicates that during the first season the frequency of agricultural drought is low (below 15%) for most parts of the Amhara Regional State, with the exception of eastern and southern parts, western parts of Oromiya, Benshangul Gumuz; Southern Nations Nationalities Peoples Region (SNNPR) and, with the exception of parts of South Omo, Gambella Regional State, the highlands of Sidammo, Arsi, Bale, and Hararghie. The frequency of drought is between 15 and 40% (moderate) for northern parts of north Gonder, southwestern parts of south Tigrai, western parts of eastern Amhara, most parts of north Shoa, parts of east Shoa, parts of east and west Hararghie, northern parts of Bale, northern and western parts of Borena, and parts of South Omo. However, drought is very frequent in most parts of Tigrai, Afar, eastern Oromiya, Somali, and southern Borena. The area with high drought frequency has decreased and is limited to the eastern parts of Tigrai, most parts of Afar, most parts of Somali, southern Borena, and parts of South Omo during the second season. Most parts of Amhara, western parts of Tigrai, most parts of Oromiya, and SNNP have a low frequency of drought during the second season, as shown in figure 17.1. The other parts of the country that experience two rainy seasons experience medium drought risk (15–40% frequency) during the second season.

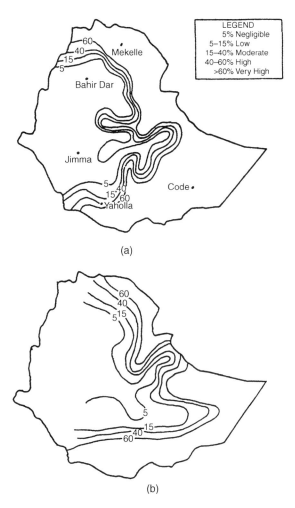

Figure 17.1 Frequency of agricultural droughts in Ethiopia during the (a) first (February–May) and (b) second (July–September) rainy season.

History of Droughts or Famines in Ethiopia

Information sources for African droughts are mainly from local records, archived data, historical texts, traveler's dairies, European settlers' notes, and folk songs. The historical reports of drought are mostly qualitative in nature. The National Meteorological Services Agency (NMSA, 1996a), Workineh (1987), Mesfin (1984), and Pankhurst (1984) attempted to collect and document the history of droughts and famine and their impact on various administrative regions of Ethiopia from different national and international documents. Analyses of the chronological events of Ethiopian droughts and famines have been divided into four parts (NMSA, 1996a). The analysis contained some interesting features. During the period from

253 BC to 1 AD, one drought or famine was reported in a seven-year period. From 1 AD to 1500 AD, there were some cruel famines that killed millions. In this period, a total of 177 droughts or famines occurred, about 1 in 9 years. From 1500 to 1950, the information is relatively based on recorded data and is more reliable. From 16th to the first half of 20th century, 69 droughts or famines were reported during a period of 450 years. That means, on average, droughts occurred once in seven years. The two notorious famines locally known as *quachine* and *kifuken* (bad days), which devastated major areas of the country, were reported during this period. The reports from 1950 onward are well documented and supported by scientific data. The analysis of the rainfall data during this period indicate 18 droughts or famines in 38 years, resulting in one drought every two years. The decadal analysis shows that the 1970–79 period was the worst, with seven disasters. The highest frequency of droughts and famines was in the 2nd century AD, followed by the first part of 20th century, and there has been an increasing trend from 16th century onward. The broad features of low rainfall and associated droughts over northern parts of Ethiopia are reflected in the Lake Tana, which feeds the Blue Nile. The water levels in the Lake Tana at Bahir Dar and the rainfall deficiency at Gonder reflected some of the major droughts such as the one in 1982. The worst period appears to be the 1980s, and the worst drought year was 1984. The areas that were affected severely by droughts/famines are mainly northern parts of the country.

Major Causes of Droughts

Atmospheric interaction, method of cultivation, selection of cropping pattern, inappropriate land use, and deforestation are important issues in the context of Ethiopian drought.

Atmospheric Interaction

Droughts in Ethiopia occur mainly due to rainfall variability. Seasonal and annual rainfall variations in Ethiopia are associated with the macro-scale pressure systems and monsoon flows related to the changes in the pressure systems (NMSA,1996b). The interactions between the pressure systems are extremely complicated and to date are poorly understood (WMO, 1975a, 1975b, 1983, 1993). However, if any one of the rain-producing systems in any season weakens, there will be an abnormal rainfall behavior during that season. A recent study (Engida, 1999) indicates that the area with stable rainfall activity has decreased, while the area with highly variable rainfall has substantially increased. As a result, the frequency of droughts has increased.

Soil Erosion

Ethiopian farmers are still continuing unsustainable methods of cultivation, and winds and rains have eroded the topsoil. Due to the erosion, water does not percolate into the soil and is wasted as runoff. Therefore, the soil cannot maintain the required amount of soil moisture. As a result of the depletion of soil moisture and soil nutrients, the soils do not sustain plant growth. The Ethiopian Highland Reclamation Study (Constable and Belshaw, 1989) estimated that more than half of the highlands (270,000 km^2) are already eroded significantly, of which about 100 tons/ha of soil are eroded every year primarily because of the erosive cropping practices. The annual soil loss due to erosion is estimated at 1.9–3.5 billion tons. According to Hurni (1986), soil loss on cultivated land is estimated to be 4–10 times higher than on grazing land, and 80% of the recorded annual soil loss occurs during the month of plowing and the following month. The reclamation study further stressed that the condition of land before sowing during the short rainy season or during the first month of growth is important for averting soil erosion.

Deforestation

Deforestation is another important factor that contributes to drought in Ethiopia. Historical sources indicate that dense forests that might have covered about 35–40% of the total area of Ethiopia have now been reduced to 2.7%. It is estimated that these resources are vanishing at an alarming rate—150,000–200,000 ha/year (IUCN, 1990; EFAP, 1994; EARO, 1999).

Overgrazing

Ethiopia has one of the largest livestock populations in Africa, with 30 million cattle, 22 million sheep, 17 million goats, 7 million equines, and 1 million camels (CSA, 1999). Approximately 70–80% of these livestock are found in the highlands (Alemneh, 1990). Ethiopian rangelands account for almost 90% of desertified lands (Mabbutt, 1984). Overgrazing of these rangelands by livestock has caused degradation of vegetation and the compaction and erosion of the soil by wind and water.

Population Growth

Growing demand of land for crop production and fuel wood, due to population growth, also contributes to land degradation and drought. Ethiopia has a total area of 1.24 million km^2 with a population of about 60 million people and an estimated 3–4% growth rate. Agriculture has always been the backbone of the country (CSA, 1998). Without a major fertility decline, Ethiopia will have to feed a population expected to double by 2030. These are frightening figures to consider because the land cannot support even the present population. Future farming practices will involve intensive cultivation, which will further result in a loss of soil fertility and drought resistance.

Land Ownership

Another important factor contributing to drought recurrence is the problem of land ownership. Investment decisions about land are affected by tenure security (Place and Hazell, 1993; Gavain and Fafchamps, 1996). Communal ownership is believed to lead to mismanagement, particularly, overgrazing and inefficient removal of wood for fuel (Hudson, 1981). The ability to transfer land sales and leasing also allows lands to be used by farmers who earn the highest return from it through mobility of draft animals, farm implements, and labor (Pender, 1998). The system of land tenure in Ethiopia has had varying and significant impacts on land management. From a historical perspective, it is believed that Ethiopia's small holders are uncertain about the security of rights to the land. This has led to cultivation for short-term needs rather than long-term yield. Accordingly, no long-term investments (e.g., soil and water conservation measures) are made that would maintain or boost yields, and this has resulted in ecological damage, which has become almost impossible to reverse (Lakew et al., 2000).

Drought Monitoring Systems

Rainfall Analysis

The NMSA of Ethiopia regularly produces a 10-day bulletin that gives analysis of rainfall based on the long-term average or normal. This bulletin is circulated to a wide range of users, ranging from local development agents to the decision-makers at national level. The rainfall analysis includes both qualitative and quantitative evaluations. The onset of a rainy season is reported as early, normal, or late onset. The rainfall distribution is expressed as whether it is deficient or excess, erratic or even. The rainfall is compared with the normal by computing the percentage deviation (D). Depending on the range of D, the area is considered to have more ($D > 125$), the same ($75 < D < 125$), insufficient ($50 < D < 75$), or below average ($D < 50\%$) rainfall. If drought continues for consecutive season or years, it signals an alarming situation.

Normalized Difference Vegetation Index

The normalized difference vegetation index (NDVI) is a satellite data-based index widely used to monitor vegetation and drought conditions. The detailed description of NDVI is provided in chapters 5 and 6. There are few organizations that provide the NDVI analysis. However, the NMSA produces a regular 10-day bulletin regarding NDVI variation that compares the current vegetation condition with normal or last-year conditions. The bulletin is distributed to higher officials and NGOs engaged in early warning activities.

Crop Monitoring

Crop performance during the growing season is monitored by the members of the National Early Warning Committee (NEWC). Land preparation, planting, and cropping pattern are monitored based on the area prepared for cultivation as compared with the normal and last-year equivalents. Moreover, area planted is compared with last year and normal year. Dates of planting are compared with the seasonal crop calendar to identify any delay in planting. Crop conditions during specific phenological phases are monitored, and poor/very poor crop conditions are also identified. Based on such monitoring, recommendations are made for alleviating any crop damage. The availability of inputs, such as fertilizers, pesticides, and improved seeds and the financial ability of the farmer to use the inputs is explained in the monitoring report. Finally, estimates on crop productions are made and compared with the last year and also with a normal year.

Water Balance Method

A crop water balance method can also be used to monitor crop performance and estimate its production. The full methodology and the table used in the computation are explained by Frere and Popov (1979). The criteria used to estimate the quality of production is presented in table 17.1. The NMSA uses this procedure to analyze the current crop situation.

Drought Early Warning System

Early warning systems are established at *woreda* (district), zonal (province), regional, and central levels. The apex body of these systems is National Committee for Early Warning that comprises representatives of the following agencies: Disaster Prevention and Preparedness Commission (DPPC), nodal officers of Ministry of Agriculture (MoA), Ministry of Health (MoH), Central Statistical Authority (CSA), Ethiopian Mapping Authority (EMA), and NMSA. The head of the Early Warning Department of DPPC serves as member/secretary. The responsibility of the committee is to coordinate the collection of information and data pertaining to weather, crops, food and nutrition status, market trends, livestock conditions, and so on, and to provide periodic information about the occurrence of disaster conditions, scale of their impact, and assessment of food availability in different parts of the country.

Drought Mitigation

Government responses to drought can be broadly classified into three types (Parry and Carter, 1987): pre-impact programs, post-impact interventions, and contingency arrangements or preparedness plans. To alleviate the root causes of drought and famines that occurred during 1972–73, 1984–85, and 1993–94 and reduce human suffering, the government of Ethiopia has

Table 17.1 Qualitative categorization of crop conditions based on the percentage to which water requirement needs are met

$I\%$[a]	% Yield	Production condition
100	≥ 100	Very good
95–99	90–100	Good
80–94	50–90	Average
60–79	20–50	Mediocre
50–90	10–20	Poor
< 50	≤ 10	Complete failure

[a] I is the cumulative water requirement satisfaction index in the crop water-balance computation.

issued a national policy on disaster prevention and management (NDPPC; Yibra, 1996). The major disaster prevention and preparedness modalities include development of an emergency food security reserve, national disaster prevention and preparedness fund, and seed reserve. The National Committee for Early Warning can recommend that the NDPPC declare the whole country or part of it as disaster areas, following which the government initiates possible actions to mitigate impacts of drought.

Potential Drought Research Needs

The complex nature of drought, coupled with its adverse consequences, still is a challenge to scientists. In Ethiopia, the need to predict drought is great because severe and prolonged droughts have multidimensional impacts on the progress and development of the country's economy, including suffering and death for thousands of humans and livestock.

Establishing Drought Prediction Methods

The existing knowledge about drought needs to be improved, and all drought detection and monitoring methods have to be upgraded. New findings are emerging for monitoring and forecasting of the global climate system in near-real time. Faster communication technology is also being developed to allow distribution and archiving of climate information. Investigations have to focus on the use of this information for better prediction of drought. There are different indices for drought monitoring. However, location- and crop-specific drought indices should be developed for Ethiopia.

Conclusions and Recommendations

The historical documentation on drought shows that droughts occurred in Ethiopia for many years, and since 1950 their occurrence has substantially increased. In drought-stricken areas of Ethiopia, it is very difficult to carry out rain-fed agriculture. Hence, it would be appropriate to introduce irrigated agriculture to reduce crop failure risk. Hence, some appropriate controlling measures against the anthropogenic causes of drought should be taken to minimize impacts of drought.

Rainfall and vegetation information obtained from satellite data has been found to be a good supplement to the conventional data sources for early warning in Ethiopia. In some remote and inaccessible areas these data are the only available source of information. The fact that these data are continuous in space and can be processed for remote areas in near-real time has made them indispensable for drought monitoring. Yet these sources have some shortcomings. They are still far from being the best data. Further efforts are required to improve the accuracy of rainfall estimates and make the best use of the vegetation index data.

Ethiopia suffers from droughts very often, and in some particular years almost the whole territory is subjected to drought (NMSA, 1996a). The seriousness of the climate-related food problem in Ethiopia requires further development of an early warning system. One of the most important aspects of the problem is to define the frequency at which drought may occur during the crop-growing season in different parts of the country. Preparation of drought probability map based on this information can help develop an effective early warning system for monitoring drought.

References

Alemneh, D. 1990. Environment, Famine, and Politics in Ethiopia: A View from the Village. Lynne Rienner Publishers, Boulder, CO.

Bekele, S., and S. Holden. 1997. Peasant agriculture and land degradation in Ethiopia: Reflection on constraints and incentives for soil conservation and food security. Forum Devel. Studies 2:277–306.

CSA. 1999. Statistical abstract. FDRE/CSA. Central Statistical Authority. Addis Ababa, Ethiopia.

CSA. 1998. The 1994 Population and housing census of Ethiopia. Results at country level, vol. 1 statistical report. Central Statistical Authority, Addis Ababa.

Constable, M., and W. Belshaw. 1989. The Ethiopian highland study. Major finding and recommendations. In: Office of the National Committee for Central Planning towards a food and nutrition strategy for Ethiopia. In: Proceedings of the National Workshop on Food Strategies for Ethiopia, December 1–12, 1986, Addis Ababa, pp. 142–179.

Doorenbos, J., and A.M. Kassam. 1979. Yield response to water. FAO Irrigation and Drainage Paper no. 33. Food and Agriculture Organization, Rome.

Engida, M. 1999. Annual rainfall and potential evapotranspiration in Ethiopia. Ethiop. J. Nat. Resources 1:137–154.

EARO. 1999. Soils and Water Conservation Program Strategic Plan. Ethiopian Agricultural Research Organisation, Addis Ababa.

EFAP. 1994. Ethiopia Forestry Action Program. Synopsis report. Ministry of Natural Resources and Environmental Protection, Addis Adaba.

FAO. 1978. Report on the agro-ecological zones project, vol. 1. Methodology and results for Africa. Food and Agriculture Organisation of the United Nations. Rome.

Frere, M., and G.F. Popov. 1979. Agrometeorological crop monitoring and forecasting. FAO Plant Production and Protection Paper no. 17. Food and Agriculture Organization, Rome, pp. 38–43.

Gavian, S., and M. Fafchamps. 1996. Land tenure and allocative efficiency in Niger. Am. J. Agric. Econ. 78:460–471.

Hurni, H. 1986. Applied soil conservation research in Ethiopia. Third National Workshop on Soil Conservation in Kenya. Department of Agricultural Engineering, Nairobi University, p. 10.

Henricksen, B.L., and J.W. Durkin. 1985. Moisture availability, cropping period and the prospects for early warning of famine in Ethiopia. ILCA bulletin 21, International Livestock Center for Africa, Addis Ababa, pp. 2–8.

Hudson, N. 1981. Soil Conservation. B.T. Batsford Ltd., London.

IGGAD. 1988. Proceedings of the Workshop on Meteorological Services and Early Warning System in the IGADD Subregion, October 31–November 3, 1988, Djibouti. Inter-Governmental Authority on Drought and Development.

IUCN. 1990. Ethiopia—National Conservation Strategy, vol. 1. International Union of Conservation of Nature and Natural Resources, Addis Ababa.

Lakew, D., K. Menale, S. Benin, and J. Pender. 2000. Land degradation and strategies for sustainable development in the Ethiopian highlands: Amhara region. Socio-economic and Policy Research Working Paper 32. Intl. Livestock Res. Inst., Nairobi, Kenya.

Mabbutt, J.A. 1984. A new global assessment of the status and trends of desertification. Environ. Conserv. 11:100–113. Vikas Publishing House, New Delhi, India.

Mesfin, W.M. 1984. Rural vulnerability to famine in Ethiopia 1958–1977.

NMSA. 1996a. Assessment of drought in Ethiopia. Meteorological Research Reports series, no. 2. National Meteorological Services Agency, Addis Ababa, Ethiopia.

NMSA. 1996b. Climatic and agroclimatic resources of Ethiopia. Meteorological Research Report Services, vol. 1, no. 1. National Meteorological Services Agency, Addis Ababa, Ethiopia.

Pankhurst, R. 1984. The History of Famine and Epidemics in Ethiopia Prior to the Twentieth Century. pp. 9–55.

Parry, M.L., and T.R. Carter. 1987. Climate impact assessment: A review of some approaches. In: D.A. Wilhite and W.E. Easterling (eds.), Planning for Drought: Towards a Reduction of Societal Vulnerability. Westview Press, Boulder, CO, pp. 165–187.

Pender, J.L. 1998. Population growth, agricultural intensification induced innovation and natural resources sustainability: An application of neo-classical growth theory. Agric. Econ. 19:99–112.

Place, F., and P.B. Hazell. 1993. Productivity effects of indigenous land tenure systems in sub-Saharan Africa. Am. J. Agric. Econ. 75:10–19.

Reddy, S.J. 1990. Methodology: Agroclimatic analogue technique and application as relevant to dry land agriculture. Agroclimatology series 3 Eth/86/021. National Meteorological Services Agency, Addis Ababa, Ethiopia.

Teshome, E. 1996. Introduction, concepts and principles of early warning system. In: Proceedings of the Workshop on Early Warning for Members of the National and Regional Early Warning Committee Members, Addis Ababa, April 15–20, 1996 and Dire Dawa, May 6–11, 1996. Disaster Prevention and Preparedness Commission, Addis Ababa, Ethiopia, pp. 23–47.

Workineh, D. 1987. Some aspects of meteorological drought in Ethiopia. In: M. Glants (ed.), Drought and Hunger in Africa. Cambridge University Press, Cambridge, pp. 23–36.

WMO. 1975a. Drought and Agriculture. Technical Note no. 138. WMO no. 392. World Meteorological Organization, Geneva.

WMO. 1975b. Drought Special Environmental Reports, no. 5. World Meteorological Organization, Geneva.

WMO. 1986. Report on drought and countries affected by drought during 1974–1985. WCP-118, WMO/TD no. 133. World Meteorological Organization, Geneva.

WMO. 1983. Report of the expert group meeting on the climatic situation and drought in Africa. WCP-61. World Meteorological Organization, Geneva.

WMO. 1993. Report of the RAII Working Group on Agricultural Meteorology CAgM Report no. 52, WMO/TD-No. 524. World Meteorological Organization.

Yibra, H. 1996. The national policy on prevention and preparedness management. In: Proceedings of the Workshop on Early Warning for Members of the National and Regional Early Warning Committee Members, Addis Ababa, April 15–20, 1996 and Dire Dawa, May 6–11, 1996. Disaster Prevention and Preparedness Commission, Addis Ababa, Ethiopia, pp. 13–20.

CHAPTER EIGHTEEN

Monitoring Agricultural Drought: The Case of Kenya

LABAN A. OGALLO, SILVERY B. OTENGI, PETER AMBENJE, WILLIAM NYAKWADA, AND FAITH GITHUI

Agriculture is the mainstay of Kenya's economic development and accounts for about 30% of the country's gross domestic product, 60% of export earnings, and 70% of the labor force. This sector is the largest source of employment (Government of Kenya, 1995). More than 85% of the population survives in one way or the other on agricultural activities (crops and livestock). Agriculture in Kenya is mainly rain-fed, with little irrigation. About 46% of the rural population live below the poverty line, with 70% of them below food poverty line.

Like many parts of the tropics, the majority of agricultural activities in Kenya are rain dependent. Small-scale farmers, pastoralists, and wildlife are most often affected by drought, with crops withering and livestock as well as wildlife dying. Drought of more than one season overwhelms the social fabric, as crops, livestock, wild animals, and humans die. Such droughts affect pastoral communities (e.g., the Masai in Kenya and Tanzania) by killing livestock and game animals, forcing these communities to invade the nearby towns and cities to find remnants of patches of grass still left there or grass growing at the roadsides. The death of game animals affects ecotourism. Interannual climate variability that often leads to the recurrence of climate extremes such as droughts has far-reaching impacts on agricultural production.

Causes of Drought

Figure 18.1 shows below-normal rainfall during different years that are often associated with droughts in Kenya. These rainfall deficits are caused by the anomalies in the circulation patterns that can extend from local or regional to very large scales. Some patterns that are responsible for spatial and temporal distribution of rainfall in Kenya include the Intertropical Convergence Zone (ITCZ), subtropical anticyclones, monsoonal wind systems, tropical cyclones, easterly/westerly wave perturbations, subtropical jet streams, East African low-level jet stream, extratropical weather systems, teleconnection with El Niño/Southern Oscillation (ENSO), and quasi-biennial oscillation (Ogallo, 1988, 1991, 1994). In addition, complex physical features such as large inland lakes, mountains, and complex orographic patterns (e.g., the Great Rift Valley) influence rainfall patterns. Lake Victoria in western Kenya is also one of the largest freshwater lakes in the world and has its own strong circulation patterns in space and time. These regional features often induce complex patterns of rainfall anomalies over the region and reduce the chances of extreme positive/negative anomalies occurring across the country.

Figure 18.2 shows that the larger amount of mean annual rainfall is concentrated over the mountain slopes and near the large bodies of water. More than 80% of Kenya may be classified as arid or semiarid lands (ASALs) with annual rainfall less than 700 mm (figure 18.3).

The arid land classification is based on the moisture index (Pratt et al., 1966), which ranges from -50 to -60 for very arid, from -40 to -50 for arid, and from -30 to -40 for semiarid land categories. Interannual rainfall fluctuations are common in all locations, but the highest variability is concentrated within the ASALs. Most of the annual rainfall is received during the two separate rainfall seasons (i.e., the March–May and October–December periods), with higher rainfall being in the first season. Agricultural practices are therefore tied to the seasonal nature of rainfall. Rainfall in many parts of Kenya is often skewed, and at times a large proportion of it falls early, even before the crop has been planted. Such rainfall is ineffective, being of little use to the young plants, which have underdeveloped root systems. Some locations to the west, however, receive a third rainfall peak during July–August, and some locations near the large bodies of water receive substantial rainfall throughout the year (figure 18.2).

Though floods occurred during 1997–98 (often attributed to El Niño), one of the longest and severest droughts in Kenya's history occurred from mid-1998 through 2001. The drought had harsh negative impacts on agriculture and livestock (figure 18.4), wildlife, tourism, water resources, and hydroelectric power generation.

The low water levels in the dams led to strict power rationing, which resulted in large losses to the economies. Water supplies for industrial and domestic consumption were not spared by the drought. There were serious water shortages both in urban and the rural areas. Lack of water and pasture led to severe conflicts between wildlife and pastoral communities. Similarly, lack of adequate power due to prolonged drought resulted in loss of employment and economic hardships. The government had to seek support from the international community to address the impacts.

Sometimes human practices contribute to droughts. For example, some farmers in the semiarid Laikipia district and Wakambas of East Province have introduced maize (a higher water-demanding crop) to replace a mark-

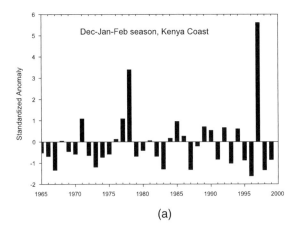

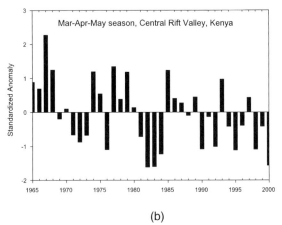

Figure 18.1 Interannual variability of rainfall for standard seasons in Kenya.

edly drought-tolerant crop such as bulrush millet. In both cases, maize was preferred to millet due to dietary habits of these people, who did not take into account the fragility of the marginal areas.

Drought Monitoring

The Drought Monitoring Center in Nairobi (DMCN) was established in 1989 to monitor drought conditions in the Greater Horn of Africa (GHA), a region comprising 10 African countries (Burundi, Djibouti, Eritrea, Ethiopia, Kenya, Rwanda, Somalia, Sudan, Tanzania, and Uganda). Some of the techniques the DMNC uses to monitor drought are described below.

Historical Climatic Data

Drought indices are derived by comparing current observations with historical records. The previous records are usually first standardized and then

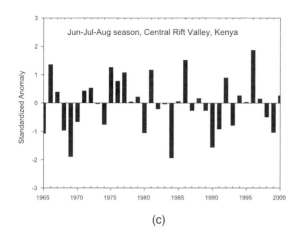

(c)

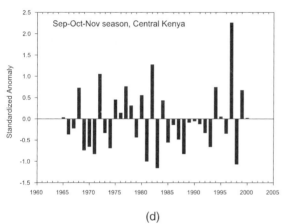

(d)

ranked in ascending or descending order. Statistical techniques are used to cluster the current observation within a group of similar past occurrences. The methods that have been used to derive such clusters range from highly sophisticated models like the Palmer drought index (PDI; Palmer, 1965; WMO, 2000) to simple indices such as the tertiles, quartiles, deciles, and percentiles. Most of the drought indices require long-term series of historical data. The choice of an index depends on the availability and extensiveness of data. For example, the use of percentiles requires about 100 years of data.

Palmer Drought Index The PDI for any individual month (X_i), may be expressed as:

$$x_i = \sum_{i=1}^{12} \frac{\{p_i - \hat{p}\}}{\left(\overline{PE} + \overline{R}\right)/\left(\overline{P} - L\right)(0.31\tau + 2.69)} \quad [18.1]$$

where $\hat{P}$ is climatically appropriate water balance for the existing condition

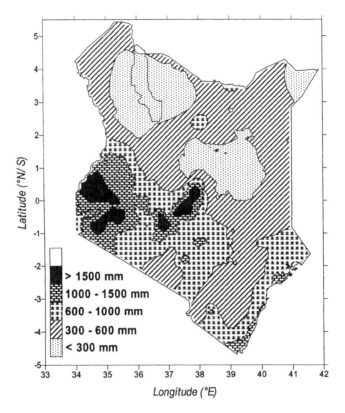

Figure 18.2 Annual rainfall distribution in Kenya.

(mm), P_i is individual monthly precipitation (mm), $\overline{P}$ is mean annual precipitation (mm), $\overline{R}$ is soil water recharge (mm), L is soil water loss (mm), $\overline{PE}$ is annual potential evapotranspiration (mm), and τ is number of months. Based on this index, one can characterize the monthly conditions as ranging from extreme wetness (index value > 4) to extreme drought (index value < −4), as shown in table 18.1.

Quartile Drought Index Using this method, historical records are first standardized (based on the specific long-term mean, standard deviation, and sometimes higher order statistics), ranked, and then divided into four groups based on quartiles, as shown in table 18.2. All new observations are then classified into one of these groups based on the magnitude of the specific observations. Assessment of drought severity in the GHA is based on the quartile index. Cumulative monthly rainfall is used to assess the persistence of drought. Figure 18.5 compares the worst drought conditions that occurred during 1984 and 2000 on the basis of cumulative rainfall.

The 2000 drought was associated with a La Niña event due to the observed slow cooling of sea-surface temperatures over much of the tropical Indian Ocean. Chapter 3 describes such atmospheric conditions in detail.

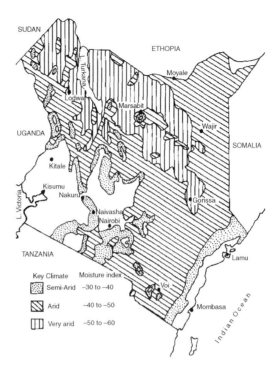

Figure 18.3 Arid and semiarid Lands of Kenya (from Ominde, 1971).

Drought indices indicated that the 2000 drought was far more severe than that of 1984 in many parts of Kenya. Many people lost their lives in 1984. Millions were displaced in search of water, food, and grazing land. Although the 2000 drought was more severe, the impacts were much less severe due to the early warning provided by DMCN and Kenya Meteorological Department (KMD).

Remote Sensing

Remotely sensed data can now be used as proxies for records. In the case of agriculture, the NDVI (chapters 5 and 6) is one such proxy for monitoring vegetation conditions, while cold cloud duration (CCD; chapter 20) and cloud temperatures are common proxies for rainfall. A comparison of the near-real time composite NDVI values with the average and any past records can help delineate areas with relatively drier and/or greener vegetation conditions. In Kenya, NDVI has been used to monitor vegetation cover, which together with the quartile indices is used to monitor drought. For example, NDVI images showed a general deterioration in vegetation conditions over northern and eastern Kenya, where some locations experienced the driest conditions since 1961 in May 2001 (DMCN, 2001).

Landsat series has also provided continuous data on the conditions of the earth's terrestrial surface for more than 25 years. These data are crucial in addressing the consequences of drought through ecosystem mapping,

(A)

(B)

Figure 18.4 Impacts of the 1999–2001 drought on (A) maize and (B) livestock in Kenya.

Table 18.1 The classification of drought or weather conditons based on Palmer drought index

Index value	Drought classification
$> +4$	Extreme wetness
$+4$ to $+3$	Severe wetness
$+3$ to $+2$	Moderate wetness
$+2$ to -2	Near normal
-2 to -3	Moderate drought
-3 to -4	Severe drought
< -4	Extreme drought

Source: Palmer (1965).

Table 18.2 The quartile drought severity index (QDSI) used by the Drought Monitoring Center, Nairobi, Kenya

Index	Range	Description
1	Rainfall $<$ min	Driest on record
2	Min $\leq$ rainfall $< Q_1$	Dry
3	$Q_1 \leq$ rainfall $< Q_3$	Near normal
4	$Q_3 \leq$ rainfall $\leq$ max	Wet
5	Rainfall $>$ max	Wettest on record

Source: Ogallo (2000).

deforestation, land-cover change, and forest and grassland fires. Many other remotely sensed data that can be used in drought monitoring can be obtained from USGCRP (1999) and WMO (2000).

Livestock Conditions

The condition of livestock (camels, cattle, goats, and sheep) can also be used to monitor drought conditions. The vegetation density of the Kenya's arid and semiarid lands is low, and the variation in forage quantity and quality is enormous. These variations and periodic lack of water for livestock due to low rainfall force pastoral communities (e.g., Maasai, Samburu, Turkana, and many others) to wander continuously with their herds of livestock. Due to drought, density as well as quality of pasture deteriorates and so do the physical conditions of the livestock that feed on these pastures. Many livestock get weaker and weaker and eventually die of starvation (figure 18.4).

Low livestock prices are usually signs of drought stress when they are being sold for money to purchase expensive food commodities or when destocking is taking place to avoid eminent death due to lack of water and forage. Livestock numbers and prices when properly monitored would give an indication of food security in a particular locality. The meat–cereal price

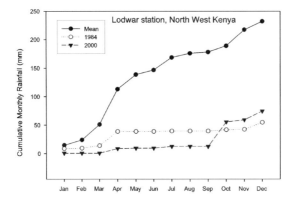

Figure 18.5 The cumulative monthly rainfall during the worst drought conditions during 1984 and 2000 in Kenya.

ratio can also be used to monitor drought. During drought, livestock are sold cheaply, and the meat price declines while the price of cereals (such as maize) increase.

The destruction of crops and livestock reduce the economic status of most rural communities. Some of the communities in marginal areas, for example, resort to survival mechanisms such as reducing the number of meals per day during periods of food shortage. Others participate in deforestation for the purpose of charcoal burning to meet income deficits (Karanja et al., 2001). Dietary/food composition also changes during drought conditions. Cereal intake by humans, which is usually higher during nondrought years, declines, and consumption of pulses (relatively drought resistant) increases during drought years.

Drought Early Warning System

The DMCN is an intergovernmental center for climate monitoring for the GHA. The DMCN and KMD play an important role in providing weather and climate advisories, including predictions and early warnings about severe climate events such as droughts. Using statistical methods, weather forecasts for 10-day, monthly, and seasonal durations are provided by the DMCN and KMD as part of their normal operation activities. (DMCN, 2001).

Drought Mitigation

Growing Drought-Resistant Crops

Timing of rainfall and choice of crop variety are critical to avoiding agricultural drought, which can occur even when the annual or seasonal rainfall is normal. Crops may resist or evade drought in two basic ways (Levitt,

1972): drought tolerance and drought escape. Drought escape is the ability to complete the plant's life cycle before serious soil and plant moisture deficits develop; plants mature early. Plant breeders in Kenya have developed the Katumani composite B and Makueni composite B maize varieties as drought-escaping genotypes that mature early. In the case of drought resistance, a plant develops the ability to advance into the next phenophase. For instance, a mild water deficit for wheat between floral initiation and anthesis phases may hasten the plants to maturity. Table 18.3 gives some of the main drought-resistant crops in the region. In modern cereal varieties and genotypes, drought resistance and yield stability in drought areas with less reliable rainfall have been achieved mainly by reducing time to maturity (Fischer and Maurer, 1978).

According to research on drought-resistance in crops (Sanchez-Diaz and Kramer, 1971; Turner, 1974; Turner and Jones, 1980; ICRISAT, 1987), the factors that confer drought resistance can generally be classified into enhancement of water acquisition from the soil by plants and restriction of transpiratory losses. At the seedling stage the aerial portions of drought-resistant crops, such as sorghum, tepary beans, bonavist beans, and bulrush millet, grow slowly until the root systems are well established. For more drought tolerance, the plants develop more lateral and adventitious roots, which are better suited to extract water from the soil (Ashley, 1993, 1999). In the semiarid areas of Laikipia in Kenya, for instance, Liniger (1991) found that maize, a less drought-resistant crop than sorghum, grew vertical roots to beyond a depth of 1.5 m and more lateral roots to adapt to the lack of water in this semiarid area. Sanchez-Diaz and Kramer (1971) established that the sorghum leaves and stems are covered with a white, waxy bloom to reduce net radiation and cuticular transpiration.

Drought-tolerant plants have high water-use efficiency. For example, sorghum requires about 20% less water than maize to produce an equivalent amount of dry matter. In drought-resistant plants, photosynthetic and growth rates under mild water stress are equal to or more than those of the nonstressed plants. Moderate stress enables continued root growth, even when aerial growth has stopped. In some grass crops (e.g., sorghum), the leaves become more erect and roll inward along their lengths to reduce energy load, which would increase respiratory losses. The stoma of the drought-tolerant crops close at relatively low water potential. The leaf stomata retain viability during periods of wilting that last about two weeks or more. Functional recovery follows the restoration of leaf turgidity. This shows a lower rate of decline in relative turgidity when subjected to increasing moisture stress. Drought-tolerant crops have a greater capacity to adjust osmotically than those that are drought susceptible, and they recover quickly and resume growth when moisture conditions become favorable. Drought-tolerant crops induce premature leaf senescence to reduce transpiratory water loss. These crops are adaptable to high temperatures. Table 18.3 shows grain crops grown in eastern Africa, including Kenya,

Table 18.3 Agroecological conditions and drought-resistant status of grain crops grown in Kenya (derived from Jaetzold and Schmidt, 1982; Ashley, 1993, 1999)

Common name	Scientific name	Range of rainfall (mm) per growing periods	Range of altitude a.m.s.l. (m) according to the growing periods	Drought-resistant status
Barley	*Hordeum vulgae*	150–650	>2100	Not drought-resistant
Kidney beans	*Phaseolus vulgaris*	230–450	>600	Not drought-resistant
Tepary beans	*Phaseolus acutifolius*	120–320	>600	Drought- and heat-resistant
Soya beans	*Flycine max*	350–750	<400	Not drought-resistant
Bonavist beans	*Lablab niger* or *Lablab purpureus*	200–2000	1200–2100	Very drought-resistant
Bulrush millet (or pearl spiked or cat-tail millet)	*Pennisetum* (syn. *P. typhoicles*, *P. typhoideum*, *P. glaucum*, *P. spicatum*)	220–800	<1200	Both drought-tolerant and drought-evading
Cowpeas	*Vigna unguiculata*	190–700	<1500	Drought-resistant
Finger millet	*Eleusine coracana*	230–900	900–2400	Drought-resistant
Green gram	*Vigna aureus*	190–400	<1500	Drought-resistant
Black gram	*Vigna mungo*	200–400	<1500	Drought-resistant
Groundnuts (or peanuts)	*Arachihypogea*	280–550	<1500	Mildly drought-resistant
Maize	*Zea mays*	240–1100	<2400	Not drought-resistant
Pigeon peas	*Cajanus cajan*	370–800	<1500	Drought-resistant
Rice	*Oryza sativa* (or *O. glaberrima*)	75–1200	<1200	Not drought-resistant
Simsim	*Sesamum indicum*	300–600	<1500	Moderately drought-resistant
Sorghum	*Sorghum bicolour* (or *Sorghum vulgare*)	200–450	900–1500	Very drought-resistant
Sunflower	*Helianthus annuus*	180–650	<2600	Very drought-resistant
Wheat	*Triticum aestivum*	350–750	1200–2900	Not drought-resistant

indicating their agroecological conditions and drought resistance status. Barley, kidney beans, soya beans, maize, and wheat can succumb to drought conditions and rainfall that is not well distributed (table 18.3). The potential effect of drought on agricultural production in Kenya has been minimized through selection of appropriate and timely agricultural practices such as Katumani composite B, Makueni composite B, and Mwezi Moja beans variety.

Conclusions

Hazards related to rainfall events such as drought have devastating impacts on almost all national socioeconomic activities in Kenya. It has been noted that a single extreme drought, like the 1999–2001 ENSO-related drought, can regress national socioeconomic growth by several decades. The best strategy to minimize drought-related impacts include drought monitoring and prediction for early warning purposes. Kenya has learned from past mistakes and has now developed a disaster management plan that has been passed to parliament for approval. The country has also established a drought monitoring system that includes climatic and rural socioeconomic indicators for 17 drought-prone areas. The program is being extended to other parts of the country. Strong support is also being provided to the national meteorological services and for some of their multidisciplinary research efforts. For example, Kenya NMS has worked with several government departments through a United Nations Development Programme project to address how climate information could be factored in the specific sectors and the national disaster management program. The NMS and DMCN have played an important and useful role in providing weather and climate advisories in Kenya and advance warnings of droughts. Some of the drought-mitigating measures that have been used in Kenya include selection of appropriate and timely agricultural practices, use of farming and water harvesting techniques to optimize production, and use of drought-resistant crops.

The challenge ahead for agricultural drought monitoring in Kenya relies heavily on the availability of skilled human resources, downscaled prediction and early warning products, timely delivery of the required information and products, and the ability to have sound drought shock-absorbers. Also, it is important that an integrated approach to drought monitoring be used that will include information about soil moisture, reservoir and groundwater levels, stream flow, and vegetation health. Some of the issues required to effectively address drought challenges in Kenya will require regional and international cooperation.

References

Ashley, J. 1993. Drought and crop adaptation. In: J.R.J. Rowland (ed.), Dryland Farming in Africa. McMillan Education Ltd., London, pp. 46–67.

Ashley, J. 1999. Food Crops and Drought. The Tropical Agriculturist Series. MacMillan Education Ltd., London.

DMCN. 2001. DMCN Monthly Climate Bulletin for May 2001. Drought Monitoring Center, Nairobi, Kenya. Available http://www.dmcn.org.

Fischer, E.A., and R. Maurer. 1978. Drought resistance in spring wheat cultivars 1: Grain yield responses. Aust. J. Agric. Res. 29:897–912.

Government of Kenya. 1995. Poverty. In: Country Position Paper at the World Summit for Social Development, Copenhagen, Denmark, March 6–12, 1995.

ICRISAT. 1987. Annual Report 1986. International Crops Research Inst. for the Semi Arid Tropics. Patancheru, A.P., India.

Jaetzold, R., and H. Schmidt. 1982. Farm management handbook of Kenya—Natural conditions and farm management information, vol. 1. Public Ministry of Agriculture of Kenya, Nairobi.

Karanja F.K., L.J. Ogallo, F.M. Mutua, C. Oludhe, and S. Kisia. 2001. Kenya country case study: impacts and responses to the 1997–98 El Niño event. In: M.H. Glantz (ed.), Once Burned Twice Shy—Lessons Learned from the 1997–98 El Niño. United Nations University Press, Tokyo, pp. 123–130.

Levitt, J. 1972. Responses of Plants to Environmental Stresses. Academic Press, New York.

Liniger, H.P. 1991. Water conservation for rain fed farming in the semi-arid footzone northwest of Mt. Kenya (Laikipia Highlands). Consequences on the water balance and the soil productivity. Laikipia Mt. Kenya Papers D-3: Discussions Papers. Laikipia Research Programme and Universities of Nairobi and Berne.

Ogallo L.J. 1988. Relationship between seasonal rainfall in East Africa and the Southern Oscillation. J. Climatol. 8:31–34.

Ogallo L.J. 1991. The Dry Land of Kenya. Drought follows Plow. Cambridge University Press, Cambridge.

Ogallo L.J. 1994. Drought and desertification: An overview. WMO Bull. 43:18–22.

Ominde, S.H. 1971. The semi-arid and arid lands (ASALs) of Kenya. In: S.H. Ominde (ed.), Studies in East African Geography and Development. Heinemann, London, pp. 146–161.

Palmer, W.C. 1965. Meteorological drought. Research Paper no. 45. U.S. Weather Bureau, Washington, DC.

Pratt, D.J., P.J. Greenway, and M.D. Gwynne. 1966. A classification of East African rangeland with an appendix on terminology. J. Appl. Ecol. 3:369–382.

Sanchez-Diaz, M.F., and P.J. Kramer. 1971. Behavior of corn and sorghum under water stress and during recovery. Plant Physiol. 48:613–616.

Turner, N.C. 1974. Stomatal behavior and water status of maize, sorghum and tobacco under field conditions 2: at low soil water potential. Plant Physiol. 53:360–365.

Turner, N.C., and M.M. Jones. 1980. Turgor maintenance by osmotic adjustment. A review and evaluation. In: N.C. Turner and P.J. Kramer (eds.), Adaptation of Plants to Water and High Temperature Stress. Wiley InterScience, New York, pp. 87–104.

USGCRP. 1999. An investment in Science for the Nation's Future. The FY 1999. Our Changing Planet. U.S. Global Change Research Program, Washington, DC.

WMO. 2000. Early warning systems for drought preparedness and drought management. In: Proceedings of an Expert Group Meeting 5–7 September 2000, Lisbon, Portugal. WMO/TD No. 1037. World Meteorological Organization, Geneva, pp. 70–87.

CHAPTER NINETEEN

Drought Monitoring Techniques for Famine Early Warning Systems in Africa

JAMES ROWLAND, JAMES VERDIN, ALKHALIL ADOUM, AND GABRIEL SENAY

Hundreds of millions of people in the world today do not enjoy food security—they do not have "access . . . at all times to enough food for an active and healthy life" (World Bank, 1986). Many of these individuals are among the quarter-billion people living in the climatically vulnerable drylands of sub-Saharan Africa (UNSO/UNDP, 1997). The drought of the early 1970s was responsible for 100,000 deaths in the Sahel and 200,000 deaths in Ethiopia (Sen, 1981) and was soon followed by another drought during 1983–85 that was responsible for 400,000 to 1 million deaths (Walker, 1989).

Famine is the most extreme food security emergency that occurs in the vulnerable areas of Africa. A practical definition of famine, offered by Cox, states that it is "the regional failure of food production or distribution systems, leading to sharply increased mortality due to starvation and associated disease" (quoted in Field, 1993). Famine's underlying cause is crop failure brought about by bad weather, armed conflict, or both (Mellor and Gavian, 1987). Famine is a slow-onset disaster, the culmination of physical and social processes occurring over two or more growing seasons. Thus, observation and detection of events leading up to famine can yield information needed to trigger preparation for famine and prevention of its worst effects. Even so, early warning of famine can come none too soon because "the time needed to get food to famine-stricken areas after an appeal for aid can stretch to six months or more" (Ulrich, 1993). Furthermore, early warning does not guarantee early response. The decision by relief organizations to commit large amounts of resources ordinarily demands clear evidence that, unfortunately, is often difficult to assemble in the early stages of the famine process.

Prevention of famine in vulnerable regions of Africa, then, requires early and unambiguous identification of unfolding food security problems so that mitigating action can be taken. Complicating the task is the fact that, over the years, the identities and numbers of those who are food insecure can shift in time and space. Relevant physical, social, and political forces are ever changing. In the face of this situation, the U.S. Agency for International Development (USAID) created the Famine Early Warning System (FEWS) to obtain the information needed to prevent widespread human suffering due to lack of food availability and access. The FEWS (1999) mission is stated as: "To provide host country and U.S. decision-makers with timely and accurate information on potential famine areas." In 2000, USAID entered a new phase of support to monitor food security in Africa with the launch of the FEWS Network (FEWS NET). FEWS NET is a USAID-funded activity that "collaborates with international, national, and regional partners to provide timely and rigorous early warning and vulnerability information on emerging or evolving food security issues. The goal of FEWS NET is to strengthen the abilities of African countries and regional organizations to manage risk of food insecurity through the provision of timely and analytical early warning and vulnerability information" (FEWS, 1999). This chapter focuses on current FEWS NET methods for monitoring drought and famine conditions in sub-Saharan Africa, in particular for the following FEWS NET countries (by region): Burkina Faso, Chad, Mali, Mauritania, and Niger; Ethiopia, Kenya, Rwanda, Somalia, southern Sudan, Uganda, and Tanzania; and Malawi, Mozambique, Zambia, and Zimbabwe.

Monitoring Drought and Famine by FEWS NET

Food security assessment in sub-Saharan Africa requires monitoring the agrophysical and socioeconomic conditions of large and spatially dispersed populations. From a physical science standpoint, food security assessment requires monitoring climatic variables and modeling their implications for rain-fed agriculture on an ongoing basis. Simultaneously, the human factors of famine vulnerability must be accounted for and mapped. These include, for example, population distribution, household income, prices of grain and cattle (McCorkle, 1987; Kinsey et al., 1998), school attendance, employment opportunities, nutritional status (Shoham, 1987; Kelly, 1993), and other variables (Reddy, 1992). Food economy analysis is used by FEWS NET to structure an understanding of livelihoods and their vulnerabilities. Joint spatial analysis of agrophysical and socioeconomic factors is used to produce an integrated picture of the food security situation, for it is the coincidence of vulnerable livelihoods and hazards that defines the level of risk. Fieldwork, interviews with local experts, consultation with national early warning committees, and professional experience build on spatial analyses to yield a synthesis of the situation presented in monthly bulletins and special reports (FEWS, 1999). Consumers of this information include the decision-makers in host country governments, USAID, donor countries, multinational organizations, and nongovernmental organizations (NGOs) having mandates and resources for response to food security emergencies. Targeting their responses directly benefits from FEWS NET analyses that identify the location and intensity of needs.

Early detection and early warning of famine must be persuasive enough to overcome the risk avoidance behaviors of decision-makers in responsible organizations—national governments, donor agencies, international organizations, and NGOs (Cutler, 1993). It is for this reason that FEWS NET food security analysts rely on a convergence of evidence to make food security assessments. No single source of information is sufficiently authoritative and comprehensive to identify potential famine areas alone (Mason et al., 1987; Shoham, 1987; Kelly, 1993). Therefore, analysts draw their conclusions most confidently when all factors indicate a certain food security status in a region. Any reduction of ambiguity associated with data or information used by FEWS NET contributes to confidence in food security assessments and an improved linkage between early warning and early response.

In subsequent sections of this chapter we explore traditional and current methods of monitoring drought and famine used by the FEWS network of scientists in the United States and Africa. We first discuss traditional use of satellite-derived vegetation index and rainfall estimates, followed by a discussion of more complex crop condition modeling using satellite-derived rainfall and vegetation information.

Remote Sensing Data for FEWS NET

In Africa, in general, sparse data observation networks resulting in inadequate spatial coverage, data quality, and timely accessibility are problems that food security analysts often face. The use of remotely sensed data offers solutions to at least part of the problem. Geographically referenced (geospatial) climate monitoring products offer food security analysts succinct and practical summaries of crop growing conditions. The products are accessible in near-real time at a continental scale and are typically the most comprehensive and up-to-date observational data available, allowing analyses to be conducted at any administrative level.

Current operational climate monitoring for FEWS NET is based primarily on maximum value composite images of the normalized difference vegetation index (NDVI; Holben, 1986; Tucker and Sellers, 1986; chapter 5) and rainfall estimate (RFE) images (Herman et al., 1997; Xie and Arkin, 1997) produced on a 10-day time step. The FEWS NDVI archive dates back to July 1981; the RFE images have only been produced since 1995. Conventional rain gauge data are also analyzed, though the availability of these data (http://edcintl.cr.usgs.gov/adds/) varies from country to country, and there is a significant delay in obtaining data for many stations. These products are the basis for analyzing climate in the past, present, and future. The variability in the historical NDVI and rainfall data is the basis for estimating predisposition to drought. NDVI and RFE are the principal tools for current seasonal monitoring.

The primary geospatial climate monitoring products used by FEWS NET are derived from remote sensing data collected by meteorological satellites (Hutchinson, 1991). NDVI images produced from Advanced Very High Resolution Radiometer (AVHRR) imagery acquired by the National Oceanic and Atmospheric Administration (NOAA) polar orbiters have the longest history of use in the project (French et al., 1996). They are prepared for FEWS NET by the Global Inventory Monitoring and Modeling Studies research unit at the NASA Goddard Space Flight Center according to techniques described by Los et al. (1994). Since the NDVI signal is approximately linearly related to the area average photosynthetic capacity of the plant canopy at a location (Tucker and Sellers, 1986), it is used as an indirect measure of the condition of rain-fed crops. Chapter 5 provides details on NDVI applications.

Exploitation of NDVI by FEWS NET for monitoring is simple and straightforward. The image for the current 10-day period (or dekad) is used to compute two difference images. The first is the difference between the NDVI for the current dekad and that of the previous dekad. This reveals areas that are greening up or drying down. The second difference is with respect to the average NDVI for the 1982–2002 historical period. This reveals areas of anomalous conditions relative to the long-term average. The other operational geospatial climate product used by FEWS NET is the RFE produced by NOAA's Climate Prediction Center. They are also compiled on a dekadal basis, with each pixel's value representing an estimate of the total millimeters of rainfall that have fallen at that location during the 10-day period. Image differencing is applied to them in much the same way that it is to the NDVI images. A difference with respect to long-term average shows wet and dry rainfall anomalies. However, since the time series of the RFE is short, a standard based on surface fitting of station data with long records is also used (Hutchinson et al., 1996).

Apart from the difference-image products, FEWS NET also produces area-average time-series traces of NDVI and RFE for key crop-growing regions. Time series traces as well as other operational monitoring products (e.g., NDVI, RFE, daily rainfall estimates, soil water index anomaly) are available on the FEWS NET Web sites (www.fews.net; edcintl.cr.usgs.gov/adds) and are described in Rowland (2001).

Water Requirement Satisfaction Index

The spatially explicit water requirement satisfaction index (WRSI) is an indicator of crop performance based on the availability of water to the crop during a growing season. Food and Agriculture Organization (FAO) studies (Doorenbos and Pruitt, 1977) have shown that WRSI can be related to crop production using a linear yield-reduction function specific to a crop. Regional implementation of WRSI has been demonstrated in a geographic information system environment (Verdin and Klaver, 2002; Senay and Verdin, 2003).

WRSI for a season is calculated as the ratio of seasonal actual evapotranspiration (AET) to the seasonal crop water requirement (WR):

$$\text{WRSI} = (\text{AET}/\text{WR})100. \qquad [19.1]$$

WR is calculated from potential evapotranspiration (PET) using the crop coefficient (K_c) to adjust for the growth stage of the crop:

$$\text{WR} = K_c(\text{PET}) \qquad [19.2]$$

AET represents the actual (as opposed to potential) amount of water withdrawn from the soil water reservoir. When soil water content is above the maximum allowable depletion (MAD) (based on crop type), the AET will remain the same as WR (i.e., no water stress). But when the soil water level is below the MAD, the AET will be lower than WR in proportion to the remaining soil water content. Soil water content is obtained through a simple mass-balance equation where soil water level is monitored by the water-holding capacity of the soil and the crop root depth; i.e.,

$$\text{SW}_i = \text{SW}_{i-1} + \text{PPT}_i - \text{AET}_i \qquad [19.3]$$

where SW is soil water content, PPT is precipitation, and i is the time step index.

The most important inputs to the model are precipitation and PET. PET values are calculated daily for Africa at 1° resolution from a 6-h numerical meteorological model output using the Penman-Monteith equation (Shuttleworth, 1992; Verdin and Klaver, 2002). Blended satellite-gauge RFE data are obtained from NOAA at 0.1° (~10 km) spatial resolution (Xie and Arkin, 1997). In addition, the model uses soil attributes from digital soils map of the world (FAO, 1988).

WRSI calculation requires start-of-season (SOS) and end-of-season (EOS) data for each modeling grid cell. The model determines SOS (or onset of rains) based on simple precipitation accounting. Figure 19.1 shows the SOS map for southern Africa during the 2001–02 growing season.

SOS is determined using a threshold amount and distribution of rainfall received in three consecutive dekads. SOS is established when there is at least 25 mm of rainfall in one dekad, followed by a total of at least 20 mm of rainfall in the next two dekads. The length-of-growing period for each pixel is determined by the persistence, on average, above a threshold value of the ratio between rainfall and PET. EOS is obtained by adding length-of-growing period to the SOS dekad. The WRSI model can be applied to different crop types (maize, sorghum, millet, etc.) for which seasonal water use patterns have been published in the form of a crop coefficient (FAO, 1998).

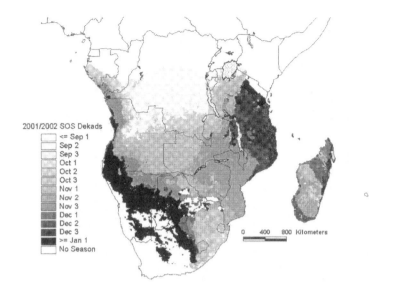

Figure 19.1 Start-of-season (SOS) map for 2001–2002 growing season in southern Africa.

At the end of the crop growth cycle, or up to a certain dekad in the cycle, cumulative AET and cumulative WR are used to calculate WRSI. A case of no deficit will result in a WRSI value of 100, which corresponds to no reduction in yield related to water stress. A seasonal WRSI value < 50 is regarded as a crop failure condition (Smith, 1992).

Yield reduction estimates based on WRSI contribute to food security preparedness and planning. As a monitoring tool, the crop performance indicator can be assessed at the end of every 10-day period during the growing season. As an early warning tool, end-of-season crop performance can be estimated by incorporating long-term average of climatological data for the period from the current dekad to EOS. Due to the different growing seasons, WRSI maps are generated and distributed on a region-by-region basis (e.g., Sahel, southern Africa, GHA). At the end of each dekad, two image products associated with the WRSI (the current WRSI and extended WRSI) are produced and disseminated by FEWS NET.

The current WRSI map portrays WRSI values for a particular crop from the onset of the growing season until the current dekad. It is based on actual estimates of meteorological data to date. For example, if the cumulative crop water requirement up to this dekad is 200 mm and only 180 mm was supplied in the form of rainfall and available soil moisture, the crop experienced a deficit of 20 mm during the period, and the WRSI value will be $(180/200)100 = 90\%$. This approach is slightly different from the traditional FAO update, where the cumulative deficit-to-date is compared to the seasonal crop water requirement instead of the requirement up to the current period. The FEWS NET WRSI may increase, decrease, or remain the same as the season progresses depending on the water supply and deficit. The FAO and FEWS NET WRSI products are mathematically equivalent when the EOS dekad becomes the current dekad.

Figures 19.2 and 19.3 demonstrate the monitoring capability of WRSI during the 2001–02 growing season in southern Africa. Figure 19.2 shows early signs of drought at the middle of the growing season (end of January 2002) in Zimbabwe and surrounding countries. Figure 19.3 shows an intensified and expanded drought by the end of the growing season (end of April 2002). The depictions in these maps were corroborated by field reports, in which 14 million people were reported to require relief assistance.

The extended WRSI map is an estimate (or forecast) of WRSI for the EOS. Long-term average rainfall and PET are used to calculate WRSI for the period between the current dekad and EOS. The calculation principles are the same as for current WRSI.

Rainfall and NDVI Combined-Departures Method

The quality of rain-fed crop production in Sahel West Africa is largely a function of the temporal and spatial distribution of seasonal rainfall. A method based on departures of NDVI and RFE data from their respective averages has been devised to monitor the annual growing season in the Sahel and to make qualitative assessments of harvest prospects up to 4–6 dekads in advance.

Food security and vulnerability analysis and mapping are often based on administrative subdivisions. This facilitates assistance to vulnerable populations in areas affected by either severe production shortfalls or other shocks to the population's livelihood. As previously mentioned, sparse data observation networks, poor quality data, and untimely accessibility are the main problems associated with ground/station data. However, neither NDVI nor RFE is problem free. Combining these measures helps decrease the uncertainty in the result. The rainfall and NDVI combined departures (RNCD) is a method based on a combination of NDVI and RFE data into an index that reflects the quality of growing season conditions, extracted for given administrative units. The growing season is divided into three distinct periods and diagnosed separately for each period. The results of the three periods are combined in a final step to diagnose the whole growing season.

Identification of Growing Season Periods   The RNCD method uses three distinct growing season periods whose lengths are agroclimatological-zone dependent: (1) the sowing period, from the dekad when generalized sowing is observed to the dekad when sowing would be considered too late; (2) the vegetative growth period, from the end of the sowing period to the maximum vegetative growth indicated by the dekad of maximum NDVI; and (3) the maturation period, from the end of the second period to complete maturation of the vegetation (pastures as well as crops). As an example, for the Sahelian zones, these periods translate to the following dates: first

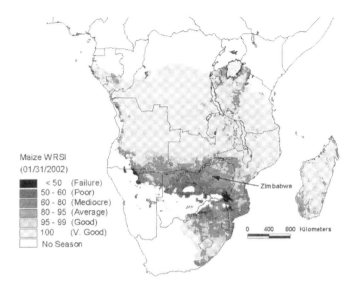

Figure 19.2 Maize WRSI at the end of January 2002 during the 2001–2002 growing season.

period, from third dekad of June to third dekad of July; second period, from first to the third dekad of August; and third period, from first to third dekad of September. Similar periods are defined for the Sahelo-Sudanian, Sudanian, and Sudano-Guinean zones.

Delineation of Base-Unit Polygons NDVI data for the period from 1982 to 1995 were used to define five agroclimatological zones in the Sahel depending on the timing of green-up. For example, zones are defined as "very early" green-up in the southernmost regions of the Sahel, all the way to "late" green-up in the northern regions of the Sahel. The agroclimatological zones have been merged with conveniently small administrative units of the Sahel, resulting in an agroclimatic–administrative unit map that consists of polygons referred to as "base units." This step is necessary because the subdivision into agroclimatological zones does not follow administrative borders, and some administrative units are large enough to straddle two or more agroclimatic zones. The base units are used to extract spatial statistics (e.g., average) for NDVI and RFE.

Computation of RNCD Indices Data used for the computation of indices are extracted from the images for all dekads during the growing season using the base units. After data are extracted for each dekad and each polygon, a dekadal index (DI) is calculated according to

$$\text{DI} = (x - \bar{x})/\sigma \qquad [19.4]$$

where x is either NDVI or RFE for a given polygon for the current season, $\bar{x}$ is the historical average, and σ is standard deviation.

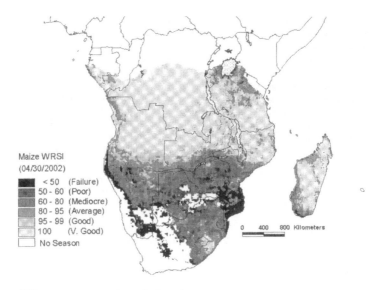

Figure 19.3 Maize WRSI at the end of April 2002 during the 2001–2002 growing season.

The average period index (PI) is obtained by averaging the DI over the period in question for NDVI and RFE. For example, if we consider the first period for those base units that fall into the very early agroclimatological zone, the averaging period will be from the second dekad of May to the third dekad of June; i.e.,

$$\text{PI} = \left(\sum \text{DI}\right)/N \qquad [19.5]$$

where DI is summed from the beginning dekad to the ending dekad, and N is the number of dekads.

PI values for NDVI and RFE are combined to obtain a growing-condition period index (GCPI) for each period. These GCPIs could be mapped to analyze the three parts of the growing season separately (the sowing period, the vegetative growth period, and the maturation period). Finally, an average of the three GCPI is calculated to obtain a growing season index called the RNCD. High values of the RNCD thus obtained are indicative of good rain-fed crop production. One should note that because of the low NDVI for water, flooded plains show a low RNCD. That is why RNCD is used to assess the seasonal outcome where rain-fed crops and pastures are predominant, as opposed to areas where recessional agriculture is predominant.

Evaluation of RNCD At the end of the growing season, the RNCD index was used to diagnose the 2001 growing season for rain-fed agricultural conditions (crops and pastures) and possible outcome. The results presented here represent the preharvest assessment for rain-fed crops and pastures and could be used for vulnerability assessment as well. Figure 19.4 shows the index based on the average of the three indices.

The analysis generally shows that good growing conditions prevailed over the Sahelian zone, with the exception of few limited areas distributed over all the countries. The largest of these areas is in eastern Chad (northeastern Biltine and southeastern Ennedi). At the western end of the Sahel, such areas are also seen in western Mali and in the border region of northern Senegal–southern Mauritania. The rest of the Sahelian zone looks very good.

The results for the Sahelian zone are in good agreement with statistical analysis. The analysis shows that 2001 growing season conditions were slightly worse than average over most of the Sudanian zone. However, the magnitude of the departures shows that the deficits are mild when compared to average conditions. The lowest negative values are around −1.0. However, over most of the Sudanian zone the index is between –0.4 and –0.1, which suggests milder departures from average conditions when considering that the highest positive value is about 3.2.

Agricultural statistics indicate that the 2001 rain-fed production was very good even in the Sudanian zone. The RNCD results are, therefore, only in partial agreement with the agricultural statistics data. Further investigation has shown that there are good reasons for this disagreement.

The RNCD index represents departures of NDVI and RFE from their respective averages. Although in the Sahelian zone any deficit in rainfall, and consequently in biomass, can have serious repercussions on production, this is not always the case in the Sudanian zone. In the Sudanian zone, where cumulative rainfall varies from 800 to 1000 mm, a 200-mm deficit may not have any ill effect on crop growth and development if rainfall is well distributed. In fact, in some cases, reasonable-length dry periods may contribute to increased yield, due to increased facility for weeding when soil moisture decreases for a period of time.

The negative departures of the RNCD index are small, since the lowest values are less than –1.0, whereas positive departures go above 3.0. Furthermore, most of the area within the Sudanian zone depicting negative departures has values ranging from –0.4 to –0.1. The area of values ranging from −1.0 to −0.4 is relatively small. The RNCD method is good for detecting anomalies in rainfall and biomass pattern. For this reason a comparison with recorded rainfall was made. Comparison of rainfall departures with respect to average was performed and good agreement was found between the spatial distribution of RNCD results and rainfall departures. Mali and Burkina Faso (figure 19.4) were chosen for this comparison because of the relatively high density of rain gauge distribution. In Mali, both RNCD and actual rainfall show some deficit in the south, especially the southeast. RNCD shows the northern part of the agricultural area of this country as above average in rainfall and rainfall records show the same area as average and slightly above average. In Burkina Faso the same is observed. RNCD shows that western and central parts of the country

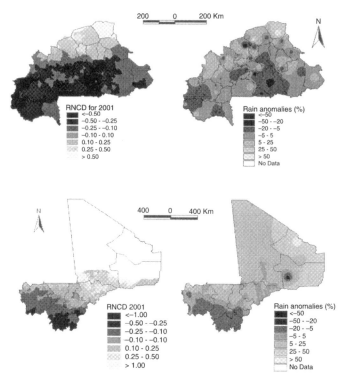

Figure 19.4 Rainfall and NDVI combined-departures versus rainfall anomalies (with respect to normal) for Burkina Faso (top) and Mali (bottom) for the year 2001.

have defcits in rainfall distribution. Here again, the RNCD method shows higher positive departure than rainfall records, which show the same area as above to slightly above average. It should be noted, however, that slight differences between the two methods may be due to the difference in the normalization of the departures from average. RNCD differences between current values and averages were normalized by the standard deviation, whereas rainfall departures were normalized by the average. In any case, the RNCD method captured the slightest anomalies in rainfall pattern.

The RNCD index is directly related to agricultural production. When these anomalies are large, its relationship with crop production is expected to be strong. However, when they are mild, as in the present case, the relationship of the index with crop production may be weak. In these conditions a crop-specific method such as the WRSI should be used for better results.

Conclusions

The FEWS and FEWS NET activities have used remote-sensing satellite data to monitor drought and famine since the late 1980s. NDVI was first used to monitor the green-up period in sub-Saharan Africa and to detect anomalous vegetation conditions. Since 1995, satellite-derived rainfall estimates have been used in a similar manner. Crop performance models have since been developed which use NDVI and RFE data as input. The WRSI crop model is applied to the Sahel, southern Africa, and Greater Horn of Africa regions for a variety of crops during their growing seasons. The RNCD method combines rainfall and vegetation conditions and is applied primarily in the Sahel region.

Remote sensing and satellite-derived products, as well as crop models based on satellite data inputs, will continue to play major roles in future FEWS NET activities. In the ongoing effort to monitor and alleviate food insecurity in Africa, FEWS NET scientists will continue to use satellite data to model and understand long-term trends exhibited in the satellite-derived vegetation, rainfall, and crop performance. Planned activities call for model validation wherever and whenever precious ground data become available. Analyses will also be undertaken to characterize teleconnections relating crop performance in Africa with El Niño and other climate indicators.

References

Cutler, P. 1993. Responses to famine: Why they are allowed to happen. In: J.O. Field (ed.), The Challenge of Famine: Recent Experience, Lessons Learned. Kumarian Press, West Hartford, CT, pp. 72–87.

Doorenbos, J., and W.O. Pruitt. 1977. Crop water requirements. FAO Irrigation and Drainage Paper no. 24. Food and Agriculture Organization, Rome.

FAO. 1988. FAO/UNESCO Soil Map of the World. Revised Legend. World Resources Report 60. Food and Agriculture Organization, Rome.

FAO. 1998. Crop Evapotranspiration: Guidelines for Computing Crop Water Requirements. FAO Irrigation and Drainage Paper 56. Food and Agriculture Organization, Rome.

FEWS. 1999. FEWS Bulletin, January 29, no. AFR/99-01. Available http://www.fews.net/.

Field, J.O. 1993. Understanding famine. In: J.O. Field (ed.), The Challenge of Famine: Recent Experience, Lessons Learned. Kumarian Press, West Hartford, CT, pp. 11–29.

French, V., N. Beninati, and S. Kish. 1996. Using Remote Sensing for Famine Early Warning. In: Proceedings of the Pecora Thirteen Symposium (CDROM). USGS EROS Data Center, Sioux Falls, SD.

Herman, A., V. Kumar, P. Arkin, and J. Kousky. 1997. Objectively determined 10-day African rainfall estimates created for famine early warning systems. Intl. J. Remote Sens. 18:2147–2159.

Holben, B. 1986. Characteristics of maximum-value composite images from temporal AVHRR data. Intl. J. Remote Sens. 7:1417–1434.

Hutchinson, C. 1991. Use of satellite data for famine early warning in sub-Saharan Africa. Intl. J. Remote Sens. 12:1405–1421.

Hutchinson, M.F., H.A. Nix, J.P. McMahon, and K.D. Ord. 1996. The development of a topographic and climate database for Africa. In: Proceedings of the Third International Conference on Integrating GIS and Environmental Modeling, Santa Fe, New Mexico (CDROM). National Center for Geographic Information and Analysis, University of California, Santa Barbara.

Kelly, M. 1993. Operational value of anthropometric surveillance in famine early warning and relief: Wollo Region, Ethiopia, 1987–88. Disasters 17:48–55.

Kinsey, B., K. Burger, and J.W. Gunning. 1998. Coping with drought in Zimbabwe: Survey evidence on responses of rural households to risk. World Development 26:89–110.

Los, S.O., C.O. Justice, and C.J. Tucker. 1994. A global 1° × 1° NDVI data set for climate studies derived from the GIMMS continental NDVI data. Intl. J. Remote Sens. 15:3493–3518.

Mason, J.B., J.G. Haaga, T.O. Maribe, G. Marks, V.J. Quinn, and K.E. Test. 1987. Using agricultural data for timely warning to prevent the effects of drought on child nutrition in Botswana. Ecol. Food Nutr. 19:169–184.

McCorkle, C.M. 1987. Foodgrain disposals as early warning signals: A case from Burkina Faso. Disasters 11:273–281.

Mellor, J., and S. Gavian. 1987. Famine: Causes, prevention, and relief. Science 235:539–545.

Reddy, A.V.S. 1992. Integrated early warning systems for disaster management and long term development. J. Rural Devel. (India) 11:265–289.

Rowland, J. 2001. Spatial databases for agroclimatic applications. In: Raymond P. Motha and M.V.K. Sivakumar (eds.), Software for Agroclimatic Data Management, Proceedings of an Expert Group Meeting, October 16–20, 2000, Washington, DC. Staff Report WAOB-2001-2. U.S. Department of Agriculture, pp. 159–169.

Sen, A. 1981. Poverty and Famines: An Essay on Entitlement and Deprivation. Clarendon Press, Oxford.

Senay, G.B., and J. Verdin. 2003. Using a GIS-based water balance model to assess regional crop performance. Can. J. Remote Sens. 29:687–692.

Shoham, J. 1987. Does nutritional surveillance have a role to play in early warning of food crisis and in the management of relief operations? Disasters 11:282–285.

Shuttleworth, J. 1992. Evaporation. In: D. Maidment (ed.), Handbook of Hydrology. McGraw-Hill, New York, pp. 4.1–4.53.

Smith, M. 1992. Expert consultation on revision of FAO methodologies for crop water requirements. Publication 73. Food and Agriculture Organization, Rome.

Tucker, C., and P. Sellers. 1986. Satellite remote sensing of primary productivity. Intl. J. Remote Sens. 7:1395–1416.

Ulrich, P. 1993. Using market prices as a guide to predict, prevent, or mitigate famine in pastoral economies. In: J.O. Field (ed.), The Challenge of Famine: Recent Experience, Lessons Learned. Kumarian Press, West Hartford, CT, pp. 239–252.

UNSO/UNDP. 1997. Aridity Zones and Dryland Populations. Office to Combat Desertification and Drought, United Nations Development Program, New York.

Verdin, J., and R. Klaver. 2002. Grid cell based crop water accounting for the Famine Early Warning System. Hydrol. Process. 16:1617–1630.

Walker, P. 1989. Famine Early Warning Systems: Victims and Destitution. Earthscan Publications Ltd., London.

World Bank. 1986. Poverty and Hunger: Issues and Options for Food Security in Developing Countries. The International Bank for Reconstruction and Development (The World Bank), Washington, DC.

Xie, P., and P.A. Arkin. 1997. A 17-year monthly analysis based on gauge observations, satellite estimates, and numerical model outputs. Bull. Am. Meteorol. Soc. 78:2539–2558.

CHAPTER TWENTY

Monitoring Agricultural Drought in Southern Africa

LEONARD S. UNGANAI AND TSITSI BANDASON

Southern Africa lies between 0°S to 35°S latitude and 10°E to 41°E longitude. In this region, annual rainfall ranges from below 20 mm along the western coastal areas of Namibia to as high as 3000 mm in some highland areas of Malawi (figure 20.1). Rainfall generally increases from south to north in response to topography and the main rain-bearing systems affecting the subregion. In the southwest sections of the sub-region, annual rainfall averages below 400 mm, whereas the high-altitude areas receive up to 3000 mm due to orographic enhancement.

Two important features that control the climate of southern Africa are the semipermanent subtropical high-pressure cells centered in the southeast Atlantic and the southwest Indian Ocean. These subtropical high pressure cells are associated with widespread and persistent subsidence (Lockwood, 1979). Part of southern Africa is under the downward leg of the Hadley Cell, superposed on the zonal Walker cell. The complex interaction of these cells, particularly during warm El Niño/Southern Oscillation (ENSO) episodes, is usually associated with drier than normal austral summers over much of southern Africa. Much of southern Africa is therefore semiarid and prone to recurrent droughts. In South Africa, for operational purposes, a drought is broadly defined as occurring when the seasonal rainfall is 70% or less of the long-term average (Bruwer, 1990; Du Pisani, 1990). It becomes a disaster or severe drought when two or more consecutive rainfall seasons experience drought.

Drought affects some part of southern Africa virtually every year. Southern Africa has suffered recurrent droughts since record keeping began (Nicholson, 1989; Unganai, 1993). Severe drought periods included 1800–30, 1840–50, 1870–90, 1910–15, 1921–25, 1930–50, 1965–75, and 1980–95. During some of these drought periods, rivers, swamps, and wells dried up and well-watered plains turned into barren lands. For Zimbabwe,

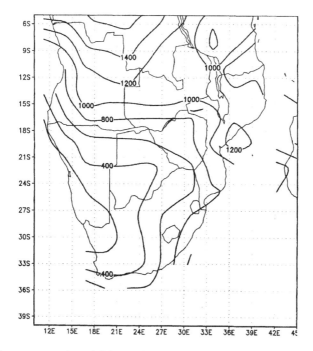

Figure 20.1 Mean annual rainfall for southern Africa.

the worst drought years were 1911–12, 1923–24, 1946–47, 1972–73, 1981–82, 1982–83, 1986–87, and 1991–92 (Zimbabwe Department of Meteorological Services, personal communication, 2002).

During the severe and recurrent droughts of the 1980s and 1990s, the impact on vulnerable communities and the environment was catastrophic. Agricultural production in the smallholder-farming sector fell by up to 70% during the 1991–92 drought in most countries in the subregion and threatened about 30 million people with starvation, requiring an estimated U.S. $4 billion in food aid (IFPRI, 1998). In Zimbabwe alone, 5.6 million people out of a population of 12 million received drought relief at a cost of about Zimbabwe $3 billion (U.S. $0.3 billion) (Ngara and Rukobo, 1999). Because of the strong dependence of Zimbabwe's economy on rain-fed agriculture, gross domestic product (GDP) for 1992 registered a negative growth rate of 6% (Franklin, 1998) as a direct result of the severe 1991–92 drought. In this chapter, the significance of drought and monitoring systems in southern Africa are reviewed with specific reference to Zimbabwe.

No objective, operational definition of drought exists in Zimbabwe. However, drought conditions are said to be occurring when rainfall is below 75% of the long-term average for a prolonged period during the rainfall season. Declaration of drought is made only when agricultural production and water supplies have been adversely affected to the extent that smallholder-farming communities cannot cope without state assistance.

Agriculture is the backbone of southern Africa's economy. In Zimbabwe it contributes about 11–18% of the GDP, 40% of annual exports, and 50% of the country's industrial raw materials. About 70% of the country's population depend directly on agriculture for their livelihood, with 30% being formally employed in that sector (Ngara and Rukobo, 1999). Agriculture also provides a significant market for products from industry. Main crops grown in the subregion include maize (staple food), tobacco, cotton (cash crops), sorghum, pearl millet, sugar, and a variety of horticultural crops. With most of the subregion's agricultural production being rain-fed, climatic extremes, if unanticipated, can produce catastrophic downstream effects on the economy, as was the case during the recent droughts of 1982–83, 1986–87, 1991–92, and 1994–95, and 2000–01.

The total rainfall and maize yield in the smallholder-farming sector in Zimbabwe are strongly correlated ($r = .72$; figure 20.2). During the severe 1991–92 drought, maize yields dropped to nearly zero in that sector. The impact of rainfall on Zimbabwe's economic performance is evident in figure 20.3. The GDP shrunk by almost 6% in 1992 following the devastating 1991–92 drought (Franklin, 1998). The strong response of the country's GDP to rainfall fluctuations is a reflection of the country's strong dependence on the performance of the agricultural sector.

For all practical purposes, southern Africa's rainy season spans November–March. December, January, and February are the peak rainfall months. Significant contributors of rainfall are the Intertropical Convergence Zone (ITCZ) and tropical–temperate troughs and their associated cloud bands (Torrance, 1981; Tyson, 1986). Some of the heaviest rains over southern Africa are associated with the infrequent passage of tropical cyclones across the coastal margins of Mozambique. Tropical–temperate troughs are frequently responsible for floods during the second half of the austral summer but are generally shorter lived during the early summer. Total rainfall from individual trough events depends on the availability of atmospheric moisture, atmospheric stability, strength of upper-level divergence, and the speed of movement of the trough. Dry summers are dominated by confluent upper winds that reduce the potential for convection over southern Africa and are often accompanied by an upsurge of tropical disturbances in the southwest Indian Ocean, representing an eastward shift in the preferred location of summer convection (Mason, 1995).

Causes of Agricultural Drought

Interannual Rainfall Variability

Southern Africa is a predominantly semiarid region with high interannual variability in rainfall (figure 20.4) and a pronounced annual cycle (Nicolson, 1986; Tyson, 1986). The coefficient of variation of annual rainfall

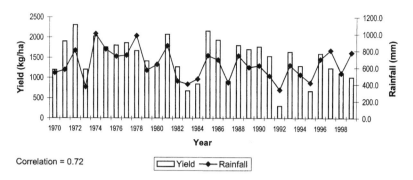

Figure 20.2 Time series of Zimbabwe smallholder-farming sector maize yields and annual rainfall from 1970 to 1999.

over Zimbabwe, for example, exceeds 40% in the drier west and southwest sections of the country and is below 25% in the north (figure 20.5). Recent studies of interannual rainfall variability over southern Africa have demonstrated their periodic nature. Historical rainfall records generally show spectral peaks in five bands, 2.2–2.4, 2.6–2.8, 3.3–3.8, 5–7, and 17–20 years (Nicholson, 1986; Makarau and Jury, 1997). It has been argued that these periodicities in annual rainfall are indicative of the influence of the Quasi-Biennial Oscillation, ENSO, periodic sea-surface temperature oscillations, and the luni-solar cycles (Nicholson, 1986; Makarau and Jury, 1997).

ENSO

Considerable evidence exists that phase shifts in the ENSO are accompanied by rainfall anomalies across southern Africa (Ropelewski and Halpert,

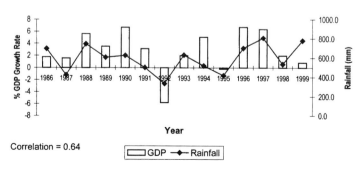

Figure 20.3 Response of Zimbabwe's GDP growth rate to rainfall.

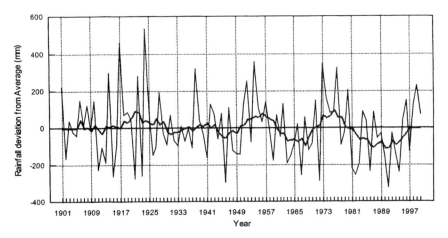

Figure 20.4 Zimbabwe's annual rainfall departures from the long-term mean from 1901 to 2000. The smooth curve represents the 10-year moving average.

1987; Matarira, 1990; Cane et al., 1994). A detailed review of ENSO can be found in chapter 3. It has been argued that the influence of ENSO events is strongest during the peak austral summer rainfall months of December–March because that is when the warm and cold events have reached maturity and when the upper westerlies have retreated significantly poleward (Mason, 2001). This delayed rainfall response has a potential value in operational long-range forecasting (Cane et al., 1994). Rainfall is reduced by 20–60% during some warm ENSO events across Zimbabwe for both the first and second part of the rainfall season. Rainfall deficits tend to be greatest in the southeast section of the country. Using maize yields from Zimbabwe's smallholder-farming sector, Cane et al. (1994) showed that more than 60% of the observed variation in yield could be predicted from Niño 3 sea-surface temperature anomalies several months in advance.

Sea-Surface Temperatures

Several studies have shown that summer rainfall in southern Africa responds to anomalous global sea-surface temperature (SST) changes (Cane et al., 1994; Makarau and Jury, 1997; Rocha and Simmonds, 1997). A warmer or cooler than normal eastern equatorial Pacific tends to be associated with dry or wet conditions, respectively, across the country (Cane et al., 1994). This response is part of the well-documented ENSO cycle (Ropelewski and Halpert, 1987). It has been shown that anomalously warm SSTs in the central equatorial Indian Ocean are usually associated with dry conditions over the central and southern parts of southern Africa. The SST anomaly pattern of the Indian Ocean could be important in the transmission of the El Niño signal to southern Africa (Rocha and Simmonds, 1997). However, it has also been observed that occasionally the Indian Ocean warm events can occur independent of ENSO events (Mason, 2001).

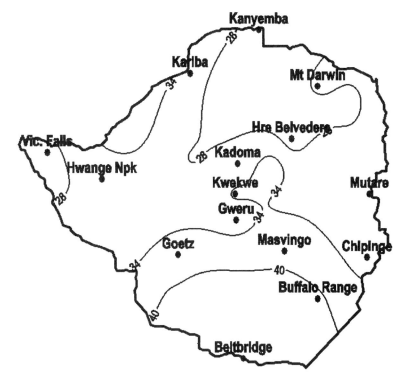

Figure 20.5 Coefficient of variation (%) of Zimbabwe's annual rainfall (1961–2000).

Associations between southern African interannual rainfall fluctuations and SSTs in the Indian Ocean appear to be complex. Although dry conditions are frequently associated with a warmer than normal western tropical Indian Ocean, this area is also an important source of atmospheric moisture for southern Africa. It becomes the dominant source during the second half of summer (Rocha and Simmonds, 1997), implying that an increase in SST here could enhance rainfall over most parts of southern Africa. A warmer tropical Indian Ocean also enhances the chances of tropical cyclone formation. Depending on the track, the tropical cyclones forming in the tropical Indian Ocean may act to dry out eastern sections of the subregion or bring floods. With a warm tropical Indian Ocean being capable of producing two opposite climatic effects, SST-based seasonal climate forecasting becomes a complex operation for much of southern Africa.

Drought Monitoring Techniques

In southern Africa most of the techniques used for drought monitoring are heavily influenced by the meteorological definition of drought. The rainfall index, the vegetation condition index, and the water requirement satisfaction index are used to monitor the progress of the rainfall season and its impact on agriculture.

Rainfall Index

Because of the difficulty and cost of monitoring parameters such as soil moisture and the lack of data to estimate potential evapotranspiration, the most widely used index for drought monitoring across southern Africa is rainfall expressed as a percentage departure from the long-term average for a given period. With this approach, cumulative seasonal rainfall is monitored and reported on weekly, monthly, and seasonal time scales during the rainfall season. A rainfall departure of 25% from the long-term average for several consecutive weeks during the rainfall season is generally classified as drought across much of southern Africa. For sub-Saharan Africa, a positive relationship between national average maize yields and rainfall has been shown for major production areas, suggesting that total rainfall may be a simple but useful indicator of drought in drought-prone areas.

Vegetation Condition Index

In southern Africa, weather data, often from a very sparse meteorological network, are incomplete or not always available to allow accurate and timely monitoring of droughts. Studies have been carried out for several parts of the world that showed the potential of satellite data for drought monitoring (van Dijk, 1985; Kogan, 1995). The vegetation condition index (VCI) has been evaluated as an alternative index for drought detection and monitoring in southern Africa (Unganai and Kogan, 1998; http://orbit-net.nesdis.noaa.gov/crad/sat/surf/vci, chapter 6).

Cold Cloud Duration

The Southern Africa Development Community Regional Remote Sensing Unit, a component of the Regional Early Warning System for Food Security, uses cloud top temperatures to monitor the progression of the rainfall season. The thermal infrared signal from the METEOSAT radiometer is used to produce statistics on the occurrence of clouds associated with rainfall over a 10-day period. Detailed description of this technique of rainfall estimation is provided in chapters 19 and 32.

Water Requirement Satisfaction Index

Crop moisture stress on grain crops across southern Africa has largely been monitored using the water requirement satisfaction index (WRSI). The WRSI indicates the extent to which the water requirements of the crop have been satisfied in a cumulative way at any stage of the crop-growing season. Details of computing the WRSI are available in chapters 19 and 32. The WRSI ranges from 0 to 100. An index below 50 indicates crop failure and a value 97–100 is associated with good crop condition.

Early Warning System

There are a couple of regional centers in southern Africa that focus on meteorological, hydrological, and food security interests related to various aspects of early warning and preparedness. These centers include the SADC Drought Monitoring Center (www.dmc.co.zw; chapter 32), and the SADC Regional Early Warning Unit for Food Security and the Regional Remote Sensing Unit (RRSU; www.sadc.fanr.org.zw). Significant progress has been made in developing an early warning system for tropical cyclones in southern Africa, but no agricultural drought early warning is widely operational in southern Africa for the benefit of both producers and policy makers. State drought declarations are largely dependent on field-crop failure reports.

Drought Preparedness and Mitigation Strategies

Maize is the staple food crop for much of southern Africa and is largely produced under rain-fed conditions. Maize requires 500–800 mm of well-distributed rainfall during the growing period. Because of its shallow rooting system that cannot draw water from soil depths greater than 80 cm, maize is highly sensitive to water stress. Drought stress and poor soil fertility are the two most important physical factors limiting rain-fed crop production in southern Africa (Heisey and Edmeades, 1999). Drought stress affects maize production at three critical stages of plant growth: during establishment, at flowering, and during middle to late grain filling. A farmer confronted with an early-season drought has several management options. The options include replanting, planting a shorter maturing cultivar, or planting a different crop that matures more rapidly or is more drought tolerant than the original crop.

A mid-season drought is more devastating than drought at the beginning or end of the season. A crop such as maize is more susceptible to drought stress during the mid-season when the plant flowers. At this stage, the farmer has no management options to respond to drought stress unless irrigation is available. To minimize water stress-related fluctuations in crop yield, a number of drought mitigation strategies have evolved across southern Africa. The most common strategies include the promotion of sustainable agricultural practices such as growing drought-tolerant crops or varieties, minimum tillage systems, water conservation, building dams for small irrigation projects, building sufficient grain reserves, and having crop mixes that minimize impact of drought on food availability.

Through financial support from the United Nations Development Programme, the Belgian government, NOAA, and the World Bank, a regional Drought Monitoring Center was set up in Harare, Zimbabwe for the purpose of providing a regionally coordinated drought early warning system in 1990. The center post-processes global climatic data sets and seasonal climate prediction products received from international climate prediction centers to produce regional forecasts for seasonal precipitation. These forecasts are disseminated to national meteorological services and other concerned agencies in the subregion.

Conclusions

Drought is endemic to southern Africa. Sustainable livelihoods across the subregion depend to a large extent on how well the subregion's economies can absorb the multiple stresses that arise from drought. The goal of sustainable livelihoods under a highly variable climatic regime can only be attained through environmental monitoring supported by a well-structured research strategy. A number of gray areas exist that limit drought mitigation strategies in southern Africa. Little is known about the physical mechanisms that trigger, prolong, and end drought episodes. Currently, no objective and scientifically sound thresholds and guidelines have been defined to provide effective drought early warning conditions beyond which nature and society cannot cope without intervention. The interface between operational drought monitoring and users of the information is currently either poorly defined or nonexistent. Research to define how operational drought monitoring systems can be better integrated to provide more useful guidance to socioeconomic activities at national level is therefore crucial.

References

Bruwer, J.J. 1990. Drought policy in the Republic of South Africa. In: Proceedings of the SARCCUS Workshop on Drought, June 1989, Pretoria. Southern African Regional Commission for Conservation and Utilization of the Soil, Pretoria, South Africa.

Cane, M.A., G. Eshel, and R.W. Buckland. 1994. Forecasting Zimbabwe maize yield using eastern equatorial Pacific sea surface temperature. Nature 370:204–205.

Du Pisani, A.L. 1990. Drought detection, monitoring and early warning. In: Proceedings of the SARCCUS Workshop on Drought, June 1989, Pretoria. Southern African Regional Commission for Conservation and Utilization of the Soil, Pretoria, South Africa.

Franklin, M. 1998. Seasonal climate forecasting and stockmarket in Zimbabwe. In: Proceedings of the Post Season National Climate Stakeholders Seminar, 29 April 1998. Department of Meteorological Services, Harare, Zimbabwe, p. 28.

Heisey, P.W., and G.O. Edmeades. 1999. Maize production in drought-stressed environments: technical options and research resource allocation. International Maize and Wheat Improvement Center Technical Report, Mexico City.

IFPRI. 1998. How will agriculture weather El Niño? In: Views and News, A 2020 Vision for Food, Agriculture and the Environment. International Food Policy Research Institute, Washington, DC.

Kogan, F.N. 1995. Application of vegetation index and brightness temperature for drought detection. Adv. Space Res. 15:91–100.

Lockwood, J.G. 1979. The Causes of Climate. John Wiley and Sons, New York.

Makarau, A., and M.R. Jury. 1997. Predictability of Zimbabwe summer rainfall. Intl. J. Climatol. 17:1421–1432.

Mason, S.J. 1995. Sea-surface temperature—South African rainfall associations, 1910–1989. Intl. J. Climatol. 15:119–135.

Mason, S.J. 2001. El Niño, climate change, and Southern African climate. Environmetrics 12:327–345.

Matarira, C.H. 1990. Drought over Zimbabwe in a regional and global context. Intl. J. Climatol. 10:609–625.

Ngara, T., and A. Rukobo. 1999. Environmental Impacts of the 1991/92 Drought in Zimbabwe. An Extreme Event. Radix Consultants, Harare, Zimbabwe.

Nicholson, S.E. 1986. The nature of rainfall variability in Africa south of the equator. J. Climatol. 6:515–530.

Nicholson, S.E. 1989. Long-term changes in African rainfall. Weather 44:46–55.

Rocha, A., and Simmonds, I. 1997. Interannual variability of southern African summer rainfall. Part I: Modelling the impact of sea surface temperatures on rainfall and circulation. Intl. J. Climatol. 17:235–265.

Ropelewski, C.F., and M.S. Halpert. 1987. Precipitation patterns associated with El Nino/Southern Oscillation. Month. Weath. Rev. 115:1606–1626.

Torrance, J.D. 1981. Climate Handbook of Zimbabwe. Department of Meteorological Services, Harare, Zimbabwe.

Tyson, P.D. 1986. Climatic Change and Variability over Southern Africa. Cambridge University Press, Cape Town, South Africa.

Unganai, L.S. 1993. Chronology of droughts in Southern Africa, the impact and future management options. In: Proceedings of the SADC-L&WMRP Fourth Annual Scientific Conference, 11–14 October 1993, SADC-Land and Water Management Research Program, Gaborone, Botswana, pp. 10–19.

Unganai, L.S., and F.N. Kogan. 1998. Southern Africa's recent droughts from space. Adv. Space Res. 21:507–511.

van Dijk, A. 1985. Crop condition and crop yield estimation method based on NOAA/AVHRR satellite data. Ph.D dissertation, University of Missouri, Columbia.

CHAPTER TWENTY-ONE

Harnessing Radio and Internet Systems to Monitor and Mitigate Agricultural Droughts in Rural African Communities

MARION PRATT, MACOL STEWART CERDA,

MOHAMMED BOULAHYA, AND KELLY SPONBERG

Humankind has not yet discovered a way to prevent drought entirely. Hence, the provision of timely and accurate climate and weather information can help rural and semiurban producers to better prepare for and mitigate the effects of insufficient precipitation (IRI, 2001). Communicating drought information to remote rural populations, however, has been a major challenge in Africa (Stern and Easterling, 1999). Seasonal rainfall forecasts, precipitation, and stream flow monitoring products, key environmental information, and even lifesaving early warnings are commonly trapped in the information bottleneck of Africa's capital cities, due to the relative lack of infrastructure in rural areas (Glantz, 2001). Without access to reliable communication networks, the majority of Africa's farmers and herders are cut off from the scientific and technological advances that support agricultural decision-making in other parts of the world.

Before the proliferation of radios, cell phones, and televisions, Africans used local methods—interpreting wind speed and direction, cloud formations, vegetation, and insect and bird migrations, for example—to predict weather patterns and the advent or cessation of precipitation. This chapter describes a Radio and Internet (RANET; http://www.ranetproject.net) system for communicating drought information to the rural communities in Niger and Uganda. This system was developed under a disaster mitigation program funded by the U.S. Agency for International Development (USAID).

The Origin of the RANET Program

The need for a drought communications system tailored to the realities of rural Africa was initially communicated to the director of the African Centre of Meteorological Applications for Development (ACMAD; http://

www.acmad.ne) by a nomad in the desert of southeastern Algeria when he declined the gift of a radio offered by the young meteorologist researching desert locusts near Djanet. The nomad did agree that information was vital to his survival. "Just tell me where it has rained. I will know where to take my flocks" (personal communication with Boulahya, Hirir, Algeria, February 1988). He explained that he was familiar with every rise and fall of the terrain and would lead his animals every rainy season to meet the water as it flowed in streams to form pools at low spots in the landscape. After watering his flocks at the pools, he would then lead them uphill to graze on the new grass. But however valuable rainfall information might have been to him, he could only receive it as long as his radio worked. A radio, carried across the isolated stretches of the Sahara, becomes little more than excess baggage once its batteries die.

Inspired by the potential that drought monitoring and prediction technologies hold for improving the quality of life in rural Africa, the meteorologist Mohammed Boulahya (one of the authors of this chapter, who later became the Founder and Director of ACMAD) worked with herders and farmers to design the RANET system (NOAA, 2003). Ten years after his encounter with the nomad, the Freeplay wind-up radio (http://www.freeplayfoundation.org) was designed to operate without batteries. Subsequently modified to incorporate a solar panel and other improvements suggested by rural listeners, the Freeplay radio was to become the front line of the RANET communications interface for remote communities. The RANET program, which soon will be established in five other African countries, is managed by ACMAD staff and faculty at the University of Oklahoma. Management of national-level content is the responsibility of each country's national meteorological service, which may in turn collaborate with other government offices or nongovernmental organizations to develop national RANET content.

Delighted with the advantages of this new technology, women in the dusty village of Bankilare in western Niger continued the trend of community-driven innovation, further challenging ACMAD to modify the technology so that they could create information as well as receive it (NOAA, 2003). ACMAD found the answer to its plea in the Wantok solar-powered FM radio transmitter. So compact that it ships in a 28-kg suitcase, this low-cost and fully portable FM radio equipment proved strikingly durable in the harsh, dry conditions of the pilot site in Bankilare. Villagers credited this new technology with having transformed Bankilare into an information oasis where they could receive broadcasts where and when they needed them—in their homes, in neighboring hamlets, and in the pastures with their flocks (Shapley, 2001).

The subsequent addition of WorldSpace Digital Satellite (WDS; http://www.firstvoiceint.org) radio provided a vital link with the outside world, permitting access not only to drought information, but also to a host of other information relevant to development. Unlike unwieldy satellite receivers with large dishes, the WDS radio receiver is comparable in size to a standard radio, and its small antenna can be easily held in one hand. The WDS radio has more than 100 channels of clear digital radio signals across the whole of Africa. Voice transmissions can be rebroadcast directly over community-owned FM radio or interpreted by community radio animators and incorporated into locally produced programing. Because the WDS radio is digital, it can broadcast data files as well as voice transmissions. When attached to the WorldSpace receiver with a special "modem" known as an adapter card, a common 486 personal computer saves the transmissions for display in a format that looks like a Web page—a format ideal for transmitting drought and environmental information, so much of which is graphical.

Technical Configuration

Using the above technologies, RANET transfers information from capital cities to rural communities in four critical steps: information gathering, transmission, reception and interpretation, and dissemination (figure 21.1). In the first step, scientists at the U.S. National Oceanic and Atmospheric Administration (NOAA), ACMAD, national meteorological services in Africa, and RANET partners gather vital information on climate, weather, and food security for drought monitoring and prediction. The information is managed and maintained by the Climate Information Project at NOAA's Office of Global Programs, in cooperation with the WorldSpace Foundation (NOAA, 2002).

In the second step, the information processed by the network of scientists is delivered to a WorldSpace uplink station via internet and loaded to the WDS radio. A partner can send a contribution by an e-mail or can post it to FTP sites on the server. At the top of every hour, the uplink station sends the most current RANET information to the WDS radio for broadcast over all of Africa. In the third step, field sites download RANET information using a WDS radio receiver, adapter card, and computer, frequently powered by solar energy. Staff at RANET field sites (including extension agents, development practitioners, and trained members of the community) interpret RANET information (including drought warnings) according to the local context and translate it into local languages.

In the fourth step, localized information is disseminated to communities by word of mouth and FM radio broadcasts. According to local priorities, communities across Niger and Uganda have devised different methods of distribution to the most vulnerable families (in particular, female-headed households). These methods include awarding radios as prizes in a neighborhood hygiene competition, or selling radios to support activities of the local RANET project (Pratt and Stewart, 2002).

Finally, feedback is generated through training workshops, site visits, and other person-to-person contacts and communication among RANET partners via e-mail and Web discussion groups. Two-way technologies such as satellite-enabled e-mail via portable ground stations are being

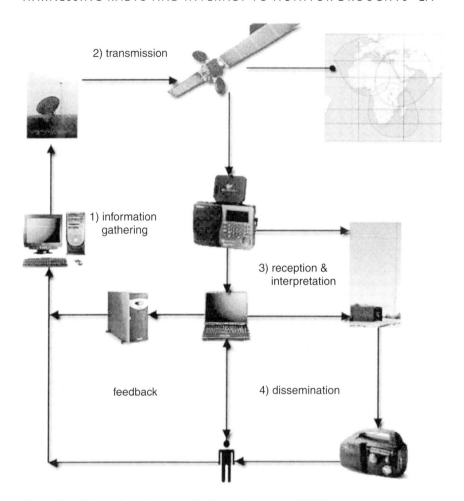

Figure 21.1 The configuration of the Radio and Internet (RANET) system for communication of drought information across African rural communities.

explored as possible avenues for facilitating communication between rural RANET sites, their national meteorological services, and the broader RANET system.

Drought Monitoring Using RANET

Rural communities in Niger and Uganda have begun to use the RANET system to improve the management of grazing and croplands, increase agricultural production, enhance food security, and reduce vulnerability to natural disasters (Shapley, 2001; Eskau, 2002; Thurow, 2002; Pratt and Stewart, 2002). The greatest benefits are realized in field sites where both the community FM radio and satellite multimedia link are functioning smoothly. Even in sites where only community radio or multimedia services were established, the communication system still has reduced vulnerability for rural populations. In all instances, a well-established radio network with full broadcast schedules attracted the attention of listeners who were then exposed to weather information on a regular basis. From the perspectives of both disaster preparedness and development, RANET was met with enthusiasm when it was initiated in Niger in 2000 and in Uganda in 2001.

Some Limitations

As the RANET system is replicated in other African countries and as potential expansion to Asia and the Pacific is considered, it is important to explore the advantages and limitations of RANET as a technology and as a drought-communicating framework. Experiences with RANET in Niger and Uganda reveal the system's successes and challenges in two very different African contexts.

RANET's powerful radio–Internet communication system risks breaking down at two critical junctures: the computer-enabled multimedia link with the outside world (the main challenge for Niger) and the dissemination of climate information by word of mouth and radio (the main challenge for Uganda). In Niger, RANET's efficiency is hindered by the difficulty of installing and maintaining computer systems in hot and dusty conditions. Technical problems frequently include insufficient knowledge to hook up parts of the solar power systems, inadequate battery storage, power surges that burn up computers, sudden losses of power that unexpectedly shut down the computers, necessitating reinstallation of the WorldSpace software, and minor malfunctions in FM station equipment such as tape recorders, microphones, and lights.

Although the RANET field sites in Uganda have had much more success in installing rural computer systems and maintaining multimedia capabilities, about 20–30% of the multimedia field sites in Uganda were sometimes not operational due to the complications resulting from solar power fluctuations (Pratt and Stewart, 2002). Other sites suffered from serious malfunctioning or damage due to improper installation of the solar and computer equipment, such as hooking up batteries in series rather than in parallel.

Technical solutions, such as new software that will not require reinstallation after power outages, are being developed to address specific technical problems. Other areas of technical improvement include reconfiguration of solar power systems, installation of relay antennas for augmenting the reach of radio programing, and strengthening the vital digital information link with the outside world by introducing voice transmission of drought information over the WorldSpace satellite system. The most pressing technical issue, however, continues to be the need for ongoing training and technical support.

Conclusions

RANET is a climate and weather information and communications support network based on the needs of remote communities and the realities of rural living in Africa. The RANET system, named for its innovative linkage of radio and Internet, brings new communications and information technologies together with the oral traditions of Africa to deliver drought information over a distributed network owned and managed by local communities.

RANET combines data from global climate data banks in the United States, seasonal rainfall predictions from the international scientific community, and forecasts generated in Africa, along with food security and agricultural information to disseminate a comprehensive information package via a network of digital satellite, receiving stations, computers, radio, and oral intermediaries. These new technologies are bringing drought and development information to rural Africa through the RANET communication network. The key to RANET's early successes lies in its dual nature as both a technological and a human communications system. Its human and technical elements depend on each other for their combined strength, but at the same time they expose the system to potential pitfalls. Because RANET is, fundamentally, a network of people supplying, interpreting, and utilizing drought and development information, the tremendous power of RANET's technology is completely dependent on the network of people that manages and maintains RANET's infrastructure and supplies, interprets, and utilizes RANET information.

References

Eskau, S. 2002. Farmers gain from satellite link. *The New Vision*, Kampala, Uganda, April 3, p. 25.

Glantz, M.H. (ed.). 2001. Lessons Learned from the 1997–98 El Niño: Once Burned, Twice Shy. UN University Press, Tokyo.

IRI. 2001. Coping with the climate: a way forward. In: Proceedings of the Regional Climate Outlook Forum, Pretoria, South Africa, October 16–20, 2000. Published on behalf of the Review Organizing Committee, Publication IRI-CW/01/2. International Research Institute for Climate Prediction, Palisades, NY.

NOAA. 2002. RANET: using science and technology for people, the environment, and development. In: Brochure for the World Summit on Sustainable Development, Johannesburg, South Africa, August 26–September 4, 2002. Office of Global Programs, Climate Information Project, U.S. National Oceanic and Atmospheric Administration. Silver Spring, MD. Available http://www.ranetproject.net/presentation_material/ranet_flyer_english.pdf.

NOAA. 2003. NOAA's Participation in Radio and Internet for the Communication of Hydro-Meteorological and Climate-Related Information Program. National Oceanic and Atmospheric Administration (NOAA) Magazine/NOAA Home Page. Available http://www.noaanews.noaa.gov/magazine/stories/mag78.htm.

Pratt, M., and M. Stewart. 2002. Field evaluation of community radio and internet (RANET) in Niger and Uganda. DCHA/OFDA-EGAT/WID trip report, January 29–February 20. U.S. Agency for International Development, Washington, DC.

Shapley, D. 2001. RANET: a weather eye on Africa. *The Financial Times*, London, June 18.

Stern, P.C., and W.E. Easterling (eds.). 1999. Making Climate Forecasts Matter. Panel on the human dimensions of seasonal-to-interannual climate variability, Committee on the Human Dimensions of Global Change, National Research Council, Washington, DC.

Thurow, R. 2002. In impoverished Niger radio provides missing links in chain of development. *The Wall Street Journal*, May 10, p. 1.

CHAPTER TWENTY-TWO

Livestock Early Warning System for Africa's Rangelands

JERRY W. STUTH, JAY ANGERER, ROBERT KAITHO, ABDI JAMA, AND RAPHAEL MARAMBII

Rangelands in Africa (i.e., grasslands, savannas, and woodlands, which contain both grasses and woody plants) cover approximately 2.1×10^9 ha. Africa's livestock population of about 184 million cattle, 3.72 million small ruminants (sheep and goats), and 17 million camels extract about 80% of their nutrition from these vast rangelands (IPCC, 1996). Rangelands have a long history of human use and are noted for great variability in climate and frequent drought events. The combination of climatic variability, low ecological resilience, and human land use make rangeland ecosystems more susceptible to rapid degeneration of ecosystems.

From a land-use perspective, there are differences between West Africa and East Africa in rangelands use. In arid and semiarid areas of West Africa (rainfall 5–600 mm), millet (or another crop) is planted over a unimodal (one peak in rainfall per year) rainy season (three to four months); then fields remain fallow during the dry season, ranging from eight to nine months. Livestock eat crop residues. Land use is dominated by cultivation, with livestock playing a subsidiary role in the village economy. In East Africa, by contrast, areas with higher rainfall (up to 600 mm) are inhabited by pastoralists rather than farmers. In dry parts, cultivation occurs mainly where irrigation is possible or where water can otherwise be sequestered and stored for cropping. Rainfall is bimodal (two peaks in rainfall per year) in most rangelands, resulting in two growing seasons. As much as 85% of the population live and depend on rangelands in a number of countries in Africa.

With emerging problems associated with the increasing population, the changes in key production areas, and the prevalence of episodic droughts and insecurity due to climatic change and ecological degradation and expansion of grazing territories, the traditional coping strategies of farmers, ranchers, and pastoralists have become inappropriate. More uncertainties require new innovations in characterizing, monitoring, analyzing, and communicating the emergence of drought to allow pastoral communities to cope with a rapidly changing environment. To this end, the United States Agency for International Development (USAID) awarded the Texas A&M University System an assessment grant to develop a Livestock Early Warning System (LEWS) as part of the Global Livestock Collaborative Research Support Program. If a drought is imminent in an area, causing scarcity of forage, pastoralists can benefit from LEWS and would be able to move their livestock to other (nondrought) areas where forage is available.

Livestock Early Warning System

The LEWS (http://cnrit.tamu.edu/lews) was designed to provide an early warning system for monitoring rangeland forage conditions, livestock nutrition, and health for maintaining food security of pastoralists. The program is an integral part of the existing framework of early warning systems for drought and famine in five countries (Tanzania, Kenya, Uganda, Ethiopia, and Eritrea) in pastoral areas of eastern Africa (figure 22.1). The development and implementation of LEWS include spatial characterization, establishment of monitoring sites, biophysical modeling, model analysis and verification, and automation of information dissemination.

Spatial Characterization

Spatial characterization first required stratification of long-term historical weather patterns in each of the five countries. Subsequently, a map showing zones of similar climate was created from a 5×5-km gridded weather surface of the entire continent of Africa developed by Corbett (1995) using the AUSPLINE algorithm of Hutchinson (1991). To conduct the climate clustering, the grid cells of the region were identified on the continental climate surface, and the primary weather attributes were queried. These attributes included maximum and minimum temperatures, annual rainfall, potential evapotranspiration (PET), and accumulated rainfall corresponding with the onset of a growing season. The attributes of the weather grid subset were then subjected to a Ward's minimum variance clustering algorithm (SAS, 1999). The resulting climatic clusters provided a mechanism to help define the boundaries for mapping the point-model output and allowed an objective mechanism to ensure that monitoring sites were located in a manner that optimized the subsequent geostatistical analysis of the model output.

Establishment of Monitoring Sites

The LEWS monitoring technology toolkit was built to serve both relief agencies and pastoral communities in the Greater Horn of Africa (Eritrea, Ethiopia, Djibouti, Somalia, Sudan, Kenya, Uganda, Tanzania, Rwanda,

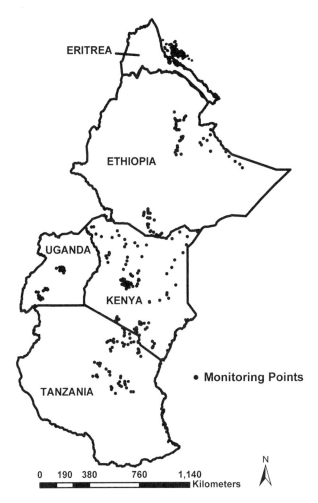

Figure 22.1 The extent of the pastoral regions in East Africa and the location of the Livestock Early Warning System drought-monitoring sites relative to climatic clusters in the region (i.e., zones of similar climatology).

and Burundi). Emphasis was placed on assisting institutions to use the system and to create educational programs to help pastoral communities better use the information to cope with drought. Keeping this in mind, the selection of the monitoring sites depended on the inherent infrastructure (roads, markets, towns, water) of the region, location of rural populations, livestock density, location of the dominant ecological sites, traditional animal movement networks, disease incidence, and conflict.

Each monitoring site was so located that the trained site monitor could easily report on-site conditions or provide the analytical team feedback on how the models were performing for that site. On average, one to three sites were selected for the site monitor to visit each month. Because the LEWS system seeks to model the forage production and availability, each site had to be surveyed to help characterize the available grazing area, water resources, animal numbers, livestock movement rules (e.g., where the animals can move during the grazing year), herd density, and estimates of other households who share the same grazing resource. This information was needed to parameterize the forage production model for those points. The households in each site were asked to designate the most probable area of migration should drought set in and force them to move. Locating both the primary and likely migration points allowed the forage conditions in both locations to be captured for the system. Both the household and the migration areas were geo-referenced using inexpensive, handheld global positioning system (GPS) units. The number of points was dictated by the analytical technique used. The minimum number of points per zone within the region was set at 30. As of this writing, there are eight 30,000–50,000-km^2 zones activated in East Africa, with 30 or more monitoring points located in each zone (figure 22.1). Five additional zones are at various stages of development.

Biophysical Modeling

Once a monitoring site was selected, the sampling protocol was largely driven by the input needs of the biophysical model used for the analysis. Point-based biophysical modeling is a mechanism to capture complex relationships at key localities (e.g., Bouman, 1995). In the case of LEWS, the Phytomass Growth Simulator model (PHYGROW; http://cnrit.tamu.edu/phygrow/; Rowan, 1995) was selected as the primary analytical engine for the monitoring program and was designed by our research team. PHYGROW is a hydrologic-based plant growth model that uses multiple plant species, soil, weather, and multiple grazer parameters to simulate daily plant production and water dynamics under grazing pressure for a specified ecological site or a designated plant community. Parameters required by the PHYGROW model include the physical and hydrologic characteristics of soils, physical response and competitive potential of plant species, and forage preferences and forage demand of grazers.

PHYGROW requires that the following major attributes be measured for each site: (1) grass basal cover by species or species group using a modified point method, (2) forb frequency, by species or species group, (3) woody plant effective canopy cover, by species or species group, (4) the name of soil and the depth and texture of each soil layer, slope, degree of rockiness, and (5) the estimates of the temporal density for each kind of livestock (this can be derived from discussions with pastoralists or individual landholders).

Model Analysis and Verification

Results obtained from the PHYGROW simulations include quantity and relative quality of daily forage production, changes in stocking rates, and a daily water balance (comprising runoff, drainage, transpiration, evaporation, interception, and soil moisture). At each monitoring site, PHYGROW simulates differences in forage production resulting from the varying plant communities, soils, grazers, and weather parameters. Each plant community is composed of its major plant species, with each plant characterized by physiological response to weather, soil conditions, and grazing pressure. Biomass production and water balance are calculated daily for each site using loops to depict natural feedback mechanisms throughout the ecosystem, as a function of the intercepted radiation, precipitation, and temperature. Plant community dynamics (growth, turnover, consumption, decay, and competition) progress with each simulated day, influencing forage production and water balance for the site.

Given the dearth of information on growth characteristics of native species, considerable effort has been made to catalog plant growth attributes (e.g., leaf area index, base/ceiling temperatures for plant growth, turnover rates for leaves, stems, litter, day length sensitivity). The Food and Agriculture Organization ECOCROP (FAO, 1998; http://pppis.fao.org) database was an excellent starting point, and our team has cataloged growth attributes of several hundred species in East Africa. A great need now exists for the ecological scientific community to design algorithms to estimate growth parameters from the known growth habits, taxonomy, and morphology of herbaceous and woody plants to accelerate the parameterization process.

Because the PHYGROW model is used in grazed environments, it was critical to have proper information on temporal changes in the animal population densities and their dietary preferences for plant species. We have collected, by interviewing experienced pastoralists, sufficient information to classify major species into the preference categories (i.e., preferred, desirable, undesirable, toxic, nonconsumed, or only used as an emergency forage by the grazing animals).

Excellent soil parameter estimators are available, which use basic information on texture and soil family class to estimate parameters such as wet bulk density, saturated hydraulic conductivity, and water holding capacity. The most useful estimators for this exercise were the Map Unit Use File (MUUF) soil attribute estimator (Soil Conservation Service, 1997) and the Washington State University hydraulic properties calculator (Saxton et al., 1986; http://www.bsyse.wsu.edu/saxton/soilwater/).

Other biophysical models that would be suitable for capturing rangeland response include the SPUR model (Wight and Skiles, 2000), the SAVANNA landscape and regional ecosystem model (Coughenour, 1992), the Erosion Productivity Impact Calculator (EPIC; Williams et al., 1984; 1997), the USDA Water Erosion Prediction Project model (WEPP; Flanagan and Nearing, 1995), and the GRASP rangeland model (Littleboy and McKeon, 1997). GRASP has been used as the basis of a prototype drought alert information system for Australia (Brook and Carter, 1996). These models have not been applied in an early warning context.

Automation of Information Dissemination

The LEWS toolkit has focused on developing fully automated computing environments that capture geo-referenced weather data, link the weather data with pre-parameterized PHYGROW files (soils, plant community, and stocking rules), and generate the necessary data files and graphics files for each monitoring site. The automation system constructs the graphs and texts and updates a Web site where the information is fully accessible to the outside world. Additional files are generated for distribution to the Arid Lands Information Network (ALIN) (http://www.alin.or.ke/), which places the data on the FTP site of the African Leaning Channel. These data are then uploaded and broadcasted as HTML files (containers) to laptop/desktop computers linked to WorldSpace satellite radios (http://www.worldspace.com/; chapter 21).

Geostatistics and Spatial Extrapolation Using Vegetation Indices

The combination of the automated modeling and mapping system based on sparse point analysis coupled with robust satellite imagery provides an excellent opportunity to interpolate results to areas not actively monitored. The geostatistical methods of ordinary kriging and co-kriging (Rossi et al., 1994) were explored for this interpolation analysis as a mechanism to make projections across large landscapes without intensive sampling.

In our case, the secondary variable is the NASA 10-day normalized difference vegetation index (NDVI) for continental Africa that provides a spatially rich data set of vegetation greenness across the landscape that has been correlated with plant biomass production (Tucker et al., 1985). A description of the index is provided at the U.S. Geological Survey Africa Data Dissemination Service Web site (http://edcsnw4.cr.usgs.gov/adds/NDVI Paper.php).

Forage production estimates from the PHYGROW biophysical simulation model for each of the monitoring sites served as the primary variable in both the kriging and co-kriging analysis. Gridded (8 × 8 km) dekadal (10-day) NDVI was used as the covariate in the co-kriging analysis. The majority of dekads analyzed have exhibited moderate to high correlations between forage production and NDVI ($r = .60$–$.86$; Angerer et al., 2002). Cross-validation indicated that the co-kriging analysis generally does a good job of estimating forage production ($r^2 = .59$–$.80$; standard error = 292–495 kg/ha). Mapped surfaces of the co-kriging output allow us to pinpoint areas of drought vulnerability (figure 22.2). During the periods of high rainfall or extended drought, we have found that the correspondence between forage production and NDVI can be low ($r < .30$), requiring that ordinary kriging be used for mapping forage production.

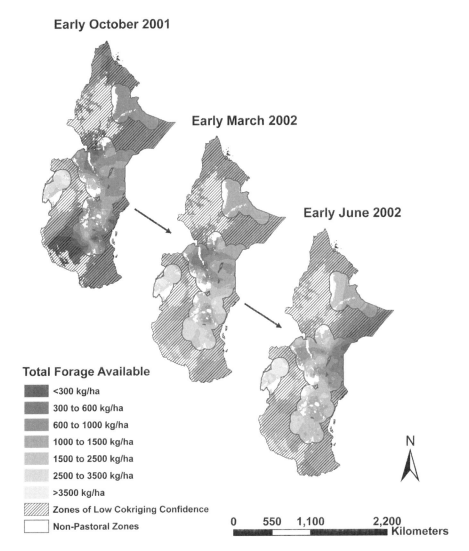

Figure 22.2 Maps of forage deviation from the long-term average that are used to pinpoint areas of drought vulnerability in the Livestock Early Warning System.

Acquiring Satellite-Based Weather Data

A technique for estimating precipitation over Africa has been developed to augment the rainfall data available from the relatively sparse observational network of rain-gauge stations over this region. For the period 1998–2001, the data used by LEWS was the rainfall estimator (RFE) version 1.0 product. The method uses METEOSAT 5 satellite data, Global Telecommunication System (GTS) rain-gauge reports, model analyses of wind and relative humidity, and orography for computing daily estimates of accumulated rainfall (Herman et al., 1997; http://edcintl.cr.usgs.gov/adds/RFEPaper.php). Since January 1, 2001, the RFE version 2.0, which has been implemented by National Oceanic and Atmospheric Administration's Climate Prediction Center (Xie and Arkin, 1997) to replace RFE 1.0, has been used by LEWS. RFE 2.0 uses additional techniques to better estimate precipitation while continuing the use of cold cloud duration (derived from cloud top temperature) and station rainfall data. The METEOSAT 7 geostationary satellite is the primary satellite data source.

The LEWS system acquires rainfall data from ftp.ncep.noaa.gov/pub/cpc/fews/newalgo_est/ site, minimum temperature from ftp.ncep.noaa.gov/pub/cpc/fews/daily_gdas_avgs/tmin/ site, and maximum temperature from ftp.ncep.noaa.gov/pub/cpc/fews/daily_gdas_avgs/tmax/ site. These data are placed on the Web (http://cnrit.tamu.edu.edu/rsg/rainfall/rainfall.cgi) for public use. These geo-referenced values of rainfall, minimum/maximum temperature, and generated radiation data are linked to the PHYGROW model to provide weather data to derive the model in the automated system. Temperature provided is skin temperature, which is not the same as the typical 2-m shaded thermometer data, reported by most standard weather status, requiring modification of model equations to accommodate this type of temperature measure. There is a need for biophysical modelers to recognize the new emerging measures of temperature, wind, and humidity from satellites and explore new algorithms to take advantages of these geographically robust data.

Using Meteorological Projections

Capturing near-real time weather data linked with automated biophysical models provides an incremental stream of analysis to the decision-maker, allowing the assessment of emerging trends. In the case of the LEWS program, information is provided in a 10-day interval. This information includes deviation from long-term average of standing crops forage availability to cattle, sheep, goats, camels, and donkeys, the percentile ranking of that response, and the estimated amount of forage available for each target herbivore. Examples of emerging trends in the state of livestock forage that can be seen in the LEWS products are shown in figures 22.3 and 22.4.

Using strong shifts in percentile ranking of current standing crops relative to 30-year averages from generated weather data and changes in NDVI satellite greenness data, it was possible to project forward 30–120 days with a relatively high confidence, allowing decision-makers time to begin planning for adjustments in livestock numbers or movements.

The LEWS program has recently developed a collaborative relationship with the Drought Monitoring Center in Nairobi, Kenya to integrate the quarterly Climate Outlook Forum (COF) 90-day projections of above, below, and average rainfall conditions as well as the 10-day and 30-day projections. Each 10-day report projects 90 days forward using the 25, 50, and

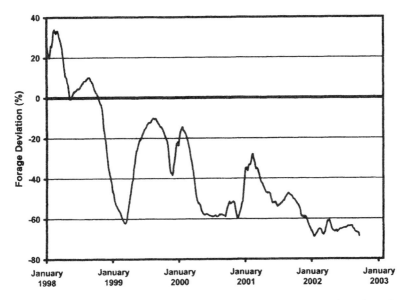

Figure 22.3 An example of one of the Livestock Early Warning System's products that reflects the percent deviation in forage standing crop for cattle compared to 30-year means of standing crop for a monitoring point in southeastern Kenya.

75% ranking of the corresponding 90-day interval from that date based on 90-day accumulated rainfall from the 30-year generated weather data.

A new promising projection technique developed by Al-Hamad (2002) has been applied to LEWS, where point-based biophysical simulation of forage production coupled with 8-km Advanced Very High Resolution Radiometer-NDVI data was used as a forecasting method for near-term forage production. NDVI data is de-noised using wavelet method (Percival and Walden, 2000) and is then co-regressed with standing crop of forage at the monitoring sites using the Box and Jenkins model or ARIMA procedure for forecasting. This methodology appears to offer stable projections of forage conditions typically from 60 to 90 days.

Community Outreach Program

Perhaps the most difficult issue to address in drought monitoring systems is devising effective communication frameworks with pastoral communities that provide information valued by community leaders and individuals in a manner that will result in rational and proactive decisions. All of the early warning institutions in East Africa working with LEWS have identified effective communication with pastoral communities as one of the major weaknesses in their information dissemination system. Currently, LEWS is exploring different models for communicating monitoring products including training local nongovernmental organizations operating in pastoral communities, placement and training of community-based site monitors,

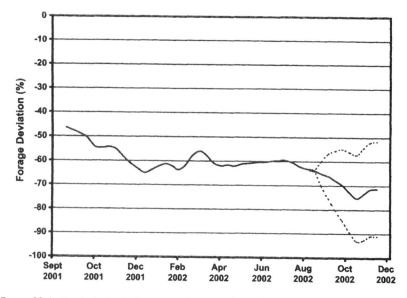

Figure 22.4 The deviation in forage standing crop for cattle compared to 30-year means of daily standing crop for a rangeland community in southeastern Kenya, as estimated by the Livestock Early Warning System. The dotted lines show the upper and lower 66% confidence limits of 10- to 90-day projections.

and improved training of local extension officers. Issues of literacy, traditional pastoral decision-making processes, water availability, disease incidence, and conflict must be considered in devising effective communication instruments in pastoral communities.

Reflections on the Future

Agricultural practitioners, regardless of the decision environment, have not had access to reliable production estimates relative to long-term data in a spatially coherent manner. The emergence of geo-referenced weather data, offered on-line in near-real time, coupled with automated modeling environments, is ushering in a new suite of tools for the agricultural and natural resource decision-maker. These tools could easily be integrated into a suite of new tools that allow the emergence of highly integrated precision landscape analysis systems, allowing decision-makers to explore issues of crop production, disease risk, drought, hydrology, and market conditions in a manner not experienced before.

Adapting models to the new realities of weather data systems (satellite-based, radar, distributed ground-satellite-web linked systems) needs to be considered in the near future. Higher resolution imagery, greater computing power, and stronger institutional collaboration on data sharing are going to make drought monitoring a more routine process that a wide array of decision-makers can access.

The key to this process is staging and positioning data for use in new and improved analytical systems and investment in access to the Internet in rural areas around the world. Improved education programs for farmers and pastoralists that focus on improved communication systems and information delivery should be a near-term priority as these technologies mature. Finally, the lack of investment in rangeland scientists over the past 15 years by universities and donor organizations has resulted in a dearth of expertise in characterizing vegetation and interpreting output of the rangeland models; resources must be prioritized to implement pastoral early warning systems.

ACKNOWLEDGMENTS The Livestock Early Warning System Project is supported by the Global Livestock Collaborative Research Support Program, funded in part by the U.S. Agency for International Development (USAID) under grant no. PCE-G-00-98-00036-00. Part of this research was supported from the Sustainable Agriculture and Natural Resource Management Collaborative Research Support Program (SANREM CRSP), funded by USAID Cooperative Agreement no. PCE-A-00-98-00019-00. The CRSP accession number is 99-GLO-004. The opinions expressed do not necessarily reflect the views of USAID. Partial funding for this program is also from the USAID Sustainable Agriculture and Resource Management CRSP subproject on Global Decision Support Systems, Association for Strengthening Agricultural Research in East and Central Africa–Crisis Mitigation Office, Kenya Agricultural Research Institute, Ethiopian Agricultural Research Organization, Uganda National Agricultural Research Organization, and Tanzanian Ministry of Livestock and Water Development.

References

Al-Hamad, M. 2002. Biophysical modeling and NDVI time series to project near-term forage supply. Ph.D. dissertation. Texas A&M University, College Station.

Angerer, J.P., J.W. Stuth, W.N. Mnene, F.P. Wandera, R.J. Kaitho, and A.A. Jama. 2002. Co-kriging of biophysical model output and NDVI to create forage production maps for livestock early warning systems. In: B. Wu (ed.), Proceedings of Landscape Analysis for Rangeland Assessment and Monitoring, Society of Range Management, Wichita, KS.

Bouman, B.A.M. 1995. Crop modeling and remote sensing for yield prediction. Neth. J. Agric. Sci. 43:143–161.

Brook, K., and J. Carter. 1996. A prototype national drought alert strategic information system for Australia. Drought Network News 8(2):2.

Corbett, J.D. 1995. Dynamic crop environment classification using interpolated climate surfaces. In: Proceedings of the Second International Conference/Workshop on Integrating GIS and Environmental Modeling, Brekenridge, Colorado, September 27–30, 1993. ESRI, Redmond, CA, pp. 23–34.

Coughenour, M.B. 1992. Spatial modeling and landscape characterization of an African pastoral ecosystem: prototype model and its potential use for monitoring drought, In: D.H. Mckenzie, D.F. Hyatt, and V.J. McDonald (eds.), Ecological Indicators. Elsevier Science, London, pp. 787–810.

FAO. 1998. ECOCROP 1 & 2—The crop environmental requirements database and the crop environment response database. Land and Water Digital Media Series no. 4. Food and Agriculture Organization, Rome.

Flanagan, D.C., and M.A. Nearing. 1995. USDA-water erosion prediction project hillslope profile and watershed model documentation. NSERL Report no. 10. National Soil Erosion Research Laboratory, U.S. Department of Agriculture, Washington, DC.

Herman, A., V.B. Kumar, P.A. Arkin, and J.V. Kousky. 1997. Objectively determined 10-day African rainfall estimates created for famine early warning systems. Intl. J. Remote Sens. 18:2147–2159.

Hutchinson, M.F. 1991. The application of the thin plate smoothing splines to continent-wide data assimilation. In: J.D. Jasper (ed.), Data Assimilation Systems, BMRC Research Report no. 27. Bureau of Meteorology, Melbourne, Australia, pp. 104–113.

IPCC, 1996. Climate Change 1995: Impacts, Adaptations, and Mitigation of Climate Change: Scientific-Technical Analyses. Contribution of Working Group II to the Second Assessment Report of the Intergovernmental Panel on Climate Change (R.T. Watson, M.C. Zinyowera, and R.H. Moss, eds.). Cambridge: Cambridge University Press, Cambridge.

Littleboy, M., and G.M. McKeon. 1997. Subroutine GRASP: Grass Production Model. In: K.A. Day, G.M. McKeon, and J.O. Carter (eds.), Evaluating the Risks of Pasture and Land Degradation in Native Pastures in Queensland. Final report for Rural Industries Research and Development Corporation project DAQ-124A. Dept. Primary Industries, Brisbane, Australia, pp. 33–91.

Percival, D.B., and A.T. Walden. 2000. Wavelet Methods for Time Series Analysis. Cambridge University Press. Cambridge.

Rossi, R.E., J.L. Dungan, and L.R. Beck. 1994. Kriging in the shadows: Geostatistical intepolation for remote sensing. Remote Sens. Environ. 49:32–40.

Rowan, R.C. 1995. PHYGROW model documentation, version 2. Texas A & M University, College Station.

SAS. 1999. SAS/STAT Users Guide: Version 8. SAS Institute, Cary, NC.

Saxton, K.E., W.J. Rawls, J.S. Romberger, and R.I. Papendick. 1986. Estimating generalized soil-water characteristics from texture. Soil Sci. Soc. Am. J. 50:1031–1036.

Soil Conservation Service. 1997. Map unit use file (MUUF). Available http://www.wcc.nrcs.usda.gov/water/quality/frame/wetdrn/Tools/tools.html.

Tucker, C.J., C.L. VanPraet, M.J. Sharman, and G. Van Ittersum. 1985. Satellite remote sensing of total herbaceous biomass production in the Senegalese Sahel: 1980–1984. Remote Sens. Environ. 17:233–249.

Wight, J.R., and J.W. Skiles. 2000. SPUR-Simulation of Production and Utilization of Rangelands: Documentation and User Guide. U.S. Department of Agriculture, Washington, DC.

Williams, J.R., P.T. Dyke, and C.A. Jones. 1984. A modeling approach to determine the relationship between erosion and soil productivity. Trans. ASAE 27:129–144.

Williams, M., E.B. Rastetter, D.N. Fernandes, M.L. Goulden, G.R. Shaver, and L.C. Johnson. 1997. Predicting gross primary production in terrestrial ecosystems. Ecol. Appl. 7:882–894.

Xie, P., and P.A. Arkin. 1997. A 17-year monthly analysis based on gauge observations, satellite estimates, and numerical model outputs. Bull. Am. Meteorol. Soc. 78:2539–2558.

PART VI

ASIA AND AUSTRALIA

CHAPTER TWENTY-THREE

Monitoring and Managing Agricultural Drought in India

AMJURI S. RAO AND VIJENDRA K. BOKEN

Agriculture is the mainstay of more than 70% of India's more than 1 billion population. Indian agriculture is predominantly rain-fed and depends on the spatial and temporal distribution of rains from southwest (June–September) and northeast (October–December) monsoons. A monsoon refers to seasonal alteration of atmospheric flow. The Indian subcontinent is predominantly characterized by a tropical monsoon climate, where climatic regimes are governed by rainfall rather than by temperature. The southwest monsoon accounts for 70–90% of the annual rainfall. The Technical Committee on Drought Prone Areas Program (DPAP) and the Desert Development Program (DDP) identified about 120 million ha of the country's land spread in 185 districts (in 13 states; figure 23.1) as drought prone (DPAP/DDP,1994).

Though the country as a whole receives an average annual rainfall of 1100 mm, the arid regions receive between 100 and 500 mm, and semiarid regions receive between 350 and 1500 mm of rainfall. Arid and semiarid regions of the country experience frequent droughts. The arid region in Rajasthan state produces 76% of pearl millet production of the country, but the average productivity is only 267 kg/ha for the whole of Rajasthan compared to 452 kg/ha for the country. Drought frequency is once in 2.5 years in arid zones and once in 4 years in semiarid regions (table 23.1).

India experienced droughts in 1792, 1804, 1812–13, 1833–34, 1838–39, 1848–49, 1850–51, 1853–54, 1868–69, 1877, 1891, 1896, 1899, 1905, 1911, 1915, 1918, 1920, 1941, 1951, 1965, 1966, 1972, 1974, 1979, 1982, 1986, 1987, 1988, 1999, and 2000. A higher frequency of drought occurred during 1891–20, 1965–90, 1997–2000, and 2002.

Figure 23.1 Drought-prone districts in India as identified under the Desert Development Program (DDP) and the Drought-Prone Areas Program (DPAP).

Impacts of Droughts

The main food crops affected by drought are rice, wheat, pearl millet, sorghum, and pegion pea that are grown during southwest monsoon season under rain-fed conditions. Legumes and pulses, which are short-duration crops, are less affected. Persistent droughts cause crop failures and lead to acute shortage of food, fodder, and water (for drinking or irrigation), affecting human and livestock health. During drought years, people and livestock in the arid parts of Rajasthan migrate to neighboring states in search of food, fodder, drinking water, and employment (figure 23.2).

Droughts that occurred in India in 1967, 1968, 1969, 1972, 1974, 1979, and 1987 (Ministry of Agriculture, 1988) had considerable impact on food grains production (figure 23.3). For example, the drought of 1966–67 reduced overall food grains production by 19%, and the loss in production of rice and other pulses was 30% in 1965–66, 66% in 1966–67, and 86%

Table 23.1 Frequency of occurrences of droughts in India

Geographical region	Drought frequency
West Rajasthan and Kutchchh	Once in 2.5 years
Telengana, Rayalaseema, Tamil Nadu, south interior Karnataka, eastern Rajasthan, and Gujarat	Once in 3 years
Vidarbha, Uttar Pradesh, north interior Karnakata, and west Madhya Pradesh	Once in 4 years
Bihar, West Bengal, Orissa, coastal Andhra Pradesh, Kerala, coastal Karnataka, Konkan, east Madhya Pradesh, and Maharastra	Once in 5 years
Northeastern regions	Once in 15 years

Source: Appa Rao (1991).

in 1967–68. The droughts of 1972–73 reduced the food grains production from 108 to 95 million tons, causing a loss of about $400 million.

In 1999, droughts occurred in 12 states, affecting about 100 million human and 60 million livestock population, mostly in Rajasthan, Gujarat, Andhra Pradesh, and Madhya Pradesh states. A survey conducted during the 1999–2000 drought in Rajasthan and Gujarat showed that the fodder deficit in the arid parts of Rajasthan was 50–75%, which resulted in 78% of the livestock migrating from the Barmer, 70% from the Jaisalmer, and 20% from the Jodhpur district in Rajasthan State. Net sowing area also declined significantly. Traditional occupations (farming, livestock husbandry) were replaced by labor for relief works during drought years. The groundwater levels declined at the rate of 0.2–0.4 m/annum. The quality of groundwater also deteriorated, and the concentration of undesirable

Figure 23.2 A family migrating from drought-prone areas (in Rajasthan, India) to adjoining areas or states in search for food, water, and employment during 1999–2000 drought year.

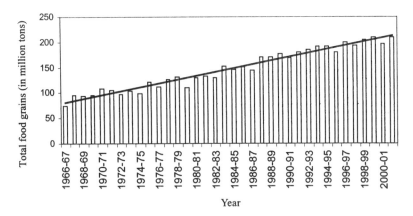

Figure 23.3 The variation in total food grain production in India for the period from 1966–67 to 1998–99 (from India, 2001).

substances such as fluoride and nitrate often reached toxic levels (Narain et al., 2000).

Venkateswarlu (1987) analyzed the yield fluctuations of different crops and found that rice was affected to a greater extent by drought because 60% of the rice crop is rain-fed, whereas wheat crop production is affected less because 65% of wheat is irrigated. During certain years, drought continued for more than one year and had multiple effects on natural resources as well as on humans and livestock. Assessment of drought effects on food grains production in the country showed that sorghum was affected only in Rajasthan and Gujarat, while pearl millet was affected in Gujarat, Uttar Pradesh, and Maharashtra states (Venkateswarlu, 1993).

During some of the worst drought years, the monsoon rainfall was significantly below normal: –26% in 1918, –25% in 1972, and –19% in 1987. In 1987, agricultural operations were affected in 43% (58.6 million ha) of cropped area in 263 districts in 15 states and 6 Union Territories. In Rajasthan and Gujarat, which were worst affected, the rainfall was less than 50% of the normal. In these states, the drought of 1987 was the third or fourth in succession and caused distress of an unprecedented level. Nearly 54,000 villages faced acute drinking water problems. India also experienced extreme drought conditions during 1998–2000, as shown in table 23.2.

Main Causes of Drought

Droughts in India occur due to late onset, early withdrawal, or failure of active monsoon, resulting in insufficient precipitation (or dry spells) with uneven temporal distribution during crop growing season (Kumar, 1986). High interannual and spatial variability in the monsoon rainfall (except in some areas, such as western coast and northeastern India) also influences crop production adversely.

Table 23.2 Impacts of extreme drought conditions in India during 1998–2000

Impact of drought	1998	1999	2000
Number of villages	20069	23406	30583
Population affected (million)	22	26	33
Cattle affected (million)	296	345	400
Damage to crop			
(Million ha)	65	78	90
Value (in million US $)	50	76	136
Rainfall deficit	−3%	−16%	−29%

Source: Narain et al. (2000).

The El Niño/Southern Oscillation (ENSO; chapter 3) weakens the summer monsoon and is related to drought occurrence in India (WMO, 1994). During the period from 1871 to 1988, 11 of the 21 droughts were attributed to ENSO phenomenon (WMO, 1994).

Subsidized electricity for irrigation systems operating in drought-prone areas has led to overexploitation of groundwater for cultivation of crops like cumin, mustard, wheat, vegetables, and fruits that earn farmers higher profits. Similarly, subsidized diesel fuel in the country has led to a spurt in tractor farming, increasing wind erosion, and deteriorating stability of natural grasslands. This has also contributed to drought and desertification phenomena.

Drought Monitoring Methods

Precipitation Deviation

The India Meteorological Department (IMD), New Delhi, monitors droughts based on the deviation of precipitation from the normal for each of 35 meteorological subdivisions in the country. The IMD issues a drought warning for a subdivision if the precipitation deviation for the subdivision is greater than 25% of the long-term average normal. Drought is considered to be moderate if the annual rainfall lies in the range of 25–50% of the normal and severe if annual rainfall is less than 50% of the normal. An area is classified as drought prone if it experiences drought conditions in 20% of the years and as a chronically drought prone if the area experiences drought conditions at least 40% of the years in a multiyear period (NCA, 1976).

Crop Watch Group

During the 1979–80 period of drought, the Ministry of Agriculture set up a Crop Watch Group that consisted of representatives from the Department of Agriculture, India Meteorological Department, Indian Council of Agricultural Research, and Ministry of Information and Broadcasting. The group provides weekly reports on rainfall, agricultural operations, water levels in 47 major reservoirs that are monitored by the Central Water Commission, employment, and other activities affected by droughts. The information thus collected by the group is sent to various agrometeorological weather advisory field units located in different agroclimatic zones of the country for preparing bulletins to farmers on various agricultural operations. Additionally, the forecasts are broadcasted through radio, television, and various newspapers. This helps monitor the development of drought conditions across the country.

Moisture Adequacy Index

The Central Arid Zone Research Institute, Jodhpur monitors agricultural drought in the Indian arid regions by using the moisture adequacy index (MAI). The MAI is equal to the ratio (expressed as a percentage) of actual evapotranspiration (AET) to the potential evapotranspiration (PET) following a soil water-balancing approach during a cropping season (Ramana Rao et al., 1981; Rao, 1997; Sastri et al., 1982).

Aridity Anomaly Index

A drought research unit was established in 1967 at the IMD, Pune. This unit prepares biweekly aridity anomaly reports for the southwest monsoon season for the entire nation and during the northeast monsoon season for five meteorological subdivisions (coastal Andhra Pradesh, Rayalaseema, south interior Karnataka, Kerala, and Tamil Nadu and Pondichery). The aridity anomalies are determined based on water balance studies (Thornthwaite and Mather, 1955) and are classified into various categories of arid conditions (e.g., mild, moderate, and severe) for monitoring and assessing agricultural droughts in the country. These reports are sent to research institutes and various agricultural authorities of state and central governments to help in agricultural planning.

Yield Modeling

Predicting yield of the major crop of a drought-prone region is a simple and practical way of drought monitoring for the region. For the Indian arid zone, pearl millet is the single most important crop. Using variables related to delay in sowing, monthly rainfall, and the number of rainy days in a month, Kumar (1998) developed early warning models for predicting pearl millet yield for Jodhpur district (in Rajasthan State), which explained up to 81% of the yield variation.

The drought research unit developed regressions equations between past yields and weather variables for forecasting rice yields during *Kharif* season for 15 states (comprising 26 meteorological subdivisions) and wheat yields during *Rabi* season for 12 states (comprising 16 meteorological subdivisions) as well as annual wheat yields for the entire nation. These forecasts are sent to the Ministry of Agriculture, and the Planning Commission, Government of India (De et al., 1997).

In addition, computer models such as the SPAW (soil-plant-air-water; Saxton, 1989) model, CERES-millet (Ramakrishna et al., 1994), and RANGETEK (Ross Wight, 1987) have been used in India to forecast crop yields and droughts. The SPAW model explained up to 89% of the variation in pearl millet yields of Jodhpur district, Rajasthan:

$$Y = -45.38 \text{ WSI} + 526.18 \quad (R^2 = 0.89) \quad [23.1]$$

where Y is pearl millet grain yields (kg/ha). The water stress index (WSI) varied from 0 (for a nonstressed crop) to 15 (for a crop with severe water stress) (Rao and Saxton, 1995; Rao et al., 2000).

The RANGETEK model, based on a yield index (ratio of transpiration to potential transpiration), could estimate moisture stress and pasture yields within 1–13% in the case of *C. ciliaris* and up to 88% in the case of *L. sindicus* (Rao et al., 1996; Singh and Rao, 1996). The CERES-millet model explained 99% of variation in pearl millet grain yield (Ramakrishna et al., 1994).

Weather Forecasting

Short range (less than 3 days), medium range (3–10 days), and long range (10–30 days) weather forecasts (DST, 1999) are issued by the IMD and the National Centre for Medium Range Weather Forecast at New Delhi, which can be used to monitor the development and persistence of drought conditions in the country.

Use of Satellite Data

Under a centrally sponsored scheme on Remote Sensing Application Mission for Agriculture, Department of Agriculture and Cooperation, in the Ministry of Agriculture, and the Indian Space Research Organization, Department of Space, initiated in 1987 (Chakraborti, 2000) a national program called the National Agricultural Drought Assessment and Monitoring System (NADAMS). The NADAMS was established at the National Remote Sensing Agency, Hyderabad, in Andhra Pradesh State. Under this system, Advanced Very High Resolution Radiometer (AVHRR) daily data have been used since 1989 to generate biweekly composites of the normalized difference vegetation index (NDVI) during the *kharif* season for 11 agriculturally sensitive states of India (Andhra Pradesh, Bihar, Gujarat, Haryana, Karanataka, Maharashtra, Madhya Pradesh, Orissa, Rajasthan, Tamil Nadu, and Uttar Pradesh).

The NDVI is a transformation of reflected radiation in the visible (0.58–0.68 μm) and near infrared (0.725–1.10 μm) bands of AVHRR data and is a function of green biomass (Thiruvengadachari, 1990; chapters 5 and 6). These composites are included in the biweekly drought bulletins that are issued to the state and central government agencies. The bulletin consists of three sections. The first section contains color-coded NDVI imageries of the state with district boundaries superimposed. The second section compares district level greenness during reporting year with the previous two years. The third section consists of a district level assessment of drought prevalence and persistence and the relative severity levels based on the extent of NDVI anomaly (Chakraborti, 2000). NDVI anomalies are interpreted in terms of moisture stress condition after verifying false alarms due to excessive rainfall or flooding, cloud contamination, poor time-composition, difference in seasonal growth cycle, and so on. Based on the relationship between NDVI and major crop yields, the early warning on relative yield assessment is provided from August onward. It is possible to further improve the performance of satellite data-based monitoring system by determining the dates of the commencement and termination of phenological phases using biometeorological time scale or crop growth models (Boken and Shaykewich, 2002; chapter 1). These models can help select the satellite data for the critical phases of the crop when yields are most significantly affected by drought conditions. In addition, Indian Remote Sensing satellite data are used for drought assessment by coupling satellite data analysis with adequate ground information, such as reduction in harvest area and water availability, reduced production of fodder, and other socio-economic factors (Chakraborti, 2000).

Drought Mitigation

The following are some strategies for drought mitigation that were developed in India since the 1970s.

Crop Planning

Based on rainfall analysis, water availability periods were identified for several locations in India in order to escape droughts and plan crop sowing according to the water availability (Srinivasamurthy, 1976). Figure 23.4 shows four different scenarios that characterize monsoon behavior in India: (1) Normal onset of monsoon followed by adequate amount of rainfall, (2) normal onset of monsoon followed by inadequate amount of rainfall, (3) late onset of monsoon followed by adequate amount of rainfall, and (4) late onset of monsoon followed by inadequate amount of rainfall. For each scenario, a different strategy is adopted to minimize the impact of drought on crop production.

In the first scenario, the crop-growing period varies from 90 to 120 days, and cereal crops such as sorghum, pearl millet, maize, pulses such as cowpea, mungbean, blackgram, and oilseeds such as sesame, castor, and groundnut with high density planting followed by leguminous/fodder crop can be grown for maximum profits. In the second scenario, mid-season corrections such as reducing plant population, weeding, and creating soil

AGRICULTURAL DROUGHT IN INDIA 305

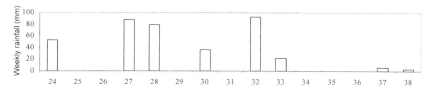
(i) Normal onset of monsoon of Jodhpur followed by adequate amount of rainfall (396mm in 2001)

(ii) Normal onset of monsoon at Jodhpur followed by inadequate amount of rainfall (64mm in 1987)

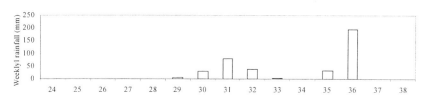
(iii) Late onset of monsoon followed by adequate amount of rainfall (384mm in 1992).

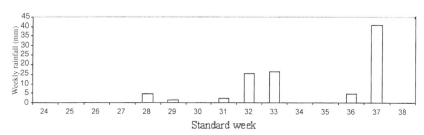
(iv) Late onset of monsoon followed by inadequate amount of rainfall (85mm in 1969)

Figure 23.4 Different monsoon rainfall situations that occur in the arid region of Jodhpur, India.

mulch can be adopted to alleviate drought effects. In the third scenario, the growing period varies from 40 to 60 days only, and therefore short-duration varieties are adopted to minimize drought impacts. Some of short-duration varieties are HHB-67, CMH-356, and RHB-30 in the case of pearl millet, S-8, K-851, and P-9075 in the case of green gram, and the *Maru* variety of clusterbean. For the last scenario, short-duration varieties of cluster bean, green gram, cowpea, moth bean, and sesame are preferred.

Crop Mixing and Integrated Farming

Crop mixing (i.e., growing multiple crops during the same growing season) is also advisable to reduce drought impacts. If the rainfall pattern is not suitable for one crop, it may be suitable for the other, and hence crop production can be maximized under drought conditions.

The Central Arid Zone Research Institute at Jodhpur, Rajasthan, developed an integrated farming systems approach for sustainable crop production in arid regions. Following this approach, crops such as pearl millet, clusterbean, mothbean, and sesame were grown in combination with traditional trees (e.g., *Prosopis cineraria, T. undulata, A. albeda*), or in combination with horticultural trees (e.g., *Zizyphus* cultvars).

Shelterbelts and Mulching

Soil evaporation losses and high temperatures can be controlled by applying mulch materials. Shelterbelts are also useful in reducing high evaporation losses. *Acacia nilotica* spp. Indica shelterbelts have been found useful in controlling wind speeds and thus reducing evaporative losses in the arid regions (Gupta et al., 1983). Shelterbelts of pearl millet provided to summer vegetables modified the crop microclimate and increased the yields of okra and cowpea by 30–40% (Ramakrishna, 1985).

Optimizing and Improving Water Resources

After the severe drought in 1987, the government of India focused more on optimizing the use of water from reservoirs and groundwater resources. An integrated watershed management approach can contribute to planning such optimization. In 1984–85, 4400 micro-watersheds covering 4.2 million ha were identified.

The flash floods that sometimes occur in arid or semiarid regions of India could also be used for augmenting groundwater sources through percolation injection wells in conjunction with subsurface barriers. Over past three decades, there has been phenomenal increase in the number of tube wells in arid regions. These wells have supplemented drinking water supply as well as irrigation for some commercial crops.

Under the auspices of the Indira Gandhi Nahar Pariyojna (i.e., the Indira Gandhi Canal Project), a canal of more than 400 km length along the western border and arid region of Rajasthan was constructed during 1961–73. This canal carries about 7.59 million acre feet (MAF) of water from Ravi-Beas rivers and irrigates about 1.1 million ha, which contributes an additional production of 3.1 million tons of food grains every year. As a result of this canal, the production of rice, groundnut, castor, rapeseed, and mustard have increased several times (table 23.3).

Table 23.3 Area and production of food grains and oil seeds in 1986–87 and 1987–88

Category	Area (million ha)			Production (m. tons)		
	1986–87	1987–88	% Difference	1986–87	1987–88	% Difference
Total *kharif* food grains	81.46	74.45	–8.6	80.20	73.89	–7
Total *rabi* food grains	45.74	44.26	–3.2	63.22	64.52	+2
Total food grains	127.20	118.71	–6.7	143.42	138.41	–3
Kharif oilseeds	11.51	11.47	–0.3	6.38	6.28	–1
Rabi oilseeds	7.12	8.53	+19.9	0.89	6.10	+24
Total oilseeds	18.63	20.00	+7.4	11.27	12.38	+10

Source: The Drought of 1987, Response and Management, vol. I, DAC, Ministry of Agriculture, New Delhi.

Soil and Water Conservation

The soil conservation department in each state has constructed contour boundaries (in areas with higher slopes) and check dams for water storage and implemented methods to control soil erosion. In addition, ridge-furrow systems, artificial microcatchments, and inter-row water harvesting is also advised to help store soil moisture throughout the growing season and thus mitigate droughts. Figure 23.5 shows a water-harvesting structure commonly used by households for storing rainwater in Rajasthan.

Relief Measures

Relief measures are undertaken mainly by governmental agencies and include supplying food, fodder, drinking water, and animal shelter (mainly for cows to avoid mortality), employment, distribution of seeds, and subsidized power for the *rabi* season to help boost production (Subbiah, 1993).

Food Security and Buffer Stock The Food Corporation of India (FCI), established in 1965, provides infrastructure for procurement, storage, and distribution of 30 million tons of food grains per year. At the time of droughts, food grains are distributed through FCI from surplus to deficit areas through a public distribution system at subsidized rates. During the 1965–66 drought, India imported 10 million tons of food grains to meet half of its food requirements for the drought-prone areas. But at the beginning of 1987 (the worst drought year in recent times) the FCI had storage (or buffer stock) of 23 million tons of food grains and also collected an additional 13 million tons from surplus areas and supplied it to 285 million people affected by drought. Although the 1987–88 drought was double in severity and affected twice the population as the 1965–66 drought, the country could meet its food requirements from its buffer stocks. In 1979,

Figure 23.5 A Tanka—a common household rainwater harvesting system in Rajasthan, India.

the drought severity was not widely felt, and no starvation deaths were reported because of the buffer stocks of food grains.

Employment Employment is generated under Food for Work program and now under *Sampoorna Grameen Rozgar Yojana* (Integrated Rural Employment Scheme introduced in 2001) in which employment is provided to one member of each family affected by drought. Employees are engaged in projects related to deepening of *Nadis* (i.e., ponds; figure 23.6), construction of medium and/or minor irrigation systems, soil conservation, pasture development and afforestation, construction of schools/governmental buildings, and road building.

Potential Drought Research or Management Needs

Water Use Estimation and Improving Storage

With increasing demand for water, the Indian arid region is likely to experience a deficit of about 2500 million m^3 by the year 2010 (Venkateswarlu et al., 1990). However, an accurate estimate of water usage (both surface and groundwater) is required to plan long-term measures to mitigate droughts, and current water-storing methods practiced in the region need to be improved. Old water-harvesting systems, such as rooftop water harvesting, *Nadis* (village pond), *Tankas* (underground cistern), *anicuts,* and *khadins*

Figure 23.6 Renovation of a village pond (*Nadi*) in Barmer district during 1999–2000 drought. As men migrate to far away places for employment, mostly the village women were seen working under Food for Work program.

need to be modernized to improve their efficiency and tackle drought situations more effectively.

Weather Network, Farmers' Participation, and Fodder Bank

There is a need to strengthen the communication network among the vast number of weather stations in India and establish a weather-data bank. Weather stations should be linked to drought advisory units to help monitor droughts on a real-time basis. Satellite remote sensing and geographic information system-based approaches for agricultural drought monitoring need standardization for precise assessment of conditions of various crops in various regions.

Early participation of local community in the implementation of a drought management scheme will make the scientific community more familiar with ground realties and make the measures more effective.

Construction of fodder-storing devices (or fodder banks) in drought-prone areas will help reduce the cost of fodder transport from surplus areas, which may be hundreds of kilometers away.

Conclusions

The commencement and termination of droughts in India are directly associated with the behavior of the southwest monsoon. India experienced severe droughts in at least 20 years of the past century and faced a drastic drop in food grains production. Due to various measures undertaken to mitigate droughts and to increase food grain production, India has not only become self-reliant but produces more than its requirements. Surplus production contributes to buffer stocks that are used to supplement food production during drought years. Nevertheless, its fodder is in short supply even in normal years. Because impact of the drought is greater on livestock than on humans, there is a need to strengthen fodder banks in rural areas to alleviate impacts on livestock.

The risk of crop production in drought-prone areas can be minimized by adopting alternate cropping strategies, efficient utilization of rain water through soil and water conservation measures, watershed development, runoff farming, and irrigation. For an early warning, the drought and crop surveillances should be carried out using short or medium range meteorological forecasts, remote sensing techniques, and computer models such as the SPAW, CERES-millet/rice/sorghum, and RANGETEK models combined with GIS capabilities for monitoring crop yields and agricultural drought. A drought monitoring system based on satellite data is currently operational in India; however, there is a need to thoroughly test its accuracy and improve its performance by including information provided by crop growth models and satellite data with a better resolution.

ACKNOWLEDGMENTS Dr. Pratap Narain, director, CAZRI, Jodhpur, provided the facilities and encouragement for writing this chapter. The assistance provided by A.S.R.'s wife, Indira, his daughter, Lakshmi Prashanthi, and his son, Murali Raj, is duly acknowledged.

References

Appa Rao, G. 1991. Droughts and southwest monsoon. In: Monsoon Meteorology. 3rd WMO Asian/African Monsoon Workshop. Pune, India.

Boken, V.K., and C.F. Shaykewich. 2002. Improving an operational wheat yield model for the Canadian Prairies using phenological-stage-based normalized difference vegetation index. Intl. J Remote Sens. 23:4157–4170.

Chakraborti, A.K. 2000. Satellite remote sensing for near-real-time food and drought impact assessment—Indian experience. In: P.S. Roy, C.J. van Westen, V.K. Jha, R.C. Lakhera, and P.K. Champati Roy (eds.), Natural Disasters and Their Mitigation. Indian Institute of Remote Sensing, Dehra Dun, India, pp. 143–150.

De, U.S., M.P. Shewale, and A.K. Sen. 1997. Role of India Meteorological Department in agrometeorological activities. Vayu Mandal, 27(3–4):7–12.

DPAP/DDP. 1994. Report of the Technical Committee on Drought Prone Areas Programme and Desert Development Programme. Ministry of Rural Development, New Delhi.

DST. 1999. Guide for Agrometeorological Advisory Service. Department of Science and Technology, New Delhi.

Gupta, J.P., G.G.S.N. Rao, G.N. Gupta, and B.V. Ramana Rao. 1983. Soil drying and wind erosion as affected by different types of shelterbelts planted in the desert region of western Rajasthan. J. Arid Environ. 6:53–58.

Ministry of Agriculture. 1988. The Drought of 1987—Response and Management, vols. I and II. Department of Agriculture Cooperation, Ministry of Agriculture, New Delhi.

(Boken) Kumar, V. 1986. National Drought Monitoring System—Status. Report RSAM-NRSA-DRM-TR-01. National Remote Sensing Agency, Balanagar, Hyderabad, India.

(Boken) Kumar, V. 1998. An early warning system for agricultural drought in an arid region using limited data. J. Arid Environ. 40:199–209.

Narain, P., K.D. Sharma, A.S. Rao, D.V. Singh, B.K. Mathur, and A. Usha Rani. 2000. Strategy to Combat Drought and Famine in the Indian Arid Region. Central Arid Zone Research Institute, Jodhpur, India.

NCA. 1976. Report on Climate and Agriculture, vol. IX. National Commission on Agriculture, New Delhi, pp. 34–39.

Ramakrishna, Y.S. 1985. Micro-crop shelter belts for improving crop productivity under irrigated conditions. In: S.D. Singh (ed.), Development and Management on Irrigated Agriculture in Arid Regions. Central Arid Zone Research Institute, Jodhpur, India, pp. 661–674.

Ramakrishna, Y.S., N.L. Joshi, A.S. Rao, and H.P. Singh. 1994. Performance of Modified CERES Millet Model under Arid and Semi-arid Conditions. Rainfed Agriculture-Research Newsletter, no. 4, Central Research Institute for Dryland Agriculture, Hyderabad, India, pp. 3–5.

Ramana Rao, B.V., A.S.R.A.S. Sastri, and Y.S. Ramakrishna. 1981. An integrated scheme of drought classification as applicable to the Indian arid zone. Idojaras, 85:317–322.

Rao, A.S. 1997. Impact of droughts on Indian arid ecosystem. In: S. Singh and A. Kar (eds.), Desertification Control in the Arid Ecosystem of India for Sustainable Development. Agro Botanical Publishers, Bikaner, India, pp. 120–130.

Rao, A.S., N.L. Joshi, and K.E. Saxton. 2000. Monitoring of productivity and crop water stress in pearl millet using the SPAW model. Ann. Arid Zone 39:151–161.

Rao, A.S., and K.E. Saxton. 1995. Analysis of soil water and water stress for pearl millet in an Indian arid region using the SPAW model. J. Arid Environ. 29:155–167.

Rao, A.S., K.C. Singh, and J.R. Ross Wight. 1996. Productivity of *C. ciliaris* in relation to rainfall, N and P fertilization. J. Range Manage. 49:143–146.

Ross Wight, J.R. 1987. ERHYM-II model description and user guide of the basic version. Bulletin 59. Agricultural Research Station, U.S. Department of Agriculture, Washington, DC.

Sastri, A.S.R.A.S., B.V. Ramana Rao, and Y.S. Ramakrishna. 1982. Agricultural droughts and crop production in the Indian arid zone. Arch. Met. Geophys. Biokl., Ser. B 31:405–411.

Saxton, K.E. 1989. User Manual of SPAW. A Soil-Plant-Atmosphere-Water Model. Agricultural Research Service, U.S. Department of Agriculture, Pullman, WA.

Singh, K.C., and A.S. Rao. 1996. A short note on water use and production potential of Sewan (*L. scindicus*) grass pasture in Thar desert region of Rajasthan, India. J. Arid Environ. 33:261–262.

Srinivasamurthy, B. 1976. Some Aspects of Water Availability to Dryland Crops in Maharashtra. Science Report no.76/14. India Meteorological Department, Pune.

Subbiah, A.R. 1993. Indian drought management: from vulnerability to resilience. In: D.W. White (ed.), Drought Assessment, Management and Planning: Theory and Case Studies. Kluwer Academic Publishers, Boston, pp. 157–179.

Thiruvengadachari, S. 1990. Satellite surveillance for improved country-wide monitoring of agricultural drought conditions. In: Proceedings of the National Symposium of Remote Sensing for Agricultural Application, New Delhi, pp. 389–407.

Thornthwaite, C.W., and J.R. Mather. 1955. The Water Balance. Publications in Climatology. Vol. 8. Laboratory of Climatology, Elmer, New Jersey, p. 104.

Venkateswarlu, J. 1987. Technological and socio-political adaptation and adjustment to drought: the Indian experience. In: D.A. Wilhite and W.E. Easterling (eds.), Planning for Drought—Towards Reduction of Societal Vulnerability. Westview Press, Boulder, CO, pp. 391–408.

Venkateswarlu, J. 1993. Effect of drought on *kharif* food grains production—a retrospect and prospect. Ann. Arid Zone 32:1–12.

Venkateswarlu, J, A.K. Sen, J.C. Dubey, N.L. Joshi, A. Kar, A.S. Kolarkar, M.L. Purohit, Y.S. Ramakrishna, A.S. Rao, and K.D. Sharma. 1990. Water 2000 A.D—The Scenario for Arid Rajasthan. CAZRI, Jodhpur, India.

WMO. 1994. Climatic Variability. Agriculture and Forestry Technical Note no. 196. World Meteorological Organization, Geneva.

CHAPTER TWENTY-FOUR

Agricultural Drought in Bangladesh
AHSAN U. AHMED, ANWAR IQBAL,
AND ABDUL M. CHOUDHURY

Bangladesh is globally known as a flood-vulnerable country—an almost flat land with too much water. In terms of annual per capita availability of water resources, it ranks among the highest in the world. But a lesser-known disaster that affects a significant proportion of its fertile land is drought. The occurrence of droughts may largely be attributed to two recent phenomena: (1) an extensive adoption of high yielding varieties (HYV) of paddy (i.e., rice) in the drier months; and (2) constraints faced in water availability during premonsoon months due to upstream water withdrawal from river systems. Up to 15% of the total cultivable land (about 0.9 million ha) now experiences droughts of moderate to very severe intensity, once in every two years (Iqbal and Ali, 2001). This chapter examines the causes of droughts in the context of the country's complex water regime, the implications of droughts, and the ways to monitor them.

About 80% of annual monsoon rainfall over the country occurs during the period from June to the first week of October. The western zones of the country receive less rainfall, averaging about 1400 mm, compared to the national average of 2150 mm, and therefore the susceptibility to droughts in the western zones of the country is higher. Table 24.1 provides a chronological overview of areas and populations in Bangladesh affected by droughts during the 1950–79 period.

The economy of Bangladesh significantly depends on agriculture. More than 63% of 130 million people, confined within a territory of 147,750 km^2, find employment in agriculture (MOF, 2003). Although the share of the crop production in the gross domestic product (GDP) has been declining steadily in recent times, dropping from 24.66% in 1990–91 to 18.58% in 2002–03, it still is the predominant economic activity of the majority of the people (Ahmad and Ahmed, 2002). More than 80% of the households in rural Bangladesh are directly dependent on the production of various crops.

Table 24.1 Area and population percentages affected by droughts in Bangladesh since 1950[a]

Year of drought	Area (%)	Population (%)
1950	13.79	14.13
1951	31.63	31.1
1952	6.57	5.95
1954	3.43	3.92
1955	9.83	15.49
1956	11.25	9.53
1957	46.54	53.03
1958	37.47	36.24
1960	2.70	3.11
1961	22.39	20.76
1962	11.30	9.74
1963	8.60	6.63
1966	18.42	16.54
1970	9.10	10.55
1971	4.80	5.66
1972	42.48	43.05
1976	5.02	5.48
1978	3.66	4.51
1979	42.04	43.90

Source: Chowdhury and Hussain (1983).

[a] Drought-free years include 1953, 1959, 1964, 1965, 1967, 1968, 1969, 1973, 1974, 1975, and 1977. Since the drought of 1979, the government has been taking proactive measures toward promoting irrigation in the drought-prone areas.

Paddy (rice) is the main crop, occupying about 80% of the cultivated land. Multiple varieties of paddy are grown in the country. In addition to the local paddy variety *Aman* (i.e., T. Aman) grown in the monsoon season, a premonsoon local variety named *Aus* is also grown during March–June. Jute, a cash crop, has also been grown in lands that were marginally suitable either for *Aus* or for *Aman*.

Because of the increasing population and food requirements, farmers first adopted HYV of paddy and began to encroach into wetlands by building embankments in the floodplains to make inundated land virtually flood free and suitable for *Aman* cropping. In addition, farmers began cultivating HYV *Boro* during the dry season with the help of irrigation. Depending on the general variability of climate and annual distribution of temperature and moisture regimes, there are three major cropping seasons in Bangladesh: pre-*Kharif*, *Kharif*, and *Rabi* (figure 24.1).

Pre-*Kharif* Drought

The pre-*Kharif* season stretches from mid-March to June. It is a period with erratic rainfall, a long period of no rainfall, high temperatures, and

AGRICULTURAL DROUGHT IN BANGLADESH 315

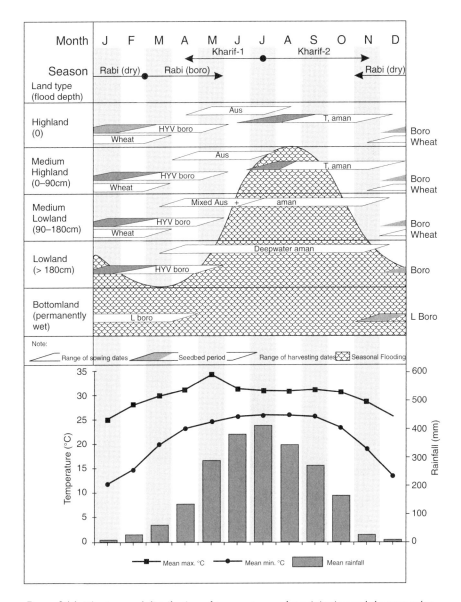

Figure 24.1 The temporal distribution of temperature and precipitation and the crop calendar for major crops in Bangladesh.

increased evapotranspiration. Jute, *Aus*, and the deep water variety of *Aman* are common crops generally grown during the pre-*Kharif* season. Because the *Aman* variety is grown over the lands subject to inundation, it is not affected by drought. In contrast, both jute and *Aus* are affected by the varying degrees of droughts depending on the field and climate conditions.

Kharif Drought

Kharif has always been the major cropping season in the country because of the abundant availability of rainfall during this season. The season usually begins in July and ends in mid-October. *Aman* is the dominant crop of this season, which is usually harvested in December. Currently, the transplanted variety of *Aman* is grown over 60% of all cultivable lands throughout the country. From the viewpoint of the livelihood strategy of the majority of the people, *Aman* is the key crop, a failure of which poses serious threats to the food security of poor households. Unfortunately, such an important crop reaches its reproductive stage when the available moisture is significantly reduced. This leads to high moisture stress for the standing crops, especially during the last one and a half months before the harvest. Intensity of drought increases with long periods without rainfall coupled with high temperatures from late October to the end of November.

Supplementary irrigation reduces the risk of *Kharif* drought significantly, especially during flowering stages. Given the intricate river networks throughout the country and the fact that most of the croplands are located in the floodplains, it is possible to cope with *Kharif* drought if there are adequate surface flows in the river systems during the postmonsoon months. In the absence of adequate surface flows, it is also possible to offer supplementary irrigation to overcome moisture stress, depending on the availability of groundwater resources.

Rabi Drought

The drought problem in Bangladesh perhaps reaches it peak during the *Rabi* season that extends from mid-October to early April. December, January, and February are the months with negligible rainfall across the 1.75 million ha of the combined catchment area of the Ganges-Brahmaputra-Meghna (GBM) river systems in the eastern Himalayan region (Ahmad et al., 1994). As a result, river flows are at their minimum. From March to early April, the temperatures shoot up rapidly, particularly during the day. These conditions lead to high potential evapotranspiration (PET), causing acute moisture stress in the topsoils. When residual moisture can no longer support the standing crops, drought develops and adversely affects crop production. The land with very poor moisture-holding capacity, found in the Barind tract, faces the worst consequences of drought during the *Rabi* season.

Drought severity during the *Rabi* season can be determined by the length of growing period (LGP). If the LGP in an area is very short (about 80–90 days), the area is likely to suffer very severe drought. With an increase of LGP by 10 days, the drought severity class changes from very severe to severe, from severe to moderate, from moderate to slight, and slight to no drought (Iqbal and Ali, 2001). Figure 24.2 presents the drought-affected regions of Bangladesh for the *Rabi* season. Iqbal and Ali (2001) categorized

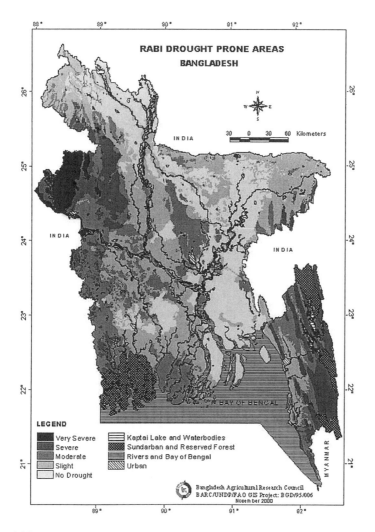

Figure 24.2 A map showing drought-affected areas in Bangladesh during the *Rabi* cropping season.

drought on the basis of yield losses. Yield losses are 45–70% for very severe drought, 35–45% for severe drought, 20–35% for moderate drought, and 15–20% for slight drought. Table 24.2 gives a summary of affected areas under drought in the three cropping seasons.

Timing of planting is also a major factor in determining yield losses. The Bangladesh Rice Research Institute (BRRI, 1999) showed different levels of yields with respect to the planting dates for the variety *Brridhan-29*. If this variety was planted on July 15, the yield was 5.40 tons/ha after 148 days to maturity. If the same variety was planted on July 30, its growing period was reduced by only 3 days, but the yield loss was about 0.3 tons/ha. If the

Table 24.2 Summary of drought-affected areas by cropping seasons in Bangladesh

Cropping season	Area under various drought severity class (million ha)					
	Very severe	Severe	Moderate	Slight	No drought	Non-T. *Aman*
Pre-*Kharif*	0.403	1.15	4.76	4.09	2.09	—
Kharif (T. *Aman* only)	0.344	0.74	3.17	2.90	0.68	4.71
Rabi	0.446	1.71	2.95	4.21	3.17	—

Source: Iqbal and Ali (2001).

planting date was August 30, the crop was forced to mature in about 128 days, resulting in a yield loss of about 28%.

Drought Monitoring

There has not been any significant step toward drought monitoring on a regular basis in Bangladesh. The country has been maintaining a fairly good research effort under the National Agricultural Research System (NARS), which includes 11 research institutes, the Bangladesh Agricultural Research Council (BARC) being the apex body of NARS.

There is a lack of coordination among institutions that could have developed a drought-monitoring mechanism. There is no clear institutional framework defining who should do drought monitoring. There is a Disaster Management Bureau, which considers drought as a form of agricultural disaster and only notes the yearly crop loss reported mostly in newspapers. The information concerning moisture availability, especially in topsoil of drought-vulnerable zones, is supposed to be collected by the Soil Resources Development Institute and by the Department of Agriculture Extension. These two organizations have personnel who can offer services at grassroot levels.

The Bangladesh Meteorological Department collects weather information, including temperature and precipitation on a daily basis. It has a good number of weather stations all over the country. But these data are seldom used for real-time monitoring of drought situations. BARC has facilities for a geographic information system (GIS) and a database equipped with models to monitor drought, but the agency does not have the scope and mandate to monitor drought on a regular basis.

Using Satellite Data

The Space Research and Remote Sensing Organization (SPARRSO) of Bangladesh is an efficient organization equipped with state-of-the-art technologies capable of monitoring drought situation. It receives Advanced Very High Resolution Radiometer (AVHRR; chapters 5 and 6) data from the U.S. National Oceanographic and Atmospheric Administration

(NOAA). The AVHRR data, received in five different channels, are digitally processed at the SPARRSO, primarily to analyze formation of cyclones and depressions in the Bay of Bengal. To analyze changes in vegetation cover, the radiometric information (% reflectance) received by the NOAA satellite is plotted against the wavelength of the radiation. From standard curves, verified by "ground truthing" with local conditions, the condition of vegetation with regard to growth can be determined. This information is then analyzed using GIS technology to get a spatial view of the analysis (SPARRSO, 1999). The technique was tried jointly by SPARRSO and BARC and has proved useful.

Unfortunately, AVHRR data can only be obtained when there are no clouds in the sky over Bangladesh (i.e., only during the winter season). Synthetic Aperture Radar (SAR) data (chapters 7 and 8) should ideally be used during the monsoon season because it can penetrate through clouds. Unfortunately, SPARRSO operates with very limited funding that does not allow it to procure SAR data from international agencies and data providers. Moreover, the organization does not have the mandate to monitor drought conditions. Its sole responsibility is to monitor cyclone formations. When there are fears of wide-scale loss of *Aman* due to the postmonsoon drought, SPARRSO is requested by the Ministry of Agriculture to analyze the situation.

Given the technological capacity of SPARRSO, it is possible to monitor drought on a regular basis for all the cropping seasons provided that the ministries involved coordinate with each other and allocate funds to procure information, engage personnel for ground-level data collection and analysis, and enhance the capacity to validate models. In the absence of a proper institutional mechanism, the country has been deprived of valuable information useful to drought monitoring.

Drought Mitigation in Bangladesh

Since ancient times, farmers have practiced irrigation using surface water from nearby sources for a variety of crops including *Aman* and a number of nongrain *Rabi* crops. There have been local *Aman* species/cultivars that required very low levels of supplementary irrigation. Traditional irrigation techniques have been used extensively. With the advent of green revolution in the mid-1960s and with the development of HYV seeds, people began to irrigate lands heavily. The initial results have been excellent in terms of grain yields. However, over the years, the requirement for fertilizers and irrigation has increased significantly due to the gradual deterioration in land quality due to the erosion of major nutrients, micronutrients, and organic carbon contents from the topsoil.

The surface water systems of the country are largely dependent on upstream countries: India, Nepal, Bhutan, and China. Increasing water withdrawal upstream and the diversion of water from the main transboundary courses reduced the flow in Bangladesh significantly during the dry months

(Rahman et al., 1990; Ahmad et al., 1994; Halcrow and Associates, 2001). Consequently, the possibility of surface water irrigation in the country has been reduced considerably.

To maintain self-sufficiency in food production, farmers have adapted to the use of modern irrigation techniques. Mechanized pumps have replaced the traditional methods of transferring water. Table 24.3 shows gradual development of various forms of irrigation with respect to time and technology. It is evident from table 24.3 that the total irrigated area more than doubled from 1985 to 2000. Moreover, the contribution of surface water and groundwater was almost equal in 1984–85. The recent expansion in total irrigated area was made possible due to about a threefold increase in groundwater irrigation, as against only about a 44% increase in surface-water irrigation during the same time. Currently, about 4.1 million ha out of 8.5 million ha of net cropped area (NCA) is being irrigated to cope with droughts. A recent study suggests that a total of 7.56 million ha could be brought under irrigation, which represents approximately 84% of NCA (Karim et al., 2001). Given the fiscal incentives provided for expansion of irrigation under the private sector, it is expected that more lands will be brought under irrigation in coming years.

Fortunately, Bangladesh has a reasonably good groundwater resource. The quaternary alluvium, the base material that raised the delta above sea level, constitutes a huge aquifer with reasonably good transmission and storage properties. Despite the growth in groundwater agriculture, heavy rainfall and annual inundation help the resources recharge substantially. It is estimated that 21 billion m^3 of groundwater is available (MPO, 1991). However, the piezometric surface of the groundwater aquifers in many areas, especially in the western parts of the country, have lowered steadily because of continued withdrawal of groundwater to offset droughts (Halcrow and Associates, 2001). In the areas dependent on the surface flows of the Ganges River, it is found that shallow tubewells have become inoperable in as many as 58 subdistricts due to the gradual lowering of aquifer surface. This is sufficient evidence that the management of drought in coming decades will be much more difficult.

Groundwater irrigation involves high production costs, especially for the poor farmers engaged in subsistence agriculture. Half of the farming community does not have any cropland; either they are sharecroppers, or they offer physical labor. For the sharecroppers, irrigation appears to be an economic burden which the owners do not share, but for which they receive the benefits.

Unfortunately, unlike flood, it is not possible to bear the loss due to drought. Often poor farmers borrow money at high interest rates, as high as 100%, to be repaid just after harvest of the standing crop (BUP, 2002). There are concerns that fighting droughts has caused high capital investment, which ultimately rendered many poor families indebted to local wealthy people. An accumulation of such debts often forces the poor farmers to migrate. Drought is therefore believed to have compounded the already endemic problem of poverty in rural Bangladesh.

Table 24.3 Area under irrigation in Bangladesh

Irrigation source	Irrigated area in different years (1000 ha)			
	1984–85	1989–90	1994–95	1999–2000
Groundwater				
Shallow tube well (STW)	586	1037	1638	2252
Deep tube well (DTW)	287	384	502	465
Manual	16	16	25	65
Subtotal for groundwater	889	1437	2165	2782
Surface water				
Low lift pumps	351	484	538	624
Canals	147	176	352	424
Traditional	384	478	250	202
Subtotal for surface water	882	1138	1140	1250
Total irrigation	1771	2576	3305	4032

Source: MOF (2000).

Conclusions

Drought affects agriculture of Bangladesh frequently. It is possible to avoid crop loss by identifying the drought-affected areas, through systematic monitoring, and motivating the farmers to take countermeasures by providing timely and adequate irrigation in the affected areas. Unfortunately, despite having the technological capacity, drought monitoring has not been given due emphasis as yet. The existing capacity needs to be enhanced through appropriate planning, and it should be brought under a proper institutional framework. Concerted efforts of all the major stakeholders will enable the country to take appropriate measures for drought monitoring and management and save the poverty-stricken farmers from losing harvest, thus seriously affecting their livelihood.

ACKNOWLEDGMENTS We thank Zahurul Karim, former Secretary to the Ministry of Fisheries and Livestock, for kindly reviewing the manuscript and providing valuable suggestions. We are indebted to Ghulam Hussain of Bangladesh Agriculture Research Council for his continued encouragement and for providing drought maps. Acknowledgment is also due to Salima Sultana, Department of Geography at Auburn University, for encouragement and support.

References

Ahmad, Q.K., and A.U. Ahmed (eds.). 2002. Bangladesh: Citizen's Perspectives on Sustainable Development. Bangladesh Unnayan Parishad, Dhaka.

Ahmad, Q.K., N. Ahmad, and K.B.S. Rasheed (eds.). 1994. Resources, Environment and Development in Bangladesh With Particular Reference to the Ganges-Brahmaputra-Meghna Basins. Bangladesh Unnayan Parishad and Academic Publishers, Dhaka.

BRRI. 1999. BRRI Annual Report. Bangladesh Rice Research Institute, Ministry of Agriculture, Joydebpur, Bangladesh.

BUP. 2002. Participatory Rapid Appraisal Report, submitted as a background paper for the study on Pro-Poor Interventions on Irrigation in Asia: Bangladesh, conducted by Bangladesh Unnayan Parishad, with assistance from the International Water Management Institute, Colombo, Sri Lanka (unpublished).

Chowdhury, M.H.K., and M.A. Hussain. 1983. On the ardity and drought conditions of Bangladesh. Musam 34(1):71–76.

Halcrow and Associates. 2001. Options for the Ganges Dependent Areas, Draft Final Report. Water Resources Planning Organization, Ministry of Water Resources, Dhaka, Bangladesh.

Iqbal, A., and M.H. Ali. 2001. Agricultural Drought: Severity and Estimation. Report prepared under project BGD/95/006, Ministry of Agriculture, Dkaha, Bangladesh.

Karim, Z., A.M. Ibrahim, A. Iqbal, and M. Ahmed, 2001. Drought in Bangladesh: Agriculture and Irrigation Schedules for Major Crops. Soils Publication no. 34, Bangladesh Agriculture Research Council, Dhaka.

MOF. 2000. Bangladesh Economic Survey. Ministry of Finance, Dhaka, Bangladesh.

MOF. 2003. Bangladesh Economic Survey. Ministry of Finance, Dhaka, Bangladesh.

MPO. 1991. National Water Plan, Phase II. Master Plan Organization, Ministry of Water Resources, Irrigation and Drainage, Dhaka, Bangladesh.

Rahman, A.A., S. Huq, and G.R. Conway (eds.). 1990. Environmental Aspects of Surface Water Systems of Bangladesh. University Press Limited, Dhaka.

SPARRSO. 1999. Annual Report: July 1998–June 1999. Bangladesh Space Research and Remote Sensing Organization, Dhaka.

CHAPTER TWENTY-FIVE

A Drought Warning System for Thailand

APISIT EIUMNOH, RAJENDRA P. SHRESTHA, AND VIJENDRA K. BOKEN

Thailand is located between 5°30' and 20°30'N latitudes and between 97°30' and 105°30'E longitudes. Geographically, the country can be divided into northern, northeastern, central, and southern regions (figure 25.1). Most of the country experiences distinct wet and dry climates, except some parts of the southern region, which experience a wet and humid climate. Of the country's total area (514,000 km^2), 41% is under agricultural use (Office of Agricultural Economics, 1999) with 92% of it being rainfed. Drought normally occurs during the hot season (March–April) and sometimes during dry season (November–April) due to inadequate rains. In recent times, the occurrence of drought has increased in Thailand, threatening sustainability of agricultural production. According to Department of Local Administration (1998), droughts of varying intensity occur in 67 out of 76 provinces of Thailand almost every year. During the period from 1987 to 1997, drought impacted a total of 5.44 million ha of agricultural land, causing $1.4 billion in losses.

Drought Monitoring

Radiative Index

Droughts of varying intensity or severity occur in different regions of Thailand. A drought is categorized as severe, moderate, slight, or none drought using a radiative index (RI) determined during the rainy season (May–October). The RI for a region is determined using the number of rainy days, percentage of irrigated area, groundwater availability, topography, land use, soil, drainage density, and watershed size. If RI ranges from 1.0 to 1.2 for 15 consecutive days for a region or area, the region is said to be affected

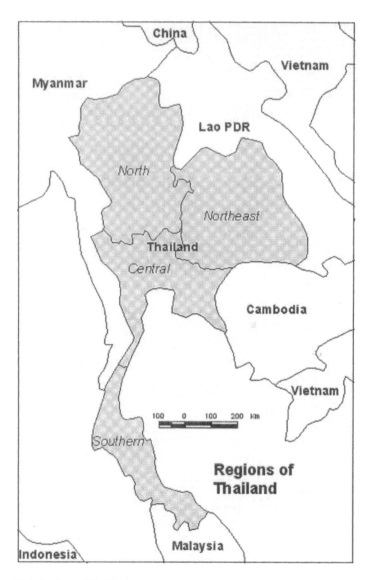

Figure 25.1 Regions of Thailand.

by slight drought. If RI exceeds 1.2 for 30 consecutive days, the region is considered to be affected by moderate drought, and if RI exceeds 1.0 for more than 30 consecutive days, severe drought is said to have occurred in the region. Using these criteria, the percentage of area affected by different drought categories has been determined in Thailand (table 25.1). It can be observed from table 25.1 that the northeastern region is the most drought-prone in Thailand.

Table 25.1 Spatial distribution of drought occurrence in Thailand

Region	Total area (km²)	Drought severity class (% area)				Source
		Severe	Moderate	Slight	None	
Northeast	168,854	68	17	9	6	OEPP/KU (1996)
North	169,756	1	7	82	10	OEPP/IWB Siam-Tech. (1998)
Central	104,319	4	57	33	6	OEPP/TRTI (1999)
South	71,040	Incidence of drought reported but not quantified				SU (1997)

OEPP = Office of Environmental Policy and Planning, TRTI = Thailand Research and Technology Institute, and SU = Songkhla University.

Drought Index

A drought index, D, is also used to monitor drought conditions in Thailand. The index is based on various variables that are given different weights and scores (table 25.2) to compute the index (Eiumnoh, 2000):

$$D = \frac{\sum_{i}^{n} S_{ij} W_i}{\sum_{i}^{n} W_i} \quad [25.1]$$

where W_i is the weight for the ith input variable, and S_{ij} is the score for the jth class of the ith variable. D ranges from 1 to 42 for slight drought, from 43 to 84 for moderate drought, and exceeds 84 for severe drought.

Agricultural Drought Warning System

Although there are numerous departments that routinely collect data relevant to drought monitoring, there is no integrated system for drought monitoring or early warning at national level. Recently, the Ministry of Agriculture and Cooperative has taken an initiative and has proposed an early warning system for agricultural drought based on a better coordination among different departments and organizations. The proposed system operates in two modes: time-warning mode and event-warning mode, as illustrated in figure 25.2.

The time-warning mode is a long-term drought warning system based on meteorological and physical data. The Department of Meteorology provides weather information throughout the year. Particularly in the dry season, the department predicts drought for different parts of the country. The Department of Land Development provides the information on current land use, land use changes, and soil types. This mode evaluates the seasonal as well as monthly weather conditions during the hot and dry season and helps identify the potential drought-prone areas and their drought sever-

Table 25.2 Input variables to compute drought severity classes for the rain-fed areas of Thailand

Variable	Weight	Score and the corresponding range of variable

Meteorological data
  Monthly rainfall, P (mm)
  Evapotranspiration, ET
	7	3, [if 0.5ET > P]
		2, [if P > 0.5ET]
		1, [if P > ET]

Ground water	6	3, 5–10m^3/h 2, 10–20m^3/h 1, 20–50m^3/h
Land use	5	3, sugarcane (high water requirement) 2, rice (moderate water requirement) 1, other crops (low water requirement)
Soil water holding capacity	4	3, sand 2, loam 1, clay
Slope	3	3, >5% 2, 2–5% 1, <2%
Density of river	2	3, <0.5 km/km^2 2, 0.5–1.0 km/km^2 1, >1.0 km/km^2
Size of basin	1	3, >3000 km^2 2, 1000–3000 km^2 1, <1000 km^2

Source: Eiumnoh (2000).

ity levels. In this mode the system basically predicts the time of drought occurrence to help farmers plan preventive measures to minimize losses.

In the case of the event mode, the farmers and agricultural extension officers at district and subdistrict levels report drought information (area of drought occurrence, pest and disease cases, rainfall amount, types of crops, etc.) to the Warning Center established in the Office of Agriculture Economics (figure 25.2). The raw data are supplied to Warning Center by different departments (irrigation, forestry, agriculture, livestock, fisheries). The information is stored and analyzed in a Geographic Information System (GIS) to plan for drought mitigation. The output of analysis is disseminated to the target areas, such as the Agricultural Technology Transfer Station (ATTS) located in each district, the Executive Information System (EIS), mass media, and the provincial government offices.

The ATTS is the lowest level administrative unit run by locally elected officials. The general public, including farmers, can access the agricultural and other related information from the ATTS. The EIS is the highest-level body that plans measures to combat drought and other disasters. Finally, based on the output, response actions in regard to drought mitigation can be planned.

The Warning Center's structure and manpower consists of three sections: system development, data analysis, and estimation of damage and

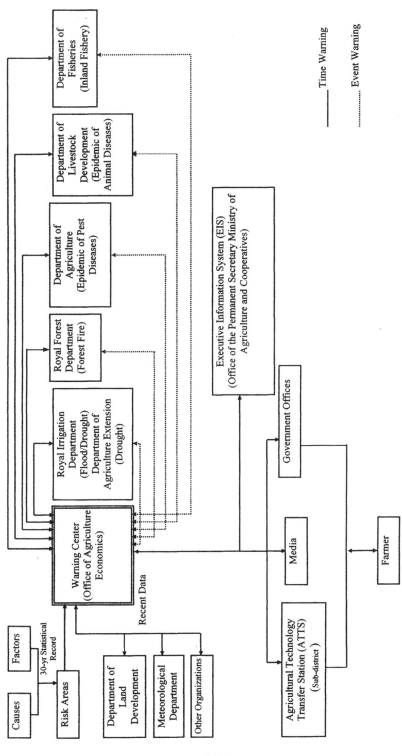

Figure 25.2 Proposed drought warning system for Thailand.

public relations. The system development section is responsible for setting up the computer system, network connection, operation, and maintenance of hardware and software and data reception and update. The data analysis section is responsible for data analysis and display of the information to the affected areas. The damage estimation and public relations section is responsible for maintaining up-to-date statistical records and performing economic analysis of the drought and reporting the outcome to the EIS and making it available to public and the media.

To perform the above task, the Warning Center is equipped, in addition to required personnel support, with the following: Remote sensing and GIS hardware and software and their specialists, relevant equipment to establish an on-line linkage with the ATTS, and public relations office. The manpower in the center is pooled from existing government organizations and consists of one director, two system engineers, two system officers, four agricultural officers, two geographers, two hydrologists, three statisticians, three economists, and two public relations officers.

Use of Satellite Data

The National Oceanic and Atmospheric Administration's (NOAA) Advanced Very High Resolution Radiometer (AVHRR) (noaasis.noaa.gov) and Moderate Imaging Scatterometer (MODIS) (modis.gsfc.nasa.gov) are suggested for drought monitoring. At this stage, the remote sensing data are suggested to generate current land-use information as required in the proposed early warning system. After the potential drought areas are identified, the use of satellite data with higher spatial resolution remote sensing data is suggested to study detailed land-use types in the identified areas.

Implementation

The proposed drought early warning system will be implemented in three phases: immediate, short term, and long term. The immediate phase will last one year, and its purpose is to collect the existing drought risk maps at various administrative levels and prepare drought maps for the whole country based on the data analysis performed at the Warning Center. The short-term phase will last two years. During this phase different drought-monitoring techniques will be examined for their suitability of use in the proposed system. In addition, the GIS database will be updated with maps of larger scales (1:50,000), replacing the existing map with scale of 1:250,000. The long-term phase will continue for four years with the objective of developing the networking facility to connect the participating organizations. Computer hardware and software will be developed during this phase, and networking and transmission facilities will be installed at all the participating organizations and field stations. Training and seminars will also be organized for all personnel with the help of external experts.

Conclusions

Drought of varying intensity occurs in almost all parts of Thailand every year due to inadequate rains during November–April. Losses due to drought during 1987–97 have been estimated at $1.44 billion. All the regions of Thailand (north, northeast, central, and southern) have been mapped with drought categories using either the radiative index or drought index by different organizations. Although there are several organizations involved in drought-related data collection and some sorts of preliminary analysis, there is no integrated system for drought monitoring or early warning. At the initiative of the Ministry of Agriculture and Cooperatives, the proposed system as explained in this chapter is a prototype early warning system for agricultural drought monitoring that integrates both vertically (all planning levels) and horizontally (all sectors). The monitoring is carried out in the time and event modes using field information, a GIS database, and remotely sensed data.

References

Department of Local Administration. 1998. National Plans to Prevent Civilian Disasters. National Civilian Disaster Prevention Authority, Department of Local Administration, Ministry of Interior, Bangkok, Thailand.

Eiumnoh, A. 2000. Final Report on Establishment of Agricultural Warning Systems Project for the Ministry of Agriculture and Cooperatives, Bangkok, Thailand. Asian Institute of Technology, Pathumthani, Thailand.

Office of Agricultural Economics. 1999. Agricultural Statistics of Thailand, Crop Year 1998–1999. Ministry of Agriculture and Cooperatives, Bangkok, Thailand.

Office of Environmental Policy and Planning and IWB-Siam Tech. 1998. Final Report on Risk Area Classification for Natural Hazards in the Northern Watersheds, Thailand. Kasetsart University Printing Office, Kamphaeng Saen, Thailand.

Office of Environmental Policy and Planning and Kasetsart University. 1996. Final Report on Risk Area Classification for Natural Hazards in the Mun and Chi Watersheds, Northeast Thailand. Kasetsart University Printing Office, Kamphaeng Saen, Thailand.

Office of Environmental Policy and Planning and Thailand Research and Technology Institute. 1999. Final Report on Risk Area Classification for Natural Hazards in the Central Watersheds, Thailand. Thailand Research and Technology Institute, Bangkok, Thailand.

Songkhla University. 1997. Final Report on Civilian Disaster Management for Southern Thailand. Songkhla University, Songkhla, Thailand.

CHAPTER TWENTY-SIX

Agricultural Drought in Indonesia
RIZALDI BOER AND ARJUNAPERMAL R. SUBBIAH

Indonesia is the largest archipelago in the world and comprises 5 main islands and about 30 smaller archipelagos. In total, there are 13,667 islands and islets, of which approximately 6,000 are inhabited. The estimated area of the Republic of Indonesia is 5,193,250 km^2, which consists of a land territory of slightly more than 2,000,000 km^2 and a sea territory of slightly more than 3,150,000 km^2. Indonesia's five main islands are Sumatra (473,606 km^2); Java and Madura (132,187 km^2), the most fertile and densely populated islands; Kalimantan or two-thirds of the island of Borneo (539,460 km^2); Sulawesi (189,216 km^2); and Irian Jaya (421,981 km^2), the least densely populated island, which forms part of the world's second largest island of New Guinea.

Of about 200 million ha of land territory, about 50 million ha area is devoted to various agricultural activities. There is nearly 20 million ha of arable land, of which about 40% is wetland (rice fields), 40% is dryland, and 15% is shifting cultivation. In the early 1970s, agriculture contributed about 33% to the gross domestic product. Its share decreased to 23% by the early 1980s and to 16.3% in 1996. However, agriculture is the most important sector in the national economy due to its capacity to employ 41% of the labor force (MoE, 1999).

Agriculture is vulnerable to drought. Ditjenbun (1995) reported that in 1994 many seedlings and young plants died due to a long dry season: about 22% of tea plants at age of 0–2 years, 4–9% of rubber plants at age of 0–1 year, 4% of cacao plants at age of 0–2 years, 1.5–11% of cashew nut plants at age of 0–2 years, 4% of coffee plants at age of 0–2 years, and 5–30% of coconut plants at age of 0–2 years. The impact of a long dry season on yields of plantation crops becomes known only a few months later. For example, oil palm production is known 6–12 months after a long dry season (Hasan et al., 1998). Rice is the main food crop severely affected by drought. On average, the total area affected per year by drought is 200,000 ha for rice, 26,000 ha for maize, 10,000 ha for soybeans, and 6,000 ha for peanuts (Ditlin, 2000).

In extreme drought years, the drought-affected area increases significantly. For rice, for example, the total affected area could go up to 900,000 ha, while for maize, soybeans, and peanuts, it could go up to 86,000, 38,000 and 23,000 ha, respectively (Ditlin, 2000). This chapter discusses briefly the causative factors of drought, drought monitoring systems, and drought mitigation for the agricultural sector in Indonesia.

Causative Factors of Drought

A significant decrease in rainfall, especially during the dry season, causes serious damage to crops. There are many factors affecting rainfall variability in Indonesia.

Types of Rainfall

There are three types of rainfall within Indonesian region (Boerema, 1938). The first type is monsoon rainfall, which peaks in December. The second is equatorial rainfall, characterized by two monthly rainfall peaks in March and October. The third is a "local" type with a monthly rainfall peak in July–August (figure 26.1).

According to Boerema (1938), monsoon rainfall can be categorized into two groups (types A and B) that are prevalent in southern Indonesia. Types A and B indicate a clear distinction between a dry and a wet season throughout the year. Type A shows a longer dry period (eastern part of Indonesia, Nusa Tenggara islands) and is drier overall compared to type B (Java, South Sumatra, and South Sulawesi). The region with type A rainfall experiences severe drought more frequently. In general, rainfall variation is larger in the dry season (April–September) compared to the wet season (October–March).

Type C (local type), which differs from the types A and B, is located mainly in the eastern equatorial part of Indonesia (e.g., Maluku and Sorong). Wet seasons of this type are between April and September, while dry seasons are between October and March. However, the dry periods of this type are not as dry as those of type A or even type B. As a result, the total rainfall in a year is relatively high.

Equatorial-type rainfall can be divided into two groups, types D and E. Type D covers the west coast of North Sumatra with no pronounced dry season, and rainfall increases slightly around March and October. Type E covers the west coast of South Sumatra with uniform distribution of rainfall throughout the year.

The length of the wet season varies from as long as 280–300 days to as short as 10–110 days. The seasonal rainfall during the wet season varies from as high as 4115 mm to as low as 640 mm. The early onset and the

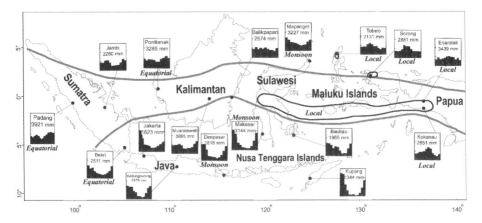

Figure 26.1 Climatology of seasonal rainfall (updated, based on Boerema, 1938, and ASEAN Secretariat, 1982).

late withdrawal of the monsoon result in a lengthy wet season, while the late onset and the early withdrawal entail a relatively short rainy season and a longer dry season.

The dry season ranges between 5 and 35 dekads (one dekad equals 10 days). Most of the area in the eastern part of Indonesia has relatively longer dry seasons. The longest and driest periods are found in central and eastern Lombok (30–35 dekads). The second longest dry season (26 dekads) occurs in Lombok North, and the third one (25 dekads) occurs in eastern Sumba, Flores, Pasuruan-Probolinggo, East Java and Subang, Indramayu, and West Java. The length of the dry season in most of East Java is between 15 and 25 dekads. The shortest periods are found in western and central Java and along the eastern coast of Sumatra. Rainfall during the dry season ranges from 250 to 500 mm in most areas of the archipelago. However, in some parts around the northern coast of West Java, the southern coast of East Java, central and northern Lampung, Sumbawa, West Lombok, the northern part of East Timor, and the northern and southern coasts of Irian Jaya, the rainfall amounts are between 500 and 750 mm.

Impact of El Niño/Southern Oscillation on Rainfall Variability

Indonesia is a tropical country influenced by the Indian and Pacific oceans. Rainfall variability is affected by at least five factors (Tjasyono, 1997): meridional circulation (Hadley), zonal circulation (Walker), monsoon activity, local effects (topography), and tropical cyclones. These five factors work simultaneously throughout the year, but in certain conditions, one factor might become more dominant than others. Table 26.1 shows that all but six drought events were associated with El Niño during the 1844–1998 period. Many studies imply that droughts in Indonesia were mostly associated with El Niño events (Braak, 1919; Berlage, 1927; Nicholls,

Table 26.1 El Niño and drought events in Indonesia, 1844–1998[a]

1844–97		1902–98	
Drought	El Niño	Drought	El Niño
1844–45	1844	1902–03	1902
1845–46	1845	1905–06	1905
1850–51	1850	1913–14	1914
1853–54	None	1918–19	1918
1855–56	1855	1923–24	1923
1857–58	1857	1925–26	1925
1864–65	1864	1929–30	1929
1873–74	1873	1932–33	1932
1875–76	1875	1835–36	None
1877–78	1877	1940–41	1939
1881–82	1880	1941–42	1941
1883–84	None	1944–45	1943
1884–85	1884	1945–46	1946
1888–89	1887	1953–54	1953
1891–92	1891	1861–62	None
1896–97	1896	1963–64	1963
		1965–66	1965
		1967–68	None
		1969–70	1969
		1972–73	1972
		1976–77	1976
		1980–81	None
		1982–83	1982
		1986–87	1986
		1991–92	1991
		1994–95	1994
		1997–98	1997

Source: Updated from Quinn et al. (1978) and ADB and BAPPENAS (1999).

[a] In drought years, about 60–90% of the Indonesian regions have rainfall below normal.

1981, 1983; Hackert and Hastenrath, 1986; Hastenrath, 1987; Malingreau, 1987; Harger, 1995; Yamanaka, 1998; chapter 3).

El Niño-related droughts occur when the Hadley cells are weak and the massive subsiding air over Indonesia feeds the increasingly westerly flow of surface air over the Pacific Ocean (Tjasyono, 1997; Kirono and Partridge, 2002). This results in the weakening or failure of the Walker circulation (Tjasyono, 1997). When the Hadley circulation is strong and the favorable local climate conditions prevail, subsidence can be localized, even during an El Niño event.

The effect of El Niño on rainfall varies between regions. It is strong in regions that are strongly influenced by monsoon systems, weak in regions that have equatorial systems, and unclear in regions that have local systems

(Tjasyono, 1997). The impact of the El Niño/Southern Oscillation (ENSO) on rainfall is more pronounced during the dry seasons than the wet season (USDA, 1984; Las et al., 1999). Various studies on the influences of ENSO on interannual rainfall variability in Indonesia reveal some seasonal patterns (USDA, 1984; ADPC, 2000; Yoshino et al., 2000; Kirono and Partridge, 2002): (1) The end of the dry season occurs later than normal during El Niño and earlier during La Niña years, (2) the onset of the wet season is delayed during El Niño and advanced during La Niña years, (3) a significant reduction of dry season rainfalls could be expected during El Niño and significant increase could be expected during La Niña years, (4) long dry spells occur during the monsoon period, particularly in eastern Indonesia. Observations at five locations during 1982–83 showed these patterns (table 26.2).

An analysis of time-series rainfall data (1951–98) for Pandeglang and Lebak (West Java) revealed that the normal onset of the dry season in this zone could be around the first week of June (ADPC, 2000). However, during extreme La Niña years, it could be delayed to the first week of August. During extreme El Niño years, the onset of the dry season could be advanced as early as mid-April. Against the normal rainfall of 379 mm during April–September, the rainfall could be as high as 979 mm during the La Niña years and as low as 15 mm during El Niño years (ADPC, 2000).

Monitoring Agricultural Drought

Decrease in rainfall below normal is a drought indicator commonly used by the Bureau of Meteorology and Geophysics (BMG), whereas the total area affected by drought is an indicator used by the Directorate of Plant Protection (Alimoeso et al., 2000, 2002). Remotely sensed data such as normalized difference vegetation index (NDVI; chapter 5) is used by Lembaga Penerbangan dan Antariksa Nasional (Kushardono et al., 1999; Heryanto et al., 2002). The following sections describe how these indicators are monitored and drought forecasted in Indonesia.

Drought Monitoring and Forecasting System

The Indonesian BMG developed an operational seasonal climate prediction scheme for Indonesia in 1993. It is a statistical-analogue scheme based on a detailed historical analysis of rainfall data for 102 meteorological regions across the whole country. The forecast is issued in early March for the dry season (April–September) and in early September for the wet season (October–March).

The seasonal forecast products are forecast of seasonal monsoon onset dates at 10-day interval; seasonal cumulative rainfall; and monthly rainfall. There is an ongoing forecast verification process undertaken by the BMG, the results of which are shown in table 26.3 for the 1993–97 period. The verification of forecasts shows a good skill in predicting the onset period

Table 26.2 Rainfall (mm) at selected stations in rice-producing areas, 1982–83[a]

	Jan	Feb	Mar	Apr	May	Jun	Jul	Aug	Sep	Oct	Nov	Dec
West Java												
Jatiwangi												
30-Year average	461	405	264	264	<u>159</u>	<u>83</u>	<u>59</u>	<u>30</u>	<u>43</u>	109	259	417
1982	538	612	245	310	<u>90</u>	<u>20</u>	<u>2</u>	<u>1</u>	<u>0</u>	55	40	379
1983	360											
Central Java												
Tegal												
30-Year average	369	307	243	<u>129</u>	<u>117</u>	<u>84</u>	<u>62</u>	<u>39</u>	<u>46</u>	<u>52</u>	120	253
1982	403	346	274	236	<u>12</u>	<u>55</u>	<u>2</u>	<u>6</u>	<u>0</u>	<u>21</u>	14	62
1983	472											
East Java												
Madiun												
30-Year average	273	271	264	232	<u>156</u>	<u>78</u>	<u>43</u>	<u>25</u>	<u>27</u>	<u>79</u>	191	243
1982	118	183	330	259	<u>0</u>	<u>0</u>	<u>3</u>	<u>0</u>	<u>0</u>	<u>0</u>	NA	196
1983	284											
Bali												
Denpasar												
30-Year average	334	276	221	<u>88</u>	<u>75</u>	<u>70</u>	<u>55</u>	<u>43</u>	<u>42</u>	<u>106</u>	168	298
1982	334	267	145	<u>53</u>	<u>2</u>	<u>1</u>	<u>0</u>	<u>3</u>	<u>0</u>	<u>1</u>	79	35
1983	<u>146</u>											
South Sulawesi												
Ujung Pandang												
30-Year average	714	515	423	<u>154</u>	<u>95</u>	<u>65</u>	<u>32</u>	<u>14</u>	<u>11</u>	<u>45</u>	183	581
1982	648	482	434	<u>109</u>	<u>77</u>	<u>6</u>	<u>0</u>	<u>0</u>	<u>0</u>	<u>0</u>	30	323
1983	348											

Source: USDA (1984).

[a] Underlining denotes dry season.

of the wet season; reasonable skill in predicting the onset period of the dry season; and relatively poor skill in predicting the dry season rainfall, particularly during ENSO years.

Despite the availability of a reliable forecast for parameters such as the onset period for the wet season, decision-makers are unable to utilize the forecast because of its coarse resolution. Onset dates are indicated for a period of one month over a large area, and rainfall distribution is presented in terms of above-normal, normal, or below-normal categories for a six-month period over a large area. Based on the climate forecast received from the BMG, national organizations such as the departments of agriculture and water resources disseminate the climate information to provincial organizations in a routine manner. At present, these forecasts are used only as a general alert.

Another problem is that there is no consensus between related agencies on the threshold onset dates for the wet and dry seasons and the distribution of rainfall during these seasons. For example, a delay of up to 20 days

Table 26.3 Verification of seasonal prediction in Indonesia (percentage of districts in the respective categories)[a]

Year	Onset of season			Rainfall classification		
	Precise	Ahead	Later	Same	Close	Different
Wet Season						
1993–94	84	5	11	52	45	3
1994–95	86	1	13	50	44	6
1995–96	83	8	9	49	42	9
1996–97	64	7	29	43	50	7
Average	79	5	16	49	45	6
Dry Season						
1993–94	74	23	3	27	59	14
1994–95	56	10	34	63	28	7
1995–96	73	20	7	51	39	9
1996–97	76	19	5	21	58	21
Average	70	18	12	41	46	13

Source: ADPC (2000).

[a] Prediction of onset for the 102 seasonal prediction areas is estimated in 10-day (dekad) periods. Rainfall is considered normal if it is 85–115% of normal, above normal if it exceeds 115% of normal, and below normal if it is below 85% of normal. Precise means that onset occurred in the 10-day period predicted; ahead means that it occurred in an earlier dekad; and later means that it occurred in a later dekad. For rainfall classification, same means that rainfall total was as predicted, and different means that it was two categories from what was predicted.

in the onset of the wet season may not upset established crop calendars at some locations because agricultural practices evolved over time are adjusted to these normal variations. Any variation beyond 20 days is likely to upset the established crop calendar, and the information on these thresholds could be of use to the end-users at the concerned locations. As another example, a 50% reduction of rainfall could still be enough to support the established crops in high rainfall zones such as Kalimantan, where average rainfall is around 3000 mm, because 1500 mm is enough to support a rice crop.

Crop Production Forecasting and Monitoring System

Forecasting rice production has long been a priority for the Indonesian government, particularly on the island of Java, which accounts for approximately 60% of the total rice production in the country. Data on the harvested area are collected on a monthly basis but, for the purpose of publication, are aggregated to a level that corresponds to time periods of recording of yields (using crop cuttings) in three sub-rounds: January–April, May–August, September–December. Official forecasts for rice production are made in mid-February, mid-June, and mid-August. All forecasts are disaggregated to provincial levels separately for *padi sawah* (wetland rice) and *padi ladang* (upland rice). Historically, the methods used to forecast harvested area have tended to rely on a combination of harvested area in the previous year and data available on planting in the current season.

The advance estimates consist of preharvest forecasts and quick estimates. Quick estimates refer to initial postharvest forecasts generally obtained through sample surveys and crop-cutting experiments. The sample surveys are conducted for rice, corn, soybeans, peanuts, cassava, and sweet potatoes. Data are collected by two field agencies, the Agricultural Extension Service (*Mantri Tani*) and the Subdistrict Statistical Office (*Mantri Statistik*). Data at the subdistrict level are collected through the discussions with village heads, the status of crops in block irrigation systems, and interviews with farmers and the sales agencies associated with the quantity of planted seed.

In the case of the crop-cutting experiments, each district is treated as a stratum, with the number of areas selected using random samples. One segment is selected randomly from each selected area. Households in each selected segment are requested to indicate whether they planted food crops and their expected harvest dates. A land parcel is selected from the household list, and one plot (2.5 × 2.5 m) is selected randomly. The crop inside the plot is harvested and weighed. The total number of sample plots for the crop-cutting experiments throughout Indonesia is 110,000. The crop cutting from 50% of the plots is done by the Sub-district Statistical Office and from the remaining 50% by the Agriculture Extension Service. The yield data thus collected are reported hierarchically (i.e., from subdistrict level to district, provincial, and national organizations).

The advance estimates from subdistrict to national levels are often tentative and subjective. No mechanism exists to validate the information flow. These estimates are revised from time to time, which poses difficulties for decision-making processes. For instance, for the 1997–98 El Niño, the initial estimates were around 45 million tons of rice, which were later revised to around 48 million tons. Therefore, decisions regarding rice imports could not be made accurately. The present system does not take into account the influence of rainfall and price variability on farmers' decisions to plant various crops. Hence, the present system has much scope for improvement, both in terms of lead time and accuracy.

Using Satellite Data   LAPAN and the Research Institute for Soils and Agroclimate have evaluated the use of satellite data and GIS technology to monitor rice crop growth to estimate rice yield (Kushardono et al., 1999; Heryanto et al., 2002). The use of this technology is promising. The yield of rice can be estimated from the normalized difference vegetation index (NDVI) during the heading phase:

$$\text{Yield(tons/ha)} = 30.9 \text{ NDVI} - 11.16 \qquad [26.1]$$

The above equation has been tested at four villages of Kronjo Subdistrict, Tanggerang District. The difference between the observed and estimated yield varied from 10% to 14% with an average of about 12% (Heryanto et al., 2002).

Using the Southern Oscillation Index Another approach to estimate the crop production in the coming season is the Southern Oscillation Index (SOI) data (Meinke and Hammer, 1997; Rahadiyan, 2002). A preliminary study indicated that the variation in national rice production could be explained partly by the SOI for April. The use of the April SOI was motivated by the evidence that after April the cumulative drought area increased rapidly, especially during El Niño years. Negative anomalies in rice production occurred mostly for El Niño years, while positive anomalies occurred for La Nina years.

Using Crop's Physical Appearance Drought severity is determined based on the proportion of the dry leaves in rice fields. If only the tip of the leaf gets dry, it is called slightly affected (*Ringan*); if one-quarter of the leaf is dry, it is called moderately affected (*Sedang*); if one-quarter to two-thirds of the leaf is dry, it is called heavily affected (*Berat*); and if all of the leaf is dry, it is called completely damaged (*Puso*). These drought symptoms are monitored every two weeks by the pest and disease observers at subdistrict levels. The data are sent to the district and then to the province and finally to the Directorate of Plant Protection. The Directorate of Plant Protection analyzes the data and provides a drought distribution map.

Drought Vulnerability Map The Directorate of Plant Protection has developed a map showing drought vulnerability in a district. Vulnerability is determined using frequency, intensity, and size of area affected by drought (Alimoeso et al., 2002). The analysis suggested that most of the vulnerable districts are located in West Java province, South Sumatra/Lampung, and South Sulawesi (figure 26.2). These three provinces are the main rice-growing areas of Indonesia. Among the three, West Java is the largest rice-producing province in the country. Any drought in West Java could have a significant influence on national rice production and food security. In West Java, districts considered to be very prone to drought are Indramayu, Bekasi, Sukabumi, Tasikmalaya, Bandung, and Cirebon. Loss of rice production in these districts increased significantly during El Niño years (figure 26.3) and could go up to 500,000 tons (Alimoeso et al., 2002).

From historical data, it was shown that, in general, the area affected by drought increased significantly during El Niño years (figure 26.4). However, from national production statistics, the impact of El Niño, apart from 1982, is not distinct, except for rice (figure 26.5). This condition appears due to a number of reasons (Suryana and Nurmalina, 2000; Meinke and Boer, 2002): (1) the statistics are based on calendar years rather than on El Niño years, (2) not all regions of the nation are affected by drought simultaneously, (3) shortage of water may force a farmer to switch from rice to secondary crops, (4) restricted water supply may reduce the area

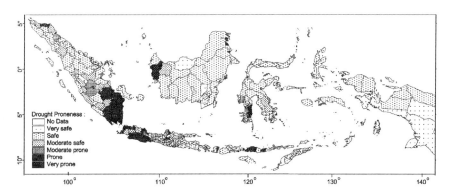

Figure 26.2 Vulnerability of rice-growing area to drought by district (from Alimoeso et al., 2002).

planted under irrigation, but yield of crops may increase due to higher solar radiation, and (5) production may be affected in the year following an El Niño event because farmers have less money to spend on fertilizers or insecticides.

Drought Mitigation

Drought mitigation strategies include identifying and mapping drought-prone areas by the Directorate of Plant Protection, improving stochastic models of climate forecasting such as time-series regression techniques (Dupe, 1999; Haryanto, 1999; Boer et al., 2000), relocation of irrigation water to drought-affected areas, adopting short-duration crops, constructing small water-harvesting structures, restoring old irrigation structures, modernizing water pumping facilities, developing drought-resistant crops, practicing mixed (cash and food crops) cropping (ADPC, 2000), and improving soil and water conservation practices.

In Java and the eastern Indonesian region, during El Niño years farmers are frequently misled by the initial rains. These rains tempt farmers to plant. However, as the rains cease later, the crops usually die during dry spells. Most farmers keep some seed reserves in case they are forced to plant for a second time, during the wet season. Rarely do farmers have sufficient seed reserves for a third attempt when there is little likelihood of crop success. In most El Niño years, the incidences of misleading rains are noticeable. A long-lead forecast could help farmers wait until regular rains set in. This would also help in advising farmers not to resort to planting crops during El Niño years (taking the misleading rains into account) but to wait for suitable dates for planting appropriate crops.

The main problems in mitigating drought in Indonesia are poor skills to forecast climate and unavailability of methods to transfer the regional climate forecasts into local climate forecasts which can be easily interpreted by farmers. Policy-makers sometimes have wrong perceptions about

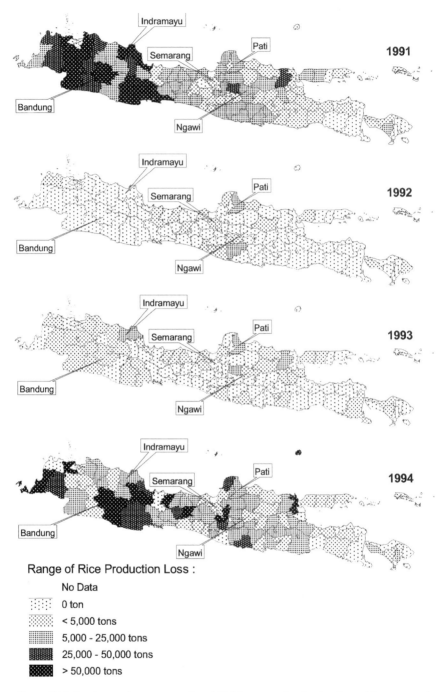

Figure 26.3 Production loss in normal and El Niño years by district (from Alimoeso et al., 2002).

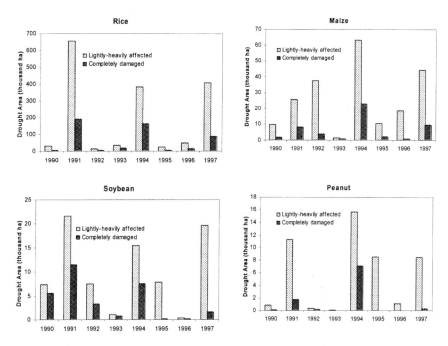

Figure 26.4 Impact of El Niño on rice and secondary crops (drawn from Ditlin, 2000).

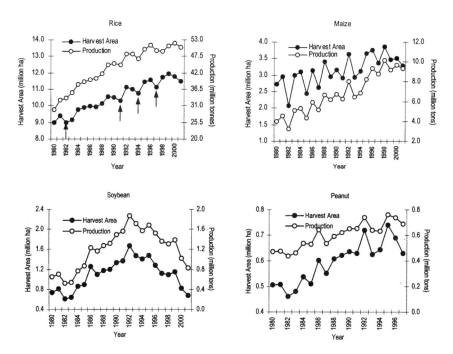

Figure 26.5 National food crops production in the period 1980–97. Arrows indicate El Niño years (drawn from BPS, 2002).

climate information. For example, based on interviews in several districts in Java, the information regarding El Niño of 2001 was given as if for 2000. All these problems need to be resolved to make the mitigation measures effective.

Conclusions

There is a need to evolve threshold dates for the onset of the wet and dry seasons and forecast the distributions of rainfall during these seasons. Efforts are needed to improve the reliability of climate forecast, accurately transfer climate forecasts from regional to local levels, and better train the policy-makers and producers about climate information to enhance the effectiveness of measures for drought mitigation in Indonesia.

ACKNOWLEDGMENTS We acknowledge input from C.J. Stigter. Ansari and Maris Karima Rahadian prepared the figures and maps.

References

ADB and BAPPENAS. 1999. Planning for the Fire Prevention and Drought Management. Final Report. Asian Development Bank and National Planning Agency, Jakarta, Indonesia.

ADPC, 2000. ENSO Impact and Potential Forecast Applications in Indonesia. Extreme Climate Events Program. Asian Disaster Preparedness Center, Bangkok, Thailand.

Alimoeso, S., R. Boer, S.W.G. Subroto, E.T. Purwani, Y. Sugiarto, R.M.K. Rahadiyan, and Suciantini. 2002. Penyebaran daerah rawan kering di wilayah pertanaman padi Indonesia. Direktorat Perlindungan Tanaman, Departemen Pertanian, Jakarta, Indonesia.

Alimoeso, S., Jasis, S.W.G Subroto, Zainita, D. Mutiawari, A. Fitri, E. Suwardiwijaya, and Issusilaningtyas. 2000. Impact of extreme climate events and food crop management in Indonesia. Project Report. Directorate of Crop Protection, Directorate General of Food Crops and Horticulture-Ministry of Agriculture Indonesia, NOAA/OFDA-USA and Asian Disaster Preparedness Center, Bangkok, Thailand.

ASEAN Secretariat. 1982. The ASEAN climatology and atlas. Asean Sub-Committee on Climatology, Asian Committee on Science and Technology, Asean Secretariate, Jakarta, Indonesia.

Berlage, H.P. 1927. East monsoon forecasting in Java. Verhandelingen no. 20. Koninklijk Magnetisch en Meterologisch Observatorium te Batavia, Jakarta, Indonesia.

Boer, R., K.A. Notodiputro, and I. Las. 2000. Prediction of daily rainfall characteristics from monthly climate indices. Paper presented at the Second International Conference on Science and Technology for the Assessment of Global Climate Change and Its Impacts on Indonesian Maritime Continent, November 29–December 1, 1999. Agency for Assessment and Application of Technology (BPPT), Jakarta, Indonesia.

Boerema, J. 1938. Rainfall Types in Nederlands Indie. Verhandelingen no. 18. Jakarta, Koninklijk Magnetisch en Meterologisch Observatorium te Batavia.

BPS. 2002. Dokumen data Biro Pusat Statistik Indonesia. Biro Pusat Statistik Indonesia, Jakarta.
Braak, C. 1919. Atmospheric variations of short and long duration in the Malay archipelago and neighbouring regions, and the possibility to forecast them. Verhandelingen No. 5. Koninklijk Magnetisch en Meterologisch Observatorium te Batavia, Jakarta.
DITJENBUN. 1995. Dampak El Niño tahun 1994 di sektor perkebunan. Bahan Diskusi Dampak El Niño 1994 di Kantor Menteri Negara Lingkungan Hidup, Direktorat Jendral Perkebunan, Jakarta, Indonesia.
DITLIN, 2000. Dokumen Data Direktorat Perlindungan Tanaman. Direktorat Perlindungan Tanaman, Departemen Pertanian, Jakarta, Indonesia.
Dupe, Z.L. 1999. Prediction Niño 3.4 SST anomaly using simple harmonic model. Paper presented at The Second International Conference on Science and Technology for the Assessment of Global Climate Change and Its Impacts on Indonesian Maritime Continent, November 29–December 1, 1999.
Hackert, E., and S. Hastenrath. 1986. Mechanisms of Java rainfall anomalies. J. Clim. Appl. Meteorol. 26:133–141.
Harger, J.R.E. 1995. ENSO variation and drought occurrence in Indonesia and the Phillipines. Atmos. Environ. 29:1943–1955.
Haryanto, U. 1999. Response to climate change: Simple rainfall prediciton based on Southern Oscillation Index. Paper presented at The Second International Conference on Science and Technology for the Assessment of Global Climate Change and Its Impacts on Indonesian Maritime Continent, November 29–December 1, 1999.
Hasan, H., H. Pawitan, R. Boer, and S. Yahya. 1998. Model simulasi produksi kelapa sawit berdasarkan karakteristik kekeringan [A simulation model of oil palm production based on drought characteristics]. J. Agromet. 13:41–54.
Hastenrath, S. 1987. Predictability of Java monsoon anomalies: A case study. J. Clim. Appl. Meteorol. 26:133–141.
Heryanto, B., S. Retno, W. Supriatna, and Wahyunto. 2002. Potential of JERS-1 satellite data for rice field estimation in Subang, West Java Province, Indonesia. Paper presented at the Symposium on Collaboration on Japanese Satelitte Data Application NASDA-LAPAN-CSARD, March 5, 2002, Jakarta, Indonesia.
Kirono, D., and I.J. Partridge. 2002. The climate and the SOI. In: I.J. Partridge and M. Ma'shum (eds.), Will It Rain? The Effect of the Southern Oscillation and El Niño in Indonesia. Department of Primary Industry, Queensland, Australia, pp. 17–24.
Kushardono, D., E.S. Adiningsih, and N.S. Haryani. 1999. Pengembangan dan pemanfaatan model sistem deteksi dini rawan pangan. Laporan Penelitian Pusat Pemanfaatan Penginderaan Jauh LAPAN, Jakarta, Indonesia.
Las, I., R. Boer, H. Syahbudin, A. Pramudia, et al. 1999. Analysis of Probability of Climate Variability and Water Availability in "300—Rice Cropping Intensity." Agriculture Research Management Project II Report. Central Research for Soil and Agroclimatology, AARD, Bogor, Indonesia.
Malingreau, J.P. 1987. The 1982–83 drought in Indonesia: Assessment and monitoring. In: M. Blantz, R. Katz, and M. Krenz (eds.), Climate Crisis: The Societal Impacts Associated with the 1982–83 Worldwide Climate Anomalies. United Nations Environment Program, Geneva, pp. 11–18.
Meinke, H., and R. Boer. 2002. Plant growth and the SOI. In: I.J. Partridge and M. Ma'shum (eds.), Will It Rain? The Effect of the Southern Oscillation and El

Niño in Indonesia. Department of Primary Industry, Queensland, Australia, pp. 25–28.

Meinke, H., and G.L. Hammer. 1997. Forecasting regional crop production using SOI phases: an example for the Australian peanut industry. Aus. J. Agric. Res. 48:789–793.

MoE. 1999. First National Communication to UNFCCC. Ministry of Environment, Jakarta, Indonesia.

Nicholls, N. 1981. Air-sea interaction and the possibility of long-range weather prediction in the Indonesian archipelago. Month. Weather Rev. 109:2435–2443.

Nicholls, N. 1983. The Southern Oscillation and Indonesian sea surface temperatures. Month. Weather Rev. 112:424–432.

Quinn, W.H., D.D. Zopf, K.S. Sort, and T.R.W. Kuo Yang. 1978. Historical trends and statistics of the southern oscillation-El-Niño and Indonesian drought. Fishery Bulletin 76 p. 3.

Rahadiyan, R.M.K. 2002. Analisis hubungan keragaman hujan dan produktivitas tanaman jagung dan kedelai dengan fenomena ENSO (El Niño Southern Oscillation). Skripsi S1, Jurusan Geofisika dan Meteorologi, Institut Pertanian Bogor, Bogor, Indonesia.

Suryana, A., and R. Nurmalina. 2000. Impact of climate change and the economic crisis on food production in Indonesia. Agriculture production and climate change in Indonesia. Global Environ. Res. 3:177–186.

Tjasyono, B. 1997. Mekanisme fisis para, selama, dan pasca El Niño. Paper presented at the Workshop Kelompok Peneliti Dinamika Atmosfer, March 13–14, 1997. Jakarta.

USDA. 1984. World Indices of Agriculture and Food Production, 1974–1983. Statistical Bulletin 710. Economic Research Service, U.S. Department of Agriculture, Washington, DC.

Yamanaka, M. (ed.). 1998. Climatology of Indonesia Maritime Continent. Kyoto University Press.

Yoshino, M., K. Urushibara-Yoshino, and W. Suratman. 2000. Agriculture production and climate change in Indonesia. Global Environ. Res. 3:187–197.

CHAPTER TWENTY-SEVEN

Agricultural Drought in Vietnam

NGUYEN VAN VIET AND VIJENDRA K. BOKEN

Vietnam is located between 8°22'N and 23°22'N latitude and between 102°10'E and 109°21'E longtitude. The country has a geographical area > 333,000 km² and a coastline > 3000 km long. Vietnam is situated in the typhoon center of the East Sea (Bien Dong), which is one of the five largest typhoon centers of the world, and it has a complicated topography ranging from narrow, low plains to steep, high mountains. Floods occur with high frequency, and drought occurs with medium frequency in Vietnam. If monthly rainfall is less than 50 mm, drought is considered to have occurred during the month.

The climate of Vietnam is strongly influenced by mountainous terrain and by the northeast and southeast monsoons. The rainfall season, which usually begins in May–June and ends in November–December, accounts for about 75–85% of the total annual rainfall. The period from November–December to April–May is usually dry and is prone to droughts.

Vietnam has been divided into seven agricultural regions (figure 27.1): (1) north mountain and midland region, (2) northern delta region (Red River delta), (3) north-central region, (4) south-central region, (5) central highland region (Taynguyen platour), (6) southeast region, and (7) southern delta region (Mekong River delta). The rainfall distribution is uneven due to complex terrain conditions. While some places (in the north mountain, central, and central highland regions) receive 3000–4000 mm of rainfall annually, other places (such as Phanthiet and Phanrang of the south-central region) receive only 750–800 mm in a year. Rice and maize are the major crops in Vietnam, whose regional distributions are given in table 27.1. Table 27.2 shows rice area lost due to droughts during 1980–99 in different regions of Vietnam.

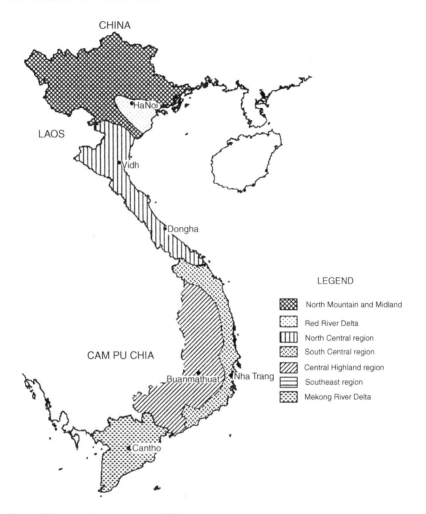

Figure 27.1 Agricultural regions of Vietnam.

General Characteristics of Regions

The north mountain and midland region is not prone to droughts. Only moderate droughts occur at some places in this region. Droughts rarely occur for consecutive years, except the droughts that occurred during the summer season of 1988 and 1989. The northeast part of this region is more prone to drought than the rest of the region. The Red River delta is relatively more drought prone, and severe droughts occurred there in 1960, 1961, 1963, and 1964.

The north-central region is a narrow belt of land by the East Sea and is covered with many forests and mountains. The geographical area of this region is about 5.1 millions ha, with nearly 1.3 millions ha being used for

Table 27.1 Regional distribution of harvested area and production of rice and maize crops in Vietnam for 1999

Region	Agricultural and forest areas (1000 ha)	Harvested area (1000 ha)		Production (1000 tons)	
		Rice	Maize	Rice	Maize
North mountain and midland	4467	669	267	2291	551
Red River delta	1479	1203	103	6354	319
North-central	5150	678	88	2653	202
South-central	4422	544	43	2236	95
Southeast	2339	390	91	1151	298
Central highland	5441	166	73	403	261
Mekong River delta	3987	3965	22	16549	50

agricultural production. This region is influenced by the eastern hot and dry winds (Laos wind) and is very hot during the dry season.

The south-central region runs along the coastline, and many streams and rivers that originate from the Truongson Mountain Range flow in this region. The geographical area of this region is about 4.4 million ha, with 640,000 ha being used for agricultural production. About 65% of the agricultural area is irrigated. Most of the population of this region (about 7 million) is engaged in agriculture or forest-related activities. During 1958–98, droughts occurred in this region during the summer season of 1962, 1969, 1993, and 1998. Relatively severe droughts occurred during winter–spring seasons of 1970 and 1984. In general, moderate and severe droughts occur in this region during the summer season.

The central highland region (Taynguyen plateau) is a mountainous plateau with average height ranging from 500 to 800 m above mean sea level. Its geographical area is 5.5 million ha, of which 112,000 ha is arable and very fertile. More than half a million hectares is hilly and is planted with commercial crops, such as coffee, tea, pepper, and fruits. Before 1980, droughts rarely occurred in consecutive seasons in this region, but since 1980, there have been cases of moderate droughts occurring for consecutive seasons.

The southeast region is an old alluvium area with a geographic area of 2.4 million ha, out of which 1.3 million ha is used for agriculture, 0.6 million ha for forestry, and about 1 million ha for commercial crops.

The Mekong River delta is the biggest rice basket of Vietnam. Its geographical area is about 4 million ha, including 2 million ha of rice and subsidiary areas. The vast majority of population of this region (16 million) lives on rice and fish production. The Tien and Hau rivers are the major irrigation sources. Water is transported to rice fields via a condensed canal network. The annual harvested area of rice is more than 1.26 million during the winter–spring season, 1.34 million ha during the summer–autumn season, and more than 0.5 million ha during other irregular periods. The

Table 27.2 Rice area (in 100 ha) lost due to droughts that occurred in different seasons and regions of Vietnam during 1980–98

Year	North mountain and midland Winter–spring	North mountain and midland Summer	Red River delta Winter–spring	Red River delta Summer	North-central Winter–spring	North-central Summer–autumn	North-central Summer	South-central Winter–spring	South-central Summer–autumn	South-central Summer	Mekong River delta Winter–spring	Mekong River delta Summer–autumn	Mekong River delta Summer
1980	945	75	30	5				7	417	279			
1981	379	81	850	13		731	1713		585	461	150	11,054	4042
1982	410	48		542		1200	2342		2440	571	150	1305	
1983	213	194	35	15		660	2464	6	1615	2060		1372	38,162
1984	690	71	470	615		659	1709	30	1998	11	200	2011	4693
1985	821	245	40	5		592	1386		2700	685		3385	4600
1986	200	25	1065	562		716			124	150		6300	109
1987	530	90	1995	1060		610	1437		1270			21,427	8357
1988	1542	2961	2036	162		680	1901		2452	450		6877	640
1989	783	112	25	12		598	162			88	5235	8340	280
1990	1542	1835	644	1200		294	105	10	277	314	524	18,103	631
1991	1676	9047	1035	685	1814	2173	264	15	257	15	50		21
1992	1538	1115	1545	620	1143	1988	3000	5	326	111	1250	39,062	7835
1993	1086	1196	752	4	2138	12,305	5500	15417	1834	7045	3180	8564	188
1994	1100	763	24	174	1222	1234		8	616	1223	771	4886	1074
1995	989	977	724	190	509	4092	2500	169	239	295	573	777	89
1996	2735	609	12	155	2815	2350			696	969	735	2735	527
1997	796	152		5	17	750	300	24	352	1067	111	14,217	
1998	1507	858		160		31,018	12,900	758	5797	657	19,414	32,109	
1999	19,482	20,454	11,282	6184	9658	62,650	37,683	16,449	23,995	16,451	32,343	182,524	71,248

weather in the Mekong River delta is moderately warm most of the year. Therefore rice can be cultivated at various times during the year, provided sufficient irrigation is available. In this region, severe drought occurred during April–June of 1983, 1992, and 1998, and during October–December of 1958 and 1992 (Bui, 2000).

Probability of Agricultural Drought

Table 27.3 presents the probability of monthly droughts in different regions of Vietnam. In general, probabilities are high during January–April, moderate during October–December, and low during June–September, except in north and south-central regions where drought probabilities are higher than other regions. In addition, drought probabilities are higher for southern regions than for northern regions.

In addition to monthly drought, studying dekadal (10-day) dry periods is important for better monitoring of crop conditions. If rainfall during a dekad is less than 30 mm, a dekadal dry period is considered to have occurred. Probability of dekadal dry periods was computed using a method proposed by Frere and Popov (1982). The probability distributions of second and third dekad dry periods for some major weather stations are shown in figure 27.2 (Nguyen, 1998, 2000, 2001).

Drought Indices

Annual and Monthly Drought Indices

Monthly drought indices are computed by Nguyen (2000):

$$K_m = E_m/P_m \qquad [27.1]$$

where K_m is monthly drought index; E_m is monthly evaporation; and P_m is monthly rainfall. The annual drought index is derived from the annual evapotranspiration and precipitation and it generally ranges from 0.3 to 3.0 (table 27.4; equation 27.3). However, there are some places (in the north mountain, central, and central highland regions) where the drought index is < 0.3.

Vietnam has been divided into three groups of regions depending on the drought characteristics (table 27.4). The first group comprises the north part of Vietnam (north mountain and midland and Red River delta regions) where the annual drought index is low and where moderate droughts occur occasionally during the dry season but severe droughts occur rarely. The second group comprises the coastal part of north-central and south-central regions, where the annual drought index is not high, but droughts occur during early- to mid-summer season. The southern part of this region is more prone to severe droughts. The third group comprises the south of Vietnam (central highland, southeast, and Mekong River delta regions),

Table 27.3 Probabilities of drought occurrence in different months for different regions of Vietnam

Region	Weather station	Probabilities (%) of drought occurrence											
		Jan.	Feb.	March	Apr.	May	June	July	Aug.	Sept.	Oct.	Nov.	Dec.
North mountain and midland	Laichau	64	47	33	6	0	0	0	0	2	29	35	34
	Sonla	50	59	57	14	2	0	0	0	4	42	54	45
	Langson	59	73	57	22	3	0	1	3	8	27	52	57
	Hongai	50	72	61	35	9	0	1	0	3	27	39	55
Red River delta	Hanoi	66	83	77	11	6	3	0	3	6	17	47	47
	Ninhbinh	64	79	63	38	0	2	6	0	0	4	29	60
North-central	Vinh	53	67	54	45	12	20	37	10	0	5	9	31
	Tuongduong	39	57	50	26	4	11	15	4	0	4	68	41
	Hue	9	45	64	54	39	39	42	31	0	2	0	4
South-central	Danang	29	79	86	82	58	50	43	28	0	0	3	7
	Bato	0	58	70	63	5	0	15	11	0	5	0	0
	Phanthiet	100	100	98	75	11	4	2	2	2	10	67	94
Central highland	Playku	100	96	86	30	2	0	0	0	0	20	59	96
	Buonmathuout	100	100	85	38	0	0	0	0	0	6	40	83
	Dalat	98	88	55	2	3	0	0	0	0	2	27	83
Southeast	Saigon	96	96	83	58	14	0	0	0	0	0	18	72
Mekong River delta	Cantho	92	99	93	65	7	3	3	4	3	3	14	71
	Phuquoc	83	79	55	17	0	0	0	0	0	0	5	5

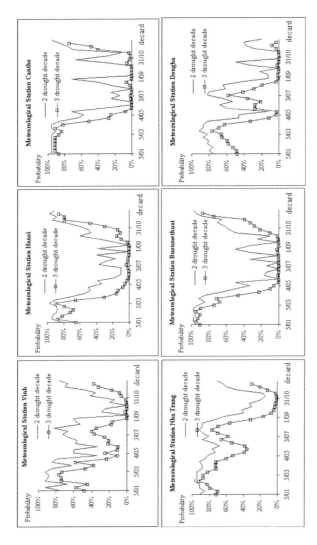

Figure 27.2 Probability distribution of two consecutive dry dekads and three consecutive dry dekads at some major weather stations in Vietnam.

Table 27.4 Agricultural regions and their drought characteristics

Group of regions	Region	Drought season	Annual drought index	Monthly drought index (K)	No. of months with $K \geq 5$ (%)
North	North mountain and midland	Nov.–April	0.3–0.8	5–8	0.5–3
	Red River delta	Nov.–March	0.3–0.8	3–5	1–3
Central coastal	North-central	April–Aug.	0.3–0.8	3–8	1–2
	South-central	Feb.–Aug.	0.3–0.8	20–60	5–10
Central highland and south	Central highland	Nov.–April	0.4–1.0	30–50	10–25
	Southeast	Nov.–April	0.4–1.0	6–50	5–10
	Mekong River delta	Dec.–April	0.3–0.9	5–50	5–10

where the annual drought index is very high. Buduko (1958) developed a drought index, K:

$$K = \frac{R}{Lr} \qquad [27.2]$$

where R is solar radiation balance, r is rainfall, and L is energy needed for 1 mm of evaporation. Using this index, percentage of months with different intensity of drought was determined for different regions of Vietnam (table 27.5).

Humidity Index

Selianinov (1958) developed the following drought index, K:

$$K = \frac{r}{0.1 \sum t > 10°C} \qquad [27.3]$$

where is r is rainfall during the crop-growing period, and $\sum t > 10°C$ is the total active daily temperature exceeding 10°C during the same period. Based on the value of K, one can categorize the intensity of drought. If K is less than 1.0, it indicates the beginning of drought. A K value between 0.4 and 0.5 means moderate drought, and a K value between 0.2 and 0.4 indicates severe drought. A value less than 0.2 means desertic conditions.

Conclusions

Every year the agricultural area (especially of rice, the main food crop of Vietnam) that is lost due to droughts exceeds 10,000 ha. The time of commencement and termination of drought is different for different regions in the country. Drought intensity is higher in southern Vietnam. While summer droughts usually occur in the central regions, winter–spring droughts occur in northern and southern Vietnam. Drought indices to

Table 27.5 Percentage of months with different intensity of drought (measured by drought index, K) for various regions in Vietnam

Agricultural regions	$K \leq 1$	$K > 1$	$K > 2$	$K > 5$
North mountain and midland	62.1	37.9	20.5	1.58
Red River delta	61.7	38.3	32.6	1.9
North-central	65.8	34.2	10.7	0.59
South-central	55.6	44.4	23.6	20.8
Southeast	63.3	36.7	30.2	16.0
Mekong River delta	65.7	34.3	30.3	18.0
Entire nation	62.0	38.0	21.1	5.6

monitor drought conditions are mainly based on weather data. There is a need to develop indices using satellite data to improve the drought-monitoring system in Vietnam.

References

Buduko, M.I. 1958. Heat Balance of the Earth's Surface. Hydrometeorological Publishing House, Leningrad.

Bui, N.L. 2000. ENSO effects on drought and inundation in agricultural production in Vietnam. Paper presented at the National Consultation Workshop on Understanding Extreme Climate Events in Vietnam, Hanoi, May 15–16, 2000.

Frere, M. and G.F. Popov. 1986. Early Agrometeorological crop yield forecasting. FAO Plant Production and Protection paper no. 73. FAO, Rome, 150 pp.

Nguyen, T.H. 2000. Distribution of drought in Vietnam. Scientific report, Agrometeorological Research Center, Hanoi.

Nguyen, V.V. 1998. Climate disaster and changing cropping pattern in central coastal region of Vietnam. Scientific report, Agrometeorological Research Center, Hanoi.

Nguyen, V.V. 2000. Impact of ENSO on climate and agriculture in Vietnam. Scientific report, Agrometeorological Research Center, Hanoi.

Nguyen, V.V. 2001. Drought and Agriculture in Vietnam. Presented at the Agrometeorological Training Workshop, Agrometeorological Research Center, Hanoi, July, 9–18, 2001, pp. 87–107.

Selianinov, G.T. 1958. Climate Zoning in Soviet Union (USSR) for Agricultural Purposes. Hydrometeorological Publishing House, Moscow.

CHAPTER TWENTY-EIGHT

Monitoring Agricultural Drought in China
GUOLIANG TIAN AND VIJENDRA K. BOKEN

Droughts account for more than half of the total number of natural disasters faced by China. Serious droughts impact industrial production, water supply, people's lives, and the ecological environment, which causes significant losses to the national economy. Because of increasing water shortages, drought has become one of the most important factors that limits agricultural production, especially in the north where droughts occur frequently. According to the Chinese terminology, if reduction in crop yields in an area is more than three-tenths of the average, the area is called a "damaged area," and if the reduction is more than eight-tenths, the area is declared a "nonyield area" (Lu and Yang, 1992; State Statistical Bureau, 1996; Chen, 2000).

History of Droughts in China

Drought has been the most serious natural disaster in Chinese history. Serious droughts occurred more than 1000 times from 206 B.C. to 1949 A.D. (Zhang, 1990; Li and Lin, 1993), or once about every two years. Over the years, the eastern part of China has become more drought prone.

Drought impacts have lessened since 1949 because the government has improved irrigation facilities. Nevertheless, agricultural production is still affected by drought because yields of most crops depend on weather conditions. Figure 28.1 shows the yearly variation in the total area affected by droughts since 1949. The most serious droughts occurred during the 1960s and the 1970s, and drought area has gradually increased (figure 28.1, table 28.1). Although serious drought occurred during 1978–79, droughts were mild during 1970–77. The droughts always damaged a greater area than did floods in a decade.

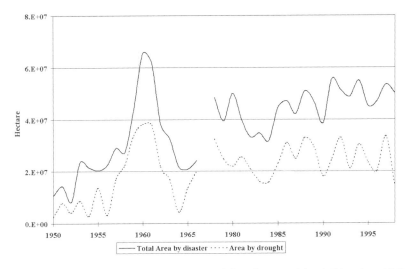

Figure 28.1 Area affected by natural disasters and droughts in mainland China since 1949.

Spatial Distribution of Droughts

Figure 28.2 presents the spatial distribution of droughts from 1951 to 1991. In particular, four regions were identified where the frequency of drought was significantly high (Li and Lin, 1993): (1) Huanghuaihai region in the North China Plain, (2) Hebei, Shanxi, Shandong, Henan, and Shaanxi provinces, (3) the coastal south China, in particular the coastal area in Guangdong and Fujian provinces, and (4) the southwestern part of southwest China, mainly the southern part of Yunnan and Sichuan provinces. In addition, there is a narrow area in the western part of northeast China and arid regions in Gansu and Xinjiang in northwest China that are prone to drought (Li and Lin, 1993).

Seasonal Characteristics of Droughts

A drought in China can be classified as a spring, summer, autumn, or winter drought on a temporal or seasonal scale. The Qin Ling Mountains and Huaihe River serve as an approximate dividing line on a spatial scale. North of the dividing line mainly spring and summer droughts occur, while south of the dividing line mainly summer, autumn, and winter droughts occur. In some years, droughts persist for two or more consecutive seasons and become highly destructive. The spring–summer and spring–summer–autumn droughts are most frequent in Huanghuaihai region (North China Plain), and spring–summer droughts are most frequent in northeast regions of China. Winter–spring droughts mainly occur in southwest and south China, while the year long and summer–autumn droughts occur in north-

Table 28.1 Area damaged by agricultural disasters in China

Decade	Total area damaged by natural hazards (10,000 ha)	Drought-damaged area (10,000 ha)	Drought-damaged area as percentage of the total
1950s	2207.3	1162.1	48
1960s	3780.4	2177.4	57
1970s	4390.5	2855.5	65
1980s	4208.6	2414.1	57
1990s	4949.7	2489.2	50

Source: State Statistical Bureau (1996).

west China and in the middle and lower reaches of Yangtze River, respectively (Li and Lin, 1993).

Main Causes of Drought

The possible factors responsible for droughts in China can be categorized as natural and, to some extent, cultural, as described in the following sections.

Natural Factors

Meteorological Factors The main agricultural region in eastern China is influenced by the East Asia monsoon. In this region, the distribution of precipitation is uneven and annual variation is high. The precipitation zone jumps between the northern and the southern parts of the country according to a tropical high-pressure concentration during the April–September period, resulting in such an uneven precipitation distribution that it causes storms and floods in some regions and droughts in other regions. In addition, movement of precipitation zones also causes droughts. Lack of rains along the Yangtze River during summer can also lead to serious droughts.

The Chinese climate is mainly controlled by four semipermanent atmospheric activities: the Mongolia high pressure and the Aleutian low pressure during winter and the India low-pressure and the Pacific tropic high pressure during summer. When the Mongolia high-pressure is strong and the Pacific tropic high pressure is weak during winter, cold air drives straight into the south from the north, and the whole country experiences a cold air mass. As a result, the weather turns cold and dry (because of no rains). In summer, when the Pacific tropic high pressure and the India high pressure are strong and active, the major part of the country is influenced by warm and wet airflow from southeast and southwest. Consequently, rains occur during summer. The rainy season in China is mainly from April to September. However, the spatial distribution of precipitation and its intensity differ from year to year depending on the timing of northward movement of subtropical high pressure, the active range of the high pressure, and the large annual variation in other weather systems.

The active intensity of the western Pacific tropical high pressure is related to ENSO (El Niño/Southern Oscillation; chapter 3). When the ENSO

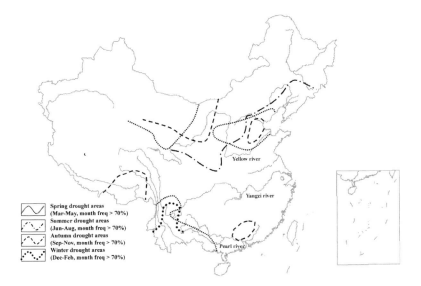

Figure 28.2 Spatial distribution of drought intensity and frequency in China, 1951–91.

develops, the western Pacific tropical high pressure is strong, and its region of influence is relatively greater toward the west and the north. This results in floods in the Yangtze and the Huaihe rivers during summer due to above-average precipitation and results in droughts along the Yellow River in northern China due to below-average precipitation. Meanwhile, the south of the lower reaches of the Yangtze River experiences drought. When the sea temperatures in the East Pacific recover and their deviations are still positive, below-average precipitation occurs along the Yangtze and Huaihe rivers during summer, leading to drought. During an ENSO year, below-average precipitation occurs along the Yellow River in southern and northern China, which also causes drought.

Distribution of Water Resources Shortage of water resources is one of the reasons for drought occurrence, particularly in the north. The national average of water availability exceeds 27000 m^3/ha, but only 4500 m^3/ha of water is available along the Yellow and the Huaihe rivers. Water availability is even lower (about 200 m^3/ha) along the Liaohe and the Hailuanhe rivers, which does not satisfy the water needs of crops at all. The national program on "water transfer from south to north" that the government is planning to undertake will alleviate the existing drought situation in northern China.

Cultural Factors

Due to the rise in industrial production and crop yields, the demand for agricultural water has increased four to five times as compared to demand in the 1950s, but the water supply has remained limited. For example, at some places two-crops-in-one-year system is transformed into three-crops-in-one-year system. This is one of the reasons that drought-damaged areas have expanded every year rather than shrunk despite the post-1949 improvement of irrigation facilities by the government. Also, sometimes farmers use water carelessly, and soil erosion occurs due to farming on steep hills. Some farmers use chemical fertilizers instead of natural fertilizers, which deteriorate soil quality and lower the water-holding capacity of soils. Such cultural activities contribute to drought occurrence, for instance, in the North China Plain.

Drought Monitoring Methods

The following models have been developed for monitoring droughts in China, and an operational drought monitoring system is also in place.

Water Balance Model

The water budget for the root zone of a soil column can be written as (Yang and Tian, 1991):

$$\frac{\partial W}{\partial t} = (P + I) - E - R + D - G \qquad [28.1]$$

where W is the water content of the root zone (cm), P is precipitation (cm), I is irrigation (cm), E is evapotranspiration (cm), R is runoff (cm), D is the discharge from groundwater (cm), and G is recharge to the groundwater (cm). Ignoring, for the moment, recharge and discharge and incorporating irrigation into rainfall, the above equation can be integrated over time t (days or weeks) to derive a simple equation for the incremental change in the soil water storage:

$$W_t - W_{t-1} = P_t - E_t - R_t \qquad [28.2]$$

The above parameters are difficult to estimate. Typically, soil moisture is estimated from the amount of rainfall and crop water demand, depending on air temperature, solar radiation, humidity, wind speed, potential evapotranspiration (E_p), and land cover conditions (Monteith and Unsworth, 1990). Simple water balancing can be performed if initial storage (W_0) is known and if operating characteristics for the area that define relationships between moisture availability [$m_a(WB)$] and runoff fraction (r_r) are also known. For example:

$$m_a(WB) = E/E_p = f[(W_t - W_w)/W_{max}] \qquad [28.3]$$

and

$$r_r = R/P = g(W_t/W_f) \qquad [28.4]$$

where W_f is the amount of water available in a soil profile at field capacity; W_w is the amount of water in the profile at which evapotranspiration stops and plant growth ceases; and W_{max} is extractable water that usually equals $W_f - W_w$. It has been sufficient (but not necessary) to use the simplest linear forms of these equations. Water availability is defined as (Jupp et al., 1997):

$$m_a(WB) = \max\{0, \min[1, \beta(W_t - W_w)/W_{max}]\} \quad [28.5]$$

Yang and Tian (1991) developed a drought index (D) following a water-balancing approach:

$$D = (W_t - W_w)/W_r \quad [28.6]$$

where W_r is the amount of water required by the vegetation for its normal growth and respiration. Derivation of D, which is closely related to m_a, needs adequate meteorological data, spatial information about the hydrological properties of soils, and the nature of the land cover. If m_a or D can be determined during various periods by remote sensing techniques (chapters 7 and 8), it is possible to validate water-balancing models.

Remote Sensing

Remote sensing provides physical measurements of components of the energy balance—daily input of solar radiation and its various forms, such as convection, evaporation, and storage in the earth or oceans. In particular, the Advanced Very High Resolution Radiometer (AVHRR) data provide information on shortwave absorption and the condition of the cover and surface temperature.

If R_n denotes net wavelength radiation (W/m^2) and G denotes the heat flux into the soil (W/m^2), the net energy available at the earth's surface will be $R_n - G$, which can be partitioned as:

$$\begin{aligned} A &= R_n - G \\ &= \lambda E + H \end{aligned} \quad [28.7]$$

where E is the evapotranspiration flux (*m/sec*) of water vapor; λ is latent heat of vaporization of water (J/m^3), H is sensible heat flux (W/m^2) or the energy involved in the movement of the air and its transfer to other objects (such as trees, grass, etc.), and λE is the energy required to transform water from liquid into vapor and can be computed as follows (Monteith and Unsworth, 1990):

$$\lambda E = (R_n - G) - H \quad [28.8]$$

The above components may be computed in various ways using resistance models (Monteith and Unsworth, 1990) of differing levels of complexity introduced by land surface structure and cover. In each case, the sensible heat flux is driven by the differences between the distribution of temperatures among the surface components and the air temperature. For simple (one-layer) or more complex models (Jupp, 1990), the remotely sensed aggregated surface temperature may be sufficient to estimate evapotranspiration (ET).

Not all energy balance models contain explicit relationships between soil moisture and surface temperature. Many of those that do are complex and may contain many parameters that are difficult to measure. Most, however, can be solved for the case when soil moisture is at field capacity and is not a limiting factor. The effective ET (λE_p, where E_p is potential ET) corresponds to a situation where soil water and vegetation condition do not limit ET while the atmospheric demand and land cover types and structures are the same. The moisture availability at other times (when soil water is below field capacity) is then defined as

$$m_a(EB) = \lambda E / \lambda E_p \qquad [28.9]$$

Crop Water Stress Index Using an integrated version of the moisture availability defined in equation 28.9, Jackson et al. (1983) defined the crop water stress index (CWSI) as

$$\begin{aligned} \text{CWSI} &= 1 - m_{ad} \\ &= 1 - \frac{E_d}{E_{pd}} \end{aligned} \qquad [28.10]$$

where the subscript d denotes that the quantity is integrated over a day. Remote sensing data can be used for computing water balancing by providing estimates for ET (Jupp et al., 1997; Tian et al., 1997). Based on the range of CWSI and W/W_f, the surface conditions can be classified as heavy drought, dry, light dry, normal, and moist conditions (table 28.2).

Normalized Difference Temperature Index Jackson et al. (1983) found that an approximation of m_a equals a normalized difference temperature index (NDTI):

$$\begin{aligned} m_a(EB) &\approx \text{NDTI} \\ &= \frac{T_\infty - T_s}{T_\infty - T_0} \end{aligned} \qquad [28.11]$$

where T_∞ is theoretical surface temperature if no water were available, and T_0 is theoretical temperature if maximum amount of water were available (i.e., the temperature corresponding to E_p).

The NDTI was successfully used to map surface temperatures for drought monitoring in the North China Plain (Tian et al., 1997). To estimate evapotranspiration, a one-layer model was used for a vegetation-covered surface (Jupp et al., 1997), and a two-layer model was used for partial vegetation-covered surface (Jupp et al., 1997). A soil-thermal inertia model was used for low vegetation-covered or bare surface for estimating soil moisture (Yu and Tian, 1997).

Table 28.2 The categories of drought

Grade	W/W_f	Crop water satisfaction index
Heavy drought	<0.4	>0.913
Dry	0.41–0.50	0.765–0.913
Light dry	0.51–0.60	0.617–0.764
Normal	0.61–0.80	0.322–0.616
Moist	>0.81	<0.321

Source: Yao (1990).

Normalized Difference Water Index

The spectral signature of vegetation canopy at 1.55–1.75 μm waveband is sensitive to water content in the canopy. Remotely sensed data at the shortwave infrared can be used to monitor water content of vegetation canopy. A normalized difference water index (NDWI) was developed to monitor drought during the cropping season (Tian et al., 2002):

$$\text{NDWI} = \frac{P_{SW} - P_R}{P_{SW} + P_R} \quad [28.12]$$

where P_{SW} is the reflectance at shortwave infrared (e.g., 1.55–1.75 μm), and P_R is the reflectance at red band (0.61–0.68 μm). The NDWI is correlated with soil moisture (Tian et al., 2002) at root zone and therefore can be used to monitor drought conditions in crops.

Anomalous NDVI Model

The anomalous NDVI (Chen, et al., 1994) can also be used to monitor drought conditions:

$$\text{ANDVI} = \frac{\text{NDVI} - \overline{\text{NDVI}}}{\overline{\text{NDVI}}} \quad [28.13]$$

where NDVI is the normalized difference vegetation index based on AVHHR data, and $\overline{\text{NDVI}}$ is an NDVI averaged for multiple (more than 10) years.

Combination Model

The combination model (CM) is defined as

$$\text{CM} = \frac{\text{NDVI}}{T_S} \quad [28.14]$$

where T_S is surface temperature that can be derived from channel 4 and channel 5 data of the AVHHR thermal data (Xin et al., 2003).

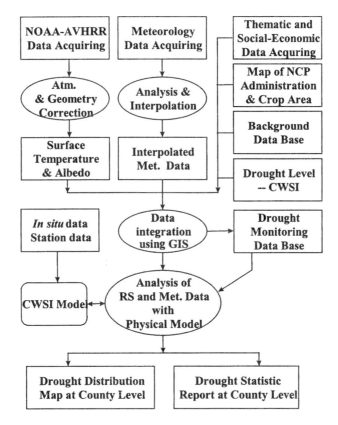

Figure 28.3 Operational system for drought early warning in the North China Plain.

Drought Early Warning System

Using some of the indices described above, an operational system for drought early warning for North China Plain (NCP) was designed (Tian et al., 1997), as illustrated in figure 28.3. Atmospheric and geometric corrections are made in the AVHRR data before estimating temperature and albedo. Meteorological data and CWSI images are appropriately registered. Finally, the system integrates drought models and remote sensing and meteorological data to produce a series of drought maps and reports.

Geographic Information System Database

The Geographic Information System (GIS) database has been set up for the management and application of historical data and real-time meteorological data for drought monitoring in NCP. The database comprises: (1) thematic, county-level, soil texture, water table, and geomorphological maps; (2) meteorological data, including rainfall data (1949–94), daily meteorological data from February to June (1992–97), and dekadal (10-day) data for agricultural season (1992–97); (3) experimental data at Yucheng Ecosystem Experimental Station, Chinese Academy of Sciences (CAS), in Shandong province, and Fenqiu Ecosystem Experimental Station and CAS in Henan province (1992–93); (4) socioeconomical data including crop productivity, population, and county area; (5) AVHRR data (1992–97); and (6) maps showing drought distribution and severity. The GIS integrates these different data sets (remote sensing data, CWSI, and NDTI models) to map soil moisture, evapotranspiration, and, finally, drought conditions.

Timeslice Products

Tian (1993) monitored droughts in the NCP region by estimating CWSI and relating it to soil moisture as follows:

$$\text{CWSI} = 1 - \frac{E}{E_p}$$

$$= 1 - m_a = 1.503 - 1.476\frac{W}{W_f} \quad (R = .90) \quad [28.15]$$

where W is the soil moisture, W_f is the field capacity, and R is the correlation coefficient. Table 28.3 indicates the accuracy of drought monitoring with the CWSI model. Figure 28.4 shows how timeslice products are used to monitor drought in NCP.

Soil Humidity Index The SHI is calculated as

$$\text{SHI}_j = (Q_j/SW_j)100 \quad [28.16]$$

where SHI_j is the ratio of soil moisture (Q_j) to field capacity (SW_j), and j varies from 1 to 10 for soil measurement. The relationship between the vegetation condition index (VCI) and SHI was established to enable the use of the AVHRR data to monitor droughts in China. The soil moisture

Table 28.3 Accuracy of drought monitoring using crop water stress index model in 1994

City/province	Accuracy (%)				
	March 4	March 26	April 13	April 30	May 8
Beijing	84.0	86.0	85.0	88.0	85.0
Tianjin	88.0	88.8	82.5	85.0	87.2
Henan	90.6	87.5	83.1	88.8	86.2
Anhui	83.3	80.6	86.6	83.3	83.3
Jiangsu	80.0	Cloud	75.0	81.5	80.0
Shandong	81.4	85.7	82.8	82.8	85.7
Hebei	85.7	80.0	84.2	87.1	85.7

Source: Tian et al. (1997).

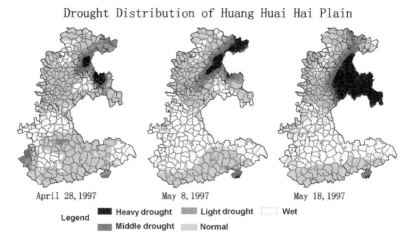

Figure 28.4 Timeslice products for drought monitoring in the North China Plain.

at 20 cm depth and the corresponding field capacity information from 102 agricultural climate observatories (covering the country) for the period from 1980 to 1994 have been collected. SHI had good correlation with VCI (table 28.4, figure 28.5).

The correlation coefficients ranged from 0.60 to 0.85. The test results of a statistical model of all 36 ten-day periods from June to December indicated that the model of each dekad passed the test at the level of $\alpha = 0.05$. Figure 28.6 illustrates an operational system of drought early warning. Some results using this system are shown in figures 28.7 and 28.8.

Table 28.4 Statistical model of drought monitoring and associated drought category

Dekad in July	Model	R	Drought categorization	
			Range of VCI	Drought category
1	VCI = 219.2832 + 83.8763 × LN(SHI)	0.846	≤90	Heavy drought
			90–109	Dry
			109–124	Light dry
			124–148	Normal
			≥148	Moist
2	VCI = 20.6373 × exp(0.0152 × SHI)	0.802	≤38	Heavy drought
			38–44	Dry
			44–51	Light dry
			51–70	Normal
			≥70	Moist
3	VCI = 12.736 × exp(0.0235 × SHI)	0.856	≤31	Heavy drought
			31–41	Dry
			41–52	Light dry
			52–83	Normal
			≥83	Moist

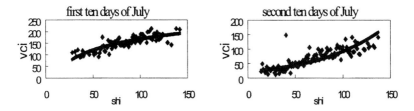

Figure 28.5 The scatter plot between measured soil humidity index (SHI) and vegetation condition index (VCI).

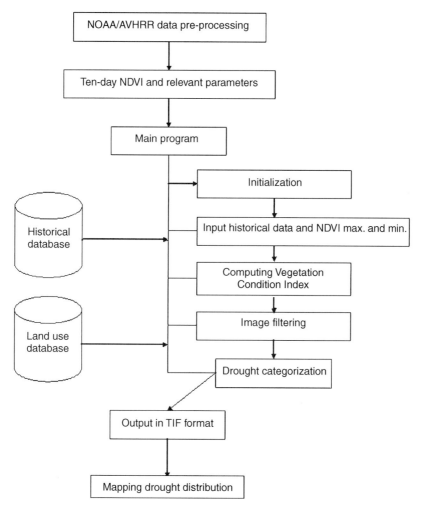

Figure 28.6 Operational system for drought early warning for China.

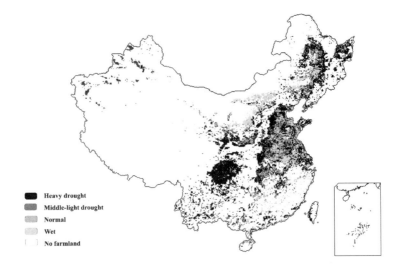

Figure 28.7 Drought distribution in China during third dekad (10 days) of May 2000.

Potential Drought Research Needs

For better monitoring of droughts in China, further research is required on linking drought monitoring models with crop yield estimation; developing models to monitor drought conditions during different phenological phases; developing new drought models by combining TCI and VCI; and combining drought models with remote sensing data and generating digital simulations of drought conditions.

Conclusions

Droughts usually last long and seriously affect large agricultural areas in China. China has paid great attention to droughts by developing drought monitoring technology to alleviate drought impacts. The combination of remote sensing techniques with GIS and some conventional methods are effective ways to monitor drought in China. However, further research is needed to improve the present setup of the drought monitoring system at the national level.

ACKNOWLEDGMENTS The drought research presented in this chapter has been supported by the State Science and Technology Commission of China, the Chinese Academy of Sciences, and the Natural Science Foundation of China. The Australia/China Joint Science and Technology Commission and Visits Program of the CSIRO/CAS Cooperative Agreement have supported the international collaboration. Contributions by Qi Yi, Feng Qiang, Li Fuqin, and Shen Guangrong are acknowledged. Yu Tao, Sui Hongzhi, and Lu Yonghong provided valuable field data.

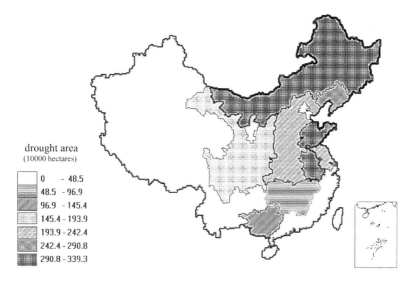

Figure 28.8 Area affected by drought at the provincial level in mainland China, 2000.

The National Climate Data Center provided meteorological and agricultural data. Xin Xiaozhou and Chen Liangfu helped translate Chinese into English.

References

Chen, W. 2000. Drought assessment and digital simulation of drought processing. M. D. dissertation. Institute of Atmospheric Physics, Chinese Academy of Sciences, Beijing.
Chen, W., Q. Xiao, Y. Sheng. 1994. Application of the anomaly vegetation index to monitoring heavy drought in 1992. Remote Sens. Environ. 9:106–112.
Jackson, R.D., J.L. Hatfield, R.J. Reginato, S.B. Idso, and P.J. Pinter Jr. 1983. Estimation of daily evapotranspiration from one time-of-day measurements. Agric. Water Manage. 7:351–362.
Jupp, D.L.B. 1990. Constrained two-layer models for estimating evapotranspiration. In: Proceedings of the 11th Asian Conference on Remote Sensing, Guangzhou, China, 15–21 November 1990. International Academic Publishers, vol. I, pp. B4-1–B4-6.
Jupp, D.L.B., G. Tian, T.R. McVicar, Q. Yi, and F. Li. 1997. Monitoring Soil Moisture Effects and Drought Using AVHRR Satellite Data I: Theory. Technical Report. CSIRO Earth Observation Centre, Canberra, Australia.
Li, K., and X. Lin. 1993. Drought in China. In: Drought Assessment, Management and Planning: Theory and Case Studies. Kluwer Academic, Dordrecht, pp. 263–269.
Lu, J. and Y. Yang. 1992. Introduction to Natural Disasters. Sichuan Science and Technology Press.
Monteith, J.L., and M.H. Unsworth. 1990. Principles of Environmental Physics, 2nd ed. Edward Arnold, London.

State Statistical Bureau. 1996. Report on Chinese Disasters. China Statistics Press, Beijing, China.

Tian, G. 1993. Estimation of evapotranspiration and soil moisture and drought monitoring using remote sensing in North China Plain. In: Proceedings of Space and Environment, 44th Congress of International Astronautical Federation. International Astronautical Federation, Graz, Austria, pp. 23–32.

Tian, G., D.L.B. Jupp, F. Qiang, W. Bingfang, and L. Qinhuo. 2002. Drought monitoring of crop using remote sensing and GIS. J. Remote Sens. 6(suppl.):145–152.

Tian, G., D.L.B. Jupp, Q. Yi, T.R. McVicar, and F. Li. 1997. Monitoring soil moisture effects and drought using AVHRR satellite data II: Applications. Technical Report. CSIRO Earth Observation Centre, Canberra, Australia.

Xin, J., G. Tian, Q. Liu, L. Chen, and X. Xin. 2003. Drought monitoring from remotely sensed temperature and vegetation index. In: Proceedings of IGARSS 2003, Toulouse, France. July 21–25.

Yang, X., and G. Tian. 1991. A remote sensing model for wheat drought monitoring. In: Proceedings of the 12th Asian Conference on Remote Sensing, Association of Asian Remote Sensing, Singapore, October 1991.

Yao, X. 1990. Physics of Soil. Science Press, Beijing, China.

Yu, T., and G. Tian. 1997. The application of thermal inertia method in the monitoring of soil moisture of North China Plain based on NOAA-AVHRR data. J. Remote Sens. 1:24–31.

CHAPTER TWENTY-NINE

Monitoring Agricultural Drought in Australia

KENNETH A. DAY, KENWYN G. RICKERT, AND GREGORY M. MCKEON

Since European settlement of Australia began in 1788, drought has been viewed as a major natural threat. Despite warnings by scientists (e.g., Ratcliffe, 1947) and many public inquiries, government policies have, in the past, encouraged closer land settlement and intensification of cropping and grazing during wetter periods. Not surprisingly, drought forms part of the Australian psyche and has been well described in poetry, literature (e.g., Ker Conway, 1993), art, and the contemporary media (newspapers and television). Droughts have resulted in social, economic, and environmental losses.

Attitudes toward drought in Australia are changing. Government policies now consider drought to be part of the natural variability of rainfall and acknowledge that drought should be better managed both by governments and by primary producers. Nonetheless, each drought serves as a reminder of the difficult challenges facing primary producers during such times.

We begin this chapter with a brief overview of drought in Australia and its impacts on agricultural production, the environment, rural communities, and the national economy. We outline some of the ways governments and primary producers plan for and respond to drought and describe in detail an operational national drought alert system.

Rainfall Variability and Agricultural Production in Australia

Australia has mainly an arid or semiarid climate. Only 22% of the country has rainfall in excess of 600 mm per annum, confined to coastal areas to the north, east, southeast, and far southwest of the country (http://www.bom.gov.au/climate/ahead/soirain.shtml). Australia also has high year-to-year and decade-to-decade variation in rainfall due, in part, to the influence of the El Niño/Southern Oscillation (ENSO) phenomenon (http://www.bom.gov.au/climate/ahead/soirain.shtml). The Interdecadal Pacific Oscillation (IPO) also contributes to the rainfall variability at annual and decadal scales and modulates ENSO impacts on rainfall (Power et al., 1999).

The current geographic boundaries of agricultural production (figure 29.1) were reached in the late 19th century, and the entire agricultural region has experienced drought, in some form, over the past 100 years. Protracted dry periods occurred during the period from late 1890s to 1902 in eastern Australia, during the mid to late 1920s and 1930s over most of the continent, during the 1940s in eastern Australia, during the 1960s over central and eastern Australia, and during 1991–95 in parts of central and northeastern Australia. During these low rainfall periods, not every year was extremely dry, but rainfall in most years was below the long-term median, and there were often runs of severe drought years. Many, but not all, droughts were associated with El Niño events (http://www.bom.gov.au/climate/ahead/soirain.shtml; chapter 3). Droughts have been more prevalent when the IPO was positive, that is, from 1896 to 1909, from 1922 to 1945, and from 1979 to 1998. During the periods when the IPO was negative, La Niña events have tended to be associated with wet conditions. During these times, Australia has largely been drought free.

Despite Australia's arid or semiarid climate, agriculture (including cropping and grazing) is practiced on 60% of Australia's land area of approximately 463 million ha (figure 29.1). Extensive grazing by beef cattle and sheep occupies approximately 90% of the agricultural land. The remaining agricultural land is equally distributed between intensively grazed, sown pastures and crops. Wheat is Australia's major crop, having the broadest geographic range (figure 29.1), and it contributes an average of 25% of total agricultural production. Only a small proportion (0.5%) of Australia's sown pastures and crops are irrigated, but these regions are nonetheless important in terms of their contribution to the value of agricultural production.

Despite the high variation in annual rainfall, cropping and grazing are geared toward years with normal rainfall, with some contingency for drought years as discussed later in this chapter. For this reason, the entire agricultural region (figure 29.1) is drought prone. In some cases, expectations of climate are biased by years with above-average rainfall, leaving a particular enterprise even more drought prone. Sensitivity to drought is therefore dependent more on an individual's experience, expectations, and management than on geographic location.

Impacts of Drought

Drought has severe impacts on agricultural production in Australia, particularly at local levels, affecting the financial and social well-being of farm families and local communities. Declining terms of trade and rising debt levels have eroded the capacity of rural families and communities to cope

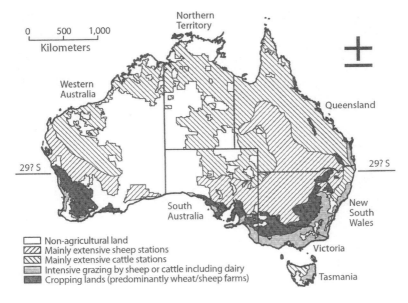

Figure 29.1 Current (2000) land-use in Australia. The entire grazed and cropped region is drought prone.

with protracted drought. Severe production losses may be tolerated for one year or longer, but when dry conditions persist for several years as they did, for example, in northeastern Australia from 1991 to 1996, rural communities deteriorate, people abandon the land, and the suicide rate in rural areas increases (Stehlik et al., 1999). Impacts are also evident at a national scale, particularly when droughts are widespread or centered on the most productive regions at critical times. For instance, the extremely dry winters of 1902, 1914, 1940, 1944, 1982, and 1994 saw large reductions in national wheat yields (figure 29.2).

One of the most serious failures of the wheat crop occurred in 1982. It is estimated that the 1982–83 drought cost Australia in excess of U.S. $1.5 billion, resulting mainly from the failure of the 1982 wheat crop. Approximately 60% of all farms in Australia involved in cropping or grazing were affected by drought during this period (Purtill et al., 1983), and the

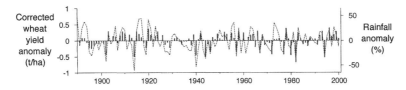

Figure 29.2 A comparison of annual anomalies in the national wheat yield corrected for a rising trend in yields (bars) and percentile growing season (May–October) rainfall for the cropping region (figure 29.1).

quantity of wheat sold per farm declined by 45% (Bureau of Agricultural Economics, 1983). The net value of rural production in Australia during 1982–83 was 50% lower than the average of the previous five years (Bureau of Agricultural Economics, 1983), and there was an estimated overall reduction in cash operating surplus of 57% across all Australian farms (Purtill et al., 1983).

In extensive grazing lands, drought has led to high mortality of livestock and widespread deterioration of the land and pasture resources. Livestock losses during drought were highest during the first half of the 20th century, as illustrated by annual changes in livestock numbers in Queensland's grazing lands (figure 29.3), a major region of beef and wool production.

The dramatic stock losses that occurred in Queensland during the protracted drought from 1895 to 1902 (figure 29.3), for example, should no longer occur due to an improvement in water supplies and transport systems for livestock, the use of more drought- (and tick-) resistant *Bos indicus* cattle, and the improved options for supplementary feeding. Through such technological advances, increased stock numbers are now possible both in dry and wet periods. High grazing pressure (before, during, and after the drought occurrence) has led to reduction in pasture cover, accelerated soil loss, decline in pasture composition, and fewer opportunities to burn pasture. Pasture burning improves pasture quality and limits the infestation and expansion of weeds, including woody plants (Tothill and Gillies, 1992).

Despite the increasing value of agricultural production in Australia, diversification of the economy has caused the importance of agricultural production to decline in relative terms, reducing the impact of drought on the Australian economy. For example, the severe drought of 1982–83 resulted in a decline in agricultural output of approximately 24%, but only accounted for about one-half of the 2% decline in economic growth recorded in that year. A 1% decline in economic growth is nonetheless significant and, if forecast, could be mitigated to some extent by relaxing monetary policy to maintain growth targets. Other ways in which governments plan for and respond to drought are considered in the following section.

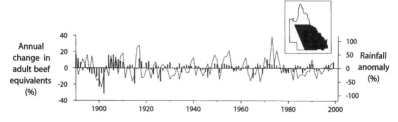

Figure 29.3 Annual change in reported livestock (sheep and cattle) numbers in Queensland (bars) and percentile summer rainfall for the region (inset) in which the majority (83%) of livestock now occur.

Government's Response to Drought

Constitutionally, Australia is a federation of effectively seven states (figure 29.1) with governments at both federal and state levels concerned with agriculture and drought. With respect to drought monitoring, the federal government is responsible for meteorological observations through the National Climate Center of the Australian Bureau of Meteorology (http://www.bom.gov.au). The Australian Bureau of Statistics (ABS; http://www.abs.gov.au) compiles, at a district level, annual statistics on crop production and livestock numbers from individual properties. Similarly, the Australian Bureau of Agriculture and Resource Economics (http://www.abare.gov.au) conducts annual surveys to obtain district-level economic information. The federal government is responsible for policy development through the portfolio of Agriculture, Fisheries and Forestry-Australia (http://www.affa.gov.au) and the Bureau of Resource Sciences (http://www.brs.gov.au), for objective analysis of claims for "exceptional circumstances" assistance. The federal government also sponsors national-scale research projects through various funding agencies (e.g., http://www.cvap.gov.au). State governments are responsible for land-use including drought management, drought monitoring, and, in some cases, administration of drought relief.

During severe droughts, the federal government provides welfare support to eligible primary producers and, together with some state governments, provides various forms of business support. The national drought policy, established in 1992, proposed that federal government support would be initiated once it was ascertained that rainfall and agronomic conditions were rare and severe enough as to be likely to occur only once in 20–25 years and to last more than 12 months. The "drought exceptional circumstances" (DEC) policy revolved around the assessment of six criteria: meteorological conditions; agronomic and stock conditions; water supplies; environmental impacts; farm income levels; and scale of the event. In 1997, the federal government broadened the concept of DEC to include exceptional events other than drought (e.g., insect plagues, disease, and water logging). Based on new declaration procedures for exceptional circumstances (EC) approved in 1999, economic criteria now take precedence in determining whether federal government assistance is warranted (White, 2000). Each Australian state has its own approach to managing droughts, but state governments are working with the federal government to align details of the drought policies and the development of objective criteria for government intervention.

Government intervention has, in the past, aimed to help primary producers survive drought. Government policies now aim to further increase the self-reliance of primary producers during drought. To this end, governments and primary producers are looking at ways to better manage and plan for rainfall variability.

Mitigating Drought at a Farm Scale

At a property level, policies for stock, pasture, and crop management tend to be based on expectations of average climate rather than on an extreme event such as drought. There is also a degree of flexibility to cope with the eventuality of extremely dry periods, although there comes a point when the most economic approach is to deal with drought when it happens (e.g., Daly, 1994). Strategic and tactical responses to drought differ according to the type of enterprise, and a variety of information is synthesized on fact sheets on various state government Web sites (e.g., http://www.longpaddock.qld.gov.au/QueenslandDroughtMonitor; http://www.dpi.qld.gov.au/drought; http://www.agric.nsw.gov.au). Strategies in nonirrigated cropping lands may include diversification of crops with varying degrees of drought resistance, sequencing of crops and fallow, or adjusting fertilizer application.

If drought results in the failure of a crop, the only alternative is to either do nothing or to plant a short-season crop. In pastoral lands, the options are also limited (Bureau of Agricultural Economics, 1969). One of the most important strategies is to maintain a conservative stocking rate in most years to reduce forced destocking, conserve fodder, make animals physically better prepared for dry conditions, and make feed available for purchase of stock from nearby areas.

A number of computer-based decision-support tools (e.g., Drought Plan; http://www.regional.org.au/au/asit/compendium/i-06.htm) have been developed to help primary producers assess options for drought management (White and Howden, 1991). Many decision-support tools are based on farm-scale agronomic models and allow alternative management strategies to be simulated and tested over a long (approximately 100-year) period. Results can be expressed as probability distributions to indicate the risk associated with specific management options and the degree to which ENSO, for instance, changes this risk. Agronomic models such as GRASP and AP-SIM (described later in this chapter) have demonstrated the usefulness of ENSO-based statistical climate forecasts for a range of decisions, including those linked with drought management in both cropping and grazing lands (Hammer et al., 2000).

The remainder of this chapter focuses on the broad-scale monitoring of drought and, in particular, the development and implementation of a national drought alert system. The system has proven valuable in assisting the process of objectively allocating drought relief and has the potential to improve preparation and planning for drought by focusing attention on emerging drought situations.

Operational Drought Monitoring in Australia

Several levels of drought monitoring exist in Australia. The Bureau of Meteorology (BoM) monitors drought across the nation from a meteorological perspective. The climate monitoring network, which is the basis for BoM's drought assessment, also serves as input to a national monitoring system operated by several state government agencies. These systems are described later in this chapter. Individual state governments also operate official drought declaration schemes that involve monitoring emerging drought situations at both individual property and district scales. Such monitoring is largely subjective and based, in the first instance, on the applications for government assistance by individual property owners. The process used in Queensland since 1964 is outlined below as an example.

Drought Monitoring at a State Level: The Queensland Example

The Queensland drought scheme primarily aims to assist primary producers with livestock management during drought. The process for declaring drought in Queensland begins with an individual primary producer approaching the local stock inspector for an "individual droughted property" (IDP) declaration. Following a property inspection by the stock inspector, a local drought committee assesses the application. If drought conditions appear widespread and a significant number of IDPs have been issued, the local drought committee may recommend that the entire district (shire or part thereof) be declared drought stricken. The state government reviews the application and, until policy changes in the early 1990s, would declare drought if summer (October–March) rainfall at official recording stations in the district was 40% below the long-term average rainfall or, for southern Queensland, if winter (April–September) rainfall was 50% below the long-term average (Daly, 1994). Since the early 1990s, drought has been defined as a once in 10- to 15-year event. From this definition, a threshold of annual rainfall below the 7th to 10th percentile has been adopted as a declaration criterion. This threshold equates to an annual rainfall of approximately 40% below the long-term average and is therefore similar to thresholds used before 1964. Drought status is revoked once it is ascertained that stock numbers can return to normal levels. Because the revocation threshold is higher than the declaration threshold, official droughts have occurred more than 10% of the time.

Rainfall and Climate Monitoring

BoM monitors rainfall across Australia through a network of approximately 7000 stations operated largely by volunteers. The National Climate Center (NCC) receives digital data via phone lines each day from approximately 300 automatic weather stations and 1500 observers. This telegraphic network provides the basis for near-real-time assessment of rainfall and climate. The remaining stations report via regular mail on a monthly basis. These written records are checked and digitized at the NCC before being placed on a national database. Data from most stations are available in digital format within three months. Maps of rainfall anomalies for various periods from 24 h to more than 12 months are updated in near-real time (http://www.bom.gov.au/cgi-bin/climate/rainmaps.cgi). Rainfall anomalies are mapped as percentiles, as absolute or percentage departures from the long-term average, and as drought severity levels.

An accepted approach to monitoring meteorological drought in Australia is to rank, as a percentile, current rainfall against historical rainfall:

$$\text{Percentile }(0-100) = [(r-1)/(n-1)]\,100 \qquad [29.1]$$

where r is a rank in a series of n values. BoM's drought analysis indicates regions in which rainfall has been below the 10th percentile (a serious rainfall deficiency) or the fifth percentile (a severe rainfall deficiency) for three months or more. The rainfall deficiency is considered removed if, for the past month alone, rainfall exceeds the 30th percentile for the three-month period commencing that month or if, for the past three months, rainfall is above the 70th percentile. This information is presented both on the Web (above) and as a monthly publication (*Drought Review-Australia*).

Use of Agronomic Models to Monitor Droughts

Anecdotal experience of graziers and the results from experimental trials show that, for different years, different amounts of pasture growth are experienced with similar amounts of seasonal rainfall (Stafford Smith and McKeon, 1998). Likewise, the relationship between crop yield and rainfall during the growing season is generally poor because yield also depends greatly on the timing of rain as shown, for example, by Nix and Fitzpatrick (1969). Thus, agronomic models are probably the best indicators of drought severity and the effectiveness of rain (White et al., 1998). The following section describes the implementation of a national-scale framework for modeling pasture and crop production in near-real time and the adaptation of model output for drought monitoring and early warning.

A National Drought Early Warning System

The development of a national drought early-warning system for Australia commenced with a Queensland prototype in 1991. The need for such a system was clear. A major land degradation episode in northeastern Queensland in the mid-1980s provided clear evidence that failure to reduce stock numbers during drought can damage the land and pasture resource. A survey across northern Australia in 1991 indicated widespread deterioration in pasture and land condition (Tothill and Gillies, 1992). An improved capacity for seasonal forecasting based on the ENSO phenomenon provided hope that the impact of future droughts could be reduced if appropriate action was taken in response to early warnings of drought. With the collaboration of several state agencies and funding support from Land and Water Australia, the Queensland Department of Natural Resources and Mines has developed a national modeling framework (figure 29.4) that is being used to monitor and forecast drought across Australia (AussieGRASS; Carter et al., 2000; http://insite.nrm.qld.gov.au/resourcenet/rsc/agrass).

Spatial Modeling Framework

The AussieGRASS spatial modeling framework (figure 29.4) allows agricultural simulation models to be run at a continental scale on a 5-km grid. The framework runs on a supercomputer and calculates daily outputs simultaneously across the continent. The framework is capable of efficiently running any daily time-step biological simulation model, provided the model is recoded to simultaneously operate across all pixels. A grazing system model, GRASP, (GRASs Production) is currently incorporated in the modeling framework to operationally monitor drought across the nation.

Pasture Model

The GRASP model was developed as a generic plant growth model and has been used to simulate growth of native pastures, sown pastures, and crops. The soil water budget is simulated using four layers (0–10 cm, 10–50 cm, 50–100 cm, and a deeper layer available only to trees). Daily calculations

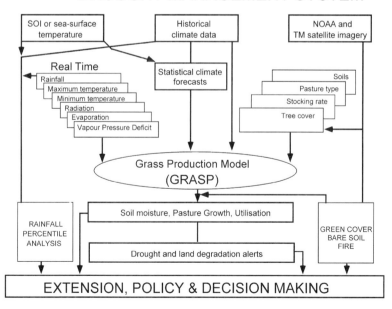

Figure 29.4 The AussieGRASS modeling framework for drought monitoring and alerts.

of runoff, drainage, soil evaporation, and transpiration are based on inputs of rainfall and pan evaporation. A daily plant-growth index is calculated from separate indices representing plant growth response to water stress (ratio of actual to potential transpiration), air temperature, vapor pressure deficit, solar radiation, and nitrogen availability. At low pasture cover, plant growth is calculated as a function of the plant growth index, plant density, and potential regrowth rate. As green cover increases, plant growth is calculated from a combination of temperature response, transpiration-efficiency, radiation-efficiency, and nitrogen limitations. For native pasture simulations, trees compete for water and nitrogen. Pasture biomass is calculated as the net result of pasture growth, detachment, and intake by grazing animals.

The GRASP model is calibrated against field data to obtain the main soil and plant parameters (McKeon et al., 1990). GRASP has been evaluated at a small plot and paddock scale for tropical pastures in northern Australia (over 100 locations, <29° S latitude; figure 29.1; Day et al., 1997) and temperate pastures in southern Australia (16 locations, >29° S latitude; figure 29.1; Tupper et al., 2001). Output from the spatial implementation of this model has been evaluated through broad-scale field surveys of pasture biomass (e.g., Carter et al., 2000; Hall et al., 2001) and, as described later in this chapter, against historical drought records.

Spatial Implementation of the Models

Rainfall and climate inputs are a generic requirement of all agronomic models. As the spatial framework runs daily time-step models, rainfall and climate data are input on individual 5-km grids (surfaces) for each day and climate element (Jeffrey et al., 2001). The surfaces are automatically created by interpolating records from individual recording stations, which are downloaded each day from the national climate database maintained by the NCC. Although rainfall is interpolated on a daily basis, the monthly rainfall totals are considered more accurate because they have been manually checked at BoM. Hence the daily rainfall surfaces are corrected such that they sum to equal the monthly surface as generated from the manually checked monthly rainfall totals. As a final check, the monthly rainfall surface is visually compared to the equivalent map on BoM's Web site. The framework also includes historical rainfall data from 1890 onward and daily climate data (temperature, humidity, and solar radiation) from 1957 onward (from 1970 for evaporation). Daily climate averages are being used for the 1890–1956 period until archival data from 1890 are added to the database.

Apart from the rainfall and climate inputs, the GRASP model also requires inputs of tree density, stock numbers, and specific parameters for different pasture types. These inputs are incorporated in the modeling framework as separate surfaces. Although important for determining the absolute amount of pasture production or biomass, these factors do not vary greatly from year to year and are therefore not critical for drought assessment purposes. Drought assessment is concerned more with relative measures of pasture or crop production (i.e., how production in the current season compares with historical production levels). Stored soil moisture, rainfall, air temperature, humidity, and evaporation vary considerably from year to year and contribute more to the year-to-year variation in pasture and crop production.

Drought Monitoring

Outputs from the GRASP model have been stored as monthly surfaces from 1890 onward and are updated each month. Various model outputs are available for drought assessment such as soil moisture, pasture growth, and pasture biomass. Pasture growth, ranked as a percentile against historical levels, is probably the most appropriate single index of drought in grazing lands because it is (1) a direct measure of rainfall effectiveness; (2) relatively insensitive to management practices except in the long term; and (3) highly correlated with carrying capacity of livestock.

Percentile rainfall and pasture growth maps are output on a monthly, seasonal, annual, and biennial basis. However, longer term maps may be required to assess protracted droughts. Twelve-month pasture growth percentiles are used as the operational basis for drought monitoring. A 12-month period is appropriate for analyzing drought in grazing lands because it corrects for the strong seasonality of rainfall and pasture growth in many parts of the country. For drought assessment purposes, the model output is aggregated to a district (e.g., shire) level. A threshold of pasture growth less than 10th percentile is adopted for triggering drought and a threshold of pasture growth more than the 30th percentile for breaking drought. It could be argued that the 30th percentile threshold for breaking drought is too low and that it takes an above-average season to break drought. While this is a common perception, a risk-averse manager is likely to gear normal stocking rates to a level that would be safe at least 70% of the time (i.e., to a level commensurate with 30th percentile pasture growth).

Based on the above criteria, the spatial model has been used to construct an historical time-sequence of drought in Queensland on a shire-by-shire basis (figure 29.5). This modeled time-series is in close agreement with the record of official droughts from the Queensland drought scheme, described earlier in this chapter. The overall close agreement between the two time series provides independent validation both of the AussieGRASS modeling framework and the criteria for monitoring drought. The close agreement also clearly dismisses any overall suggestion that official droughts in Queensland were declared too often in terms of frequency and duration (e.g., Daly, 1994). The major difference between the two time series occurred in the late 1960s and late 1980s, when the model calculates a higher proportion of land stricken by drought than evidenced by official drought declarations (figure 29.5). The far-north of the state is a region of major

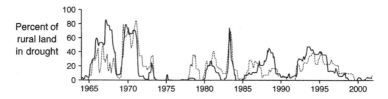

Figure 29.5 Percentage of rural holdings officially drought declared in Queensland and the percentage of land calculated to be in drought (line) based on thresholds of simulated pasture growth.

discrepancy in both periods. In this region pastures are of poor quality due to high rainfall and low soil nutrient status. As a result, cattle numbers are generally low in relation to the amount of forage available, and animal condition is influenced more by the length of the growing season than by the amount of pasture grown per se. Thus, a model based on pasture growth alone may not adequately represent drought for this region.

In the late 1960s, model calculations also indicated more droughts in far-western Queensland than were officially declared. In this region pasture growth is highly variable from one year to the next (coefficient of variation > 100%), and extremely low pasture growth is more the norm than the exception. Stocking rates are, in general, extremely light, and some larger companies have the capacity to move livestock in and out of the region as seasonal conditions dictate. Hence, industry is more adapted to drought in this region than in regions with more reliable pasture growth.

Drought and Pasture Condition Alerts

The GRASP model is run forward in time to estimate the probability of future drought. The probability is dependent both on current conditions (e.g., soil moisture) and the likelihood of future rainfall. Current conditions are determined by operationally running the model up to the current month. The model is then run forward for several months using rainfall and climate for each past year for which historical rainfall surfaces are available. A subset of historical years are objectively chosen as being analogous to the current year in terms of the Southern Oscillation Index (SOI) or global sea-surface temperatures (SSTs). The likelihood of drought is based on the proportion of model runs based on these analogue years for which projected pasture growth for a 12-month period is less than the 10th percentile. When combined with calculations of pasture utilization by livestock and kangaroos, the drought alert can be modified to provide pasture condition alerts as first envisaged by Pressland and McKeon (1990). The alert is triggered when high grazing pressure (i.e., >30% utilization of pasture grown over a 12-month period) is likely to occur during the periods of low pasture growth (i.e., less than the 30th percentile). Such circumstances are likely to cause losses of perennial grasses and pasture cover.

Currently the method for selecting analogue years is based on phases of the Southern Oscillation (Stone et al., 1996). Alternative approaches are also available, such as the SST scheme developed by BoM (Drosdowsky, 2002). Seasonal climate forecasting is a rapidly evolving field, and the spatial modeling framework is flexible enough to incorporate new statistical forecast schemes or downscaled outputs from global circulation models.

Use of Satellite Imagery

Satellite imagery was initially envisaged as a means of improving the spatial resolution of the modeling framework and of providing an independent drought assessment. However, the role of satellite imagery has fallen short of initial expectations, particularly in grazing lands, due to (1) the unreliability of the signal, (2) the short historical record against which to rank current conditions, (3) an inability to project forward from the current situation to provide warnings, (4) the tree cover confounding the pasture signal, and (5) the difficulty in distinguishing bare ground from dry pasture. Current studies are addressing such difficulties using Landsat data (e.g., Taube, 1999) and normalized difference vegetation index (NDVI) data (Carter et al., 2000). NDVI data are also used to monitor wheat yields (e.g., Smith et al., 1995), particularly in western Australia, jointly through Agriculture Western Australia (http://www.agric.wa.gov.au) and Department of Land Administration (http://www.dola.wa.gov.au).

Incorporation of Crop Models in the Modeling Framework

Apart from the GRASP model, the Agricultural Production Systems Simulator (APSIM) model has been tested within the national modeling framework for calculating district crop yields. APSIM is a detailed modeling framework developed for farm-scale simulations of a range of crop and farming systems. However, Hammer et al. (1996) found that for modeling wheat at regional scales, simpler approaches tailored to that the scale (e.g., the stress index [STIN] model, Stephens, 1998) were more accurate, robust, and easier to implement. STIN incorporates a daily soil water balance and calculates an accumulated crop moisture-stress index (SI) from a nominal sowing date. SI is sensitive both to moisture deficits and moisture excesses through the growing season. Inputs for the model include sowing date, daily rainfall, and average daily climate data (maximum and minimum temperature and solar radiation). SI is transformed to the district wheat yields through regression relationships between SI and historical shire yield data from the Australian Bureau of Statistics.

Yield is generally considered the best measure of drought in cropping lands (e.g., Stephens, 1998), and calculated district wheat yields from STIN are used as the basis for a drought alert issued by the Queensland Department of Primary Industries for Queensland wheat-growing shires (http://www.dpi.qld.gov.au/climate). Although alerts are only issued in

Queensland, calculations are made for all wheat-growing districts in Australia. The alert is based on the likelihood of shire yields falling below the 10th percentile. Projected yields are calculated using rainfall from analogue years based on phases of the SOI as described previously for pastures. The calculations are made in near-real time using climate and rainfall from the spatial modeling framework. Similar calculations based on STIN are also made by Agriculture Western Australia (http://www.agric.wa.gov.au/climate).

Operational Reports

An operational monthly report, "A Summary of Seasonal Conditions in Queensland," has been developed for grazing lands in northeastern Australia (Day and Paull, 2001; http://www.LongPaddock.qld.gov.au/AboutUs/Publications/ByAuthor/KenDay). This report combines a range of information including rainfall, pasture growth, remote sensing, and forecasts of rainfall and pasture growth into a four-page color leaflet. A similar product, "Regional Crop Outlook: Wheat," has now been developed for wheat in northeastern Australia (http://www.dpi.qld.gov.au/climate). Each of these products can be adapted to a national basis. Drought alerts for pastures and crops, as well as pasture condition alerts, have been incorporated into a more comprehensive prototype booklet (Day and Paull, 2001). A formal survey of the recipients of the seasonal conditions leaflet endorsed the usefulness of these reports. However, the production of these reports has been suspended due to the lack of demand and funding.

Conclusions

The impacts of drought in Australia have been considerable, particularly at local scales, and it is important to mitigate them. In the past, governments have responded both during and after drought occurrence, in most cases prompted by the applications for assistance from primary producers. However, government policies are now directed toward a more proactive response to drought to enhance the self-reliance of primary producers. The monitoring system, described in this chapter, has the capacity to focus attention on regions experiencing drought and those most likely to experience drought within a few months. What is required is to link the national-scale drought monitoring to a focused and timely campaign to encourage primary producers to prepare for drought forecasts. To this end, a variety of tools are available to assess management options at a farm scale.

It is important to continue to evaluate the accuracy of the modeling framework and, in particular, the drought monitoring and alert components. Apart from the real-time evaluation of the operational system, the most valuable evaluation will come through case studies of historical droughts. Currently not all states are financially committed to the ongoing development of the AussieGRASS alert system.

In Queensland, where most progress has been made to convert output from the spatial modeling framework to operational drought monitoring, it has been difficult to maintain support for drought alert products. Interest in monitoring has naturally been highest during dry periods, and it has proven difficult to maintain the same level of support during recent wet periods. In this regard it is important that the drought monitoring system does not fall victim to the very "drought hydro-illogical cycle" (Wilhite, 1993) it aims to address.

ACKNOWLEDGMENTS We thank David White, Bill Parton, Wayne Hall, Beverley Henry, and an anonymous reviewer for their advice and comments in the preparation of the manuscript. The assistance of David Ahrens, Belinda Cox, Grant Stone, and Michael Gutteridge in preparation of data and figures is also appreciated. The views expressed herein are those of the authors and do not represent a policy position of the Queensland Government or the Department of Natural Resources and Mines.

References

Bureau of Agricultural Economics. 1969. An economic survey of drought affected pastoral properties in New South Wales and Queensland 1964–1965 to 1965–1966. Report 15. Wool Economics Research, Bureau of Agricultural Economics, Canberra, Australia.

Bureau of Agricultural Economics. 1983. Overview as at January. Q. Rev. Rural Econ. 5:3–11.

Carter, J.O., W.B. Hall, K.D. Brook, G.M. McKeon, K.A. Day, and C.J. Paull. 2000. AussieGRASS: Australian grassland and rangeland assessment by spatial simulation. In: G.L. Hammer, N. Nicholls, and C. Mitchell (eds.), Applications of Seasonal Climate Forecasting in Agricultural and Natural Ecosystems—The Australian Experience. Kluwer Academic, Dordrecht, The Netherlands, pp. 329–349.

Daly, J.J. 1994. Wet as a shag, dry as a bone; drought in a variable climate. Information Ser. QI93028. Qld. Dept. Primary Industries, Brisbane, Australia.

Day, K.A., G.M. McKeon, and J.O. Carter. 1997. Evaluating the risks of pasture and land degradation in native pasture in Queensland. Project Report DAQ124A. Rural Industries Research and Development Corporation, Canberra, Australia.

Day, K.A., and C.J. Paull. 2001. Development and evaluation of seasonal conditions reports for Queensland. Rural Risks Strategy Unit Report, Qld. Dept. Primary Industries, Brisbane, Australia.

Drosdowsky, W., 2002. SST phases and Australian rainfall. Aust. Meteorol. Mag. 51:1–12.

Hall, W.B., D. Bruget, J.O. Carter, G.M. McKeon, J. Yee Yet, A. Peacock, R. Hassett, and K.D. Brook. 2001. Australian grassland and rangeland assessment by spatial simulation AussieGRASS. Project Report Land and Water Australia, Canberra.

Hammer, G.L., N. Nicholls, and C. Mitchell (eds.). 2000. Applications of Seasonal Climate Forecasting in Agricultural and Natural Ecosystems—The Australian Experience. Kluwer Academic, Dordrecht, The Netherlands.

Hammer, G.L., D. Stephens, and D. Butler. 1996. Development of a national drought alert strategic information system, vol. 6. Wheat modeling sub-project: Development of predictive models of wheat production. Project Report QP120. Land and Water Resources Research and Development Corporation, Canberra, Australia.

Jeffrey, S.J., J.O. Carter, K.B. Moodie, and A.R. Beswick. 2001. Using spatial interpolation to construct a comprehensive archive of Australian climate data. Environ. Model. Software 16:309–330.

Ker Conway, J. 1993. The Road from Coorain. William Heinemann Ltd., Melbourne, Australia.

McKeon, G.M., K.A. Day, S.M. Howden, J.J. Mott, D.M. Orr, W.J. Scattini, and E.J. Weston. 1990. Northern Australian savannas: management for pastoral production. J. Biogeogr. 17:355–372.

Nix, H.A., and E.A. Fitzpatrick. 1969. An index of crop water stress related to wheat and grain sorghum yields. Agric. Metereol. 6:321–327.

Power, S., T. Casey, C. Folland, A. Colman, and V. Mehta. 1999. Inter-decadal modulation of the impact of ENSO on Australia. Clim. Dynam. 15:319–324.

Pressland, A.T. and G.M. McKeon. 1990. Monitoring animal numbers and pasture condition for drought administration: an approach. In: Proceedings of the Australian Soil Conservation Conference, Perth, May 1990, Department of Agriculture, Canberra, Australia, pp. 17–27.

Purtill, A., M. Backhouse, A. Abey, and S. Davenport. 1983. A study of the drought. Q. Rev. Rural Econ. (suppl.) 5:3–11.

Ratcliffe, F.N. 1947. Flying Fox and Drifting Sand. Angus and Robertson, Sydney, Australia.

Smith, R.C.G., J. Adams, D.J. Stephens, and P.T. Hick. 1995. Forecasting wheat yield in a mediterranean-type environment from NOAA satellite. Aust. J. Agric. Res. 46:113–125.

Stafford Smith, D.M., and G.M. McKeon. 1998. Assessing the historical frequency of drought events on grazing properties in Australian rangelands. Agric. Syst. 57:271–299.

Stehlik, D., Gray, I., and G. Lawrence. 1999. Drought in the 1990s: Australian farm families' experiences. RIRDC Pub. no. 99/14. Rural Industries Research and Development Corporation, Canberra, Australia.

Stephens, D.J. 1998. Objective criteria for estimating the severity of drought in the wheat cropping areas of Australia. Agric. Syst. 57:333–350.

Stone, R.C., G.L. Hammer, and T. Marcussen. 1996. Prediction of global rainfall probabilities using phases of the Southern Oscillation Index. Nature 384:252–255.

Taube, C.A. 1999. Estimating ground cover in northern Australian rangelands using Landsat TM imagery. In: D. Elridge and D. Freudenberger (eds.), Proceedings of the Sixth International Rangelands Congress, Townsville, Australia, July 19–23, 1999, ITRC, Inc. pp. 757–758.

Tothill, J.C., and C. Gillies. 1992. The pasture lands of northern Australia. Trop. Grassl. Soc. Aust., Occasional Publication no. 5.

Tupper, G., J. Crichton, D. Alcock, and H. Mavi. 2001. The AussieGRASS high rainfall zone temperate pastures sub-project final report. Project Report. Land and Water Australia, Canberra.

White, D.H. 2000. Drought policy, monitoring and management in arid lands. Ann. Arid Zone 39:105–129.
White, D.H., and S.M. Howden. 1991. The emerging role of decision support systems in managing for drought on Australian farms. In: J. Stuth and B. Lyons (eds.), Decision Support Systems for Resource Management. Texas A&M University, College Station, pp. 129–132.
White, D.H., S.M. Howden, J.J. Walcott, and R.M. Cannon. 1998. A framework for estimating the extent and severity of drought, based on a grazing system in south-eastern Australia. Agric. Syst. 57:259–270.
Wilhite, D.A. 1993. Planning for drought: A methodology. In: D.A. Wilhite (ed.), Drought Assessment, Management, and Planning: Theory and Case Studies. Kluwer Academic, Boston, pp. 87–108.

CHAPTER THIRTY

Monitoring Agricultural Drought in South Korea

HI-RYONG BYUN, SUK-YOUNG HONG,
AND VIJENDRA K. BOKEN

South Korea (hereinafter referred to as Korea) lies in the middle latitudes of the Northern Hemisphere. Until the 1960s, Korea was a typical agrarian country, with agriculture generating roughly half of its gross national product (GNP) and employing more than half of the labor force. Agriculture still plays an important role in the Korean national economy, but it accounts for a relatively much lower share of the GNP (5.3% in 1997) and engages much less of the population (11.0%). The agricultural share of the national economy is declining continuously. Farms in Korea, as in many other Asian countries, have traditionally been small. Average farm size has been growing slowly from 0.86 ha in 1960 to 1.39 ha in 2001, despite a significant reduction in the average number of persons per household engaged in farming—from 6.20 persons to 2.91 persons. As a result, agriculture has become more intensive.

The country has four distinct seasons: summer, fall, winter, and spring. Summer and winter have a longer duration than spring or fall. The summer rainy season (*Changma*) in the Korean Peninsula includes the period from late June to late July. About three quarters of the annual precipitation falls during the summer season. The average annual precipitation in Korea is 1,274 mm, which is about 1.3 times the world average (973 mm). The variation in annual precipitation is larger, with an annual minimum of 784 mm and an annual maximum of 2675 mm in Seoul. Heavy rains fall during the *Changma* season, which is influenced by monsoons.

The National Institute of Agricultural Science and Technology (NIAST) classified Korea (except Jeju Island) into 19 climate zones (figure 30.1) to efficiently use agricultural resources for wetland rice production. Among the 19 zones, zone 14, which is the Southern Charyeong Plain, yields the best harvest and the most stable rice production. Zones 11 (Yeongnam Basin), 17 (the northeastern coast), and 18 (the mid-eastern coast) are

Figure 30.1 Nineteen climate zones of South Korea and seven climate zones of North Korea (Choi and Yun, 1989).

categorized as drought-risk areas at transplanting stage based on the ratio of evaporation to precipitation (Choi and Yun, 1989).

About 157,000 ha (8% of Korea's arable land) is damaged by climate disasters every year, costing about half a billion dollars for mitigation. Table 30.1 shows areas affected by severe droughts during the 1960s. The most severe drought occurred in 1968. Drought also occurred in 2001. In spring, even a small rain deficit can cause agricultural damage due to the high demand for water for rice transplanting. Every spring, there is at least some level of water shortage. However, the most severe drought generally occurs during summer.

Causes of Drought Occurrence

Drought occurs due to restricted irrigation, lack and uneven distribution of precipitation in any season, late onset of rainy season, and occurrence of dry spells. Agricultural water use is the largest component (45%) of total direct water use in Korea (MOCT, 2001). Sometimes agricultural water usage is curtailed to meet the urban and industrial water demands, and as a result drought may occur.

The pressure distribution of the marine tropical air mass, polar air mass, or continental air mass determines the amount of precipitation in summer season. When one of the air masses is excessively strong or weak such that the high pressure system is centered over the Sea of Ohkotsk and not over the northern Pacific Ocean, a drought begins (Park and Schubert, 1997).

Table 30.1 The main drought years during 1960s–90s

Year	Rainfall (mm) (May–July)	Rainy days		Rate of reservoir storage		No. of dried reservoirs	Drought area (ha)	Amount of damage (million $)	Budget for drought mitigation (million $)
		No.	Frequency	%	Frequency				
1967	307.4	56	7	5	25	—	402,547	522	5
1968	122.2	72	50	4	30	—	470,422	584	5
1981	658.2	50	5	46	2	5,306	145,547	181	43
1982	300.8	54	7	27	7	13,593	231,244	287	40
1994	231.3	68	30	15	15	6,728	113,300	—	57

Source: Kim (1995).

Low Precipitation during the Spring Season   The period from April 1 to May 15 is the spring rainy season (Tian and Yasunari, 1998; Byun and Lee, 2002). The warm and humid southerly airflow in the layer below the 700 hPa level combines with the upper-level, cold northerly current (that is still dominant in the layer over 500 hPa) and causes rains with intensified hydrostatic instability. However, sometimes, for reasons not adequately understood at present, the precipitation is low and evaporation is high in spite of the gradual rise in temperature and humidity. Recently Byun and Lee (2003) found that the low precipitation is caused partly due to the decrease in the meridional gradient of the adjacent sea-surface temperature. In March, the groundwater level rises due to the melting of snow and ice. However, this has only a small effect on the replenishment of water resources. As a result, drought occurs and hampers the process of germination and sprouting of seeds, thus damaging the crops.

Late Onset of the Summer Rainy Season   Influenced by the Asian monsoon, the summer rainy season in Korea begins in the second half of June and continues until the end of July. The precipitation during this period makes up more than 50% of the annual precipitation. Therefore, if the summer rainy season begins late, nonirrigated land is affected first, but occasionally irrigated farming also suffers. The main reason for a late onset of the summer rainy season is that the marine tropical air mass (mT; Oliver, 1987) called the Northern Pacific High becomes weak and develops late due to low heating of the Asian continent. Hence, the warm southerly flow does not reach Korea. For example, when snow is widespread on the Tibetan Plateau, it reflects more solar rays and retards the heating and melting of snow. Further, if there is a significant amount of ice in the Okhotsk Sea, a marine polar air mass (mP; Oliver, 1987) develops due to snow thawing in May and June. When the mP is strong, the frontal system that forms the rainy season is obstructed from proceeding northward, and there is a delay in onset of the rainy season (Byun et al., 1992a, 1992b).

Low Precipitation in Summer Rainy Season   Low precipitation during summer seasons is attributed not only to the absence or infrequence of low-pressure systems, but also to lower temperature, a lower moisture gradient, and the persistence of upper level high-pressure systems (Byun and Han, 1994; Byun, 1996).

Low Precipitation during Winter   Kaul-Changma (the early fall rainy season in Korea) begins about the middle of August and ends in the middle of September. This phenomenon is dominant in the southern half of the Korean Peninsula, and the precipitation during this period is stored in reservoirs or dams for irrigating crop in the following year. In this case, the precipitation is supplied mainly by the lows that pass near the Korean Peninsula. At times, the weakened tropical cyclones pass the peninsula and cause heavy rains. Also, an early-winter downpour occurs between late

October and mid-December and replenishes reservoirs. But such rains occur only once in 2.5 years (Jeong, 2002). Without heavy rains, shortage of water resources occurs in the following spring.

Drought Monitoring Methods

Drought Indices

The Korea Meteorological Administration (KMA) has 509 automated raingauge stations, which record precipitation on hourly or daily basis. Total precipitation from September to the next May and the mean and standard deviation of monthly precipitation from June to August, as well as monthly precipitation deviation from the long-term average are simple parameters used for drought monitoring. In addition, one-month forecasts of the temperature and precipitation are made and broadcasted every 10 days, and the outlook of temperature and precipitation for the next three and even six months is also announced.

The KMA determines the Palmer drought severity index (PDSI; Palmer, 1965; chapters 9 and 12). The Korean Institute of Construction Technology (KICT) also determines drought indices such as standard precipitation index (SPI; chapter 9), PDSI, and surface water supply index (SWSI). These indices are reported online (http://apply1.kma.go.kr/home/service/drought/dro_home.htm; http://www.kict.re.kr/wed/researches/wres/DroughtWeb/).

A drought index, EDI, developed by Byun and Wilhite (1999), is the standardized value of the available water resources index (AWRI; Byun and Lee, 2002), which is the accumulated precipitation with a time dependent

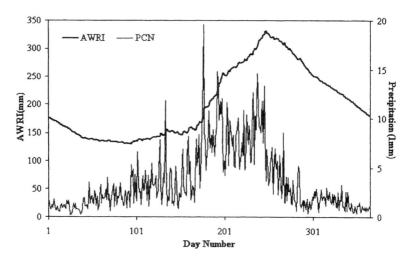

Figure 30.2 Daily precipitation and available water resources index (AWRI) averaged for 60 stations across Korea.

weighting function. The difference between AWRI and precipitation that was averaged for 60 stations from 1974 through 1998 is shown in figure 30.2. It can be seen in figure 30.2 that the AWRI curve shows the concentration of water resources better than the precipitation curve. These indices and software for computing these indices are currently being verified and are available on a Web site (http://atmos.pknu.ac.kr/~mdr).

Figure 30.3 shows the spatial distributions of four drought indices for August 31, 2001, when a severe drought occurred. The EDI is different from the PDSI, SPI, and other indices (McKee et al., 1993; Hayes et al., 1999) because it can precisely capture the daily variation in drought intensity. The most important characteristic of the EDI is that the EDI shows both drought duration and severity. An EDI of less than −0.7, between −0.8 and −1.5, and between −1.6 and −2.5, means mild, moderate, and severe

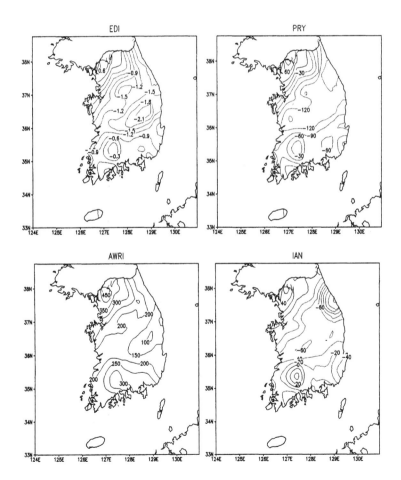

Figure 30.3 Four indices that show the drought severity on August 31, 2001. Detailed information on these indices are available online (http://atmos.pknu.ac.kr/ mdr) and in Byun and Wilhite (1999).

drought, respectively. A severe drought with EDI of −2.1 and AWRI of 100 mm is seen in the central area. Figure 30.4 shows the variation in drought days (EDI less than −1.5) for a 25-year period (1974–98). It is evident from figure 30.4 that severe droughts occurred in 1978, 1982, 1988, and 1994.

Empirical Orthogonal Function

Figure 30.5 shows the empirical orthogonal function (EOF) analysis of the EDI. The first eigen mode explained only 55.9% of the total variance. Most parts of the peninsula show a positive value, with a center over southern inland area. Its time series shows four negative peaks near pentad (5-day period) number 13, 26, 44, and 61 that coincide with the four drought periods in early March, middle May, early August, and October, as shown by Byun and Han (1994) and Byun and Lee (2002). The second eigen mode includes 13.5% of the total variance shown by east–west contours. The time series shows drought concentration at mid-September. The third eigen mode shows 7.9% of the total variance with the meridional contours. In the eastern part of the peninsula, with negative values in contours, the drought occurs from March till July, as shown by positive values in time series.

Growing Degree Days

The Rural Development Administration (RDA) provides agricultural weather information to public on a Web site (http://weather.rda.go.kr/). The daily, monthly, and yearly data in time and space are available for temperature, humidity, radiation, wind speed, precipitation, and so on.

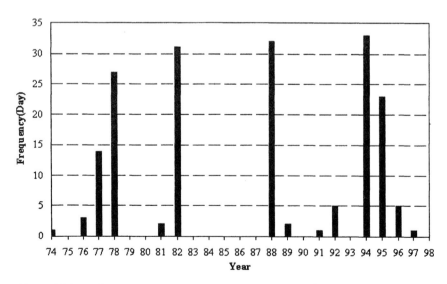

Figure 30.4 Yearly frequency, i.e. the number of drought days, of drought index (EDI < −1.5) based on data for 60 weather stations across Korea.

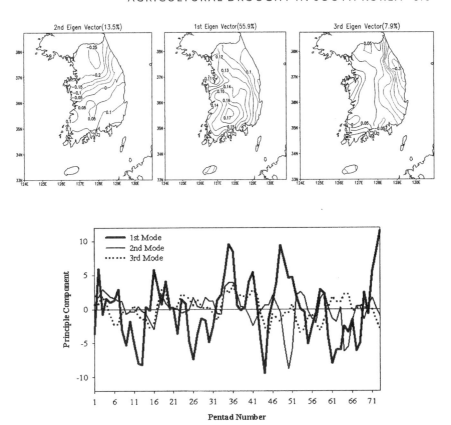

Figure 30.5 Spatial structure of the first (55.9%) and second (13.5%) eigen vectors of 25-year mean drought index and the time series of the eigen vectors.

A drought index based on growing degree day (GDD) and soil moisture can be computed by either the Penman-Monteith equation or the Priestly-Taylor equation (Priestly and Taylor, 1972) for different soil types.

Use of Satellite Data

Combining the drought index with the vegetation index derived from satellite data can be a practical approach for agricultural drought monitoring. Normalized difference vegetation index (NDVI) composites with half-degree spatial resolution over the growing season in Korea (Szilagyi, 2002) were used to estimate areal evapotranspiration (AET) across Korea using the biosphere model (SiB2). Soil type, canopy structure, and phenology were taken into account to estimate the AET.

NDVI maps at a 1.1 km × 1.1 km grid interval were derived from the Advanced Very High Resolution Radiometer (AVHRR) onboard the U.S. National Oceanic and Atmospheric Administration's (NOAA) polar-orbiting satellites to monitor vegetation conditions in Korea in 1999 (figure 30.6). These maps were used to classify the land cover of rice at a 1-km grid spacing for land surface parameterization of the biosphere model (Koo et al., 2001). Ha et al. (2001) analyzed the temporal variability in the NDVI, leaf area index (LAI), and surface temperature (Ts) estimated from AVHRR data collected from Korean Peninsula during 1981–1994. These products can be applied to estimate AET (Szilagyi, 2002):

$$\text{AET}_{\text{est}} = \text{NDVI}(\sigma_{\text{gs}}) + \bar{E}_{\text{gs}} \quad [30.1]$$

where AET_{est} is the estimated AET, σ_{gs} is the standard deviation of monthly AET during the growing season (mm/day), and $\bar{E}_{\text{gs}}$ is mean growing season AET. The estimated AET can be used for monitoring drought conditions.

Standardized Vegetation Index

A standardized vegetation index (SVI) based on calculation of a z score of NDVI distribution can also be produced for drought monitoring, as reported by Peters et al. (2002):

$$z_{ijk} = \left(\text{NDVI}_{ijk} - w\text{NDVI}_{ij}\right)/\sigma_{ij} \quad [30.2]$$

where z_{ijk} is z value for pixel i during week j for year k, $z_{ijk} \sim N(0,1)$, NDVI_{ijk} is weekly NDVI value for pixel i during week j for year k, $w\text{NDVI}_{ij}$ is the mean NDVI for pixel i during week j over n years, and σ_{ij} is the standard deviation of pixel i during week j over n years. This per-pixel proba-

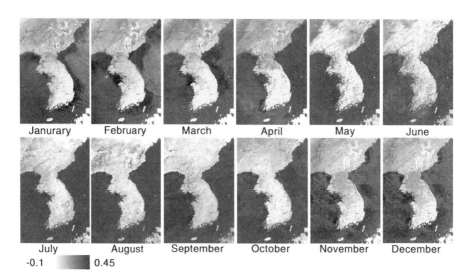

Figure 30.6 Monthly maximum-value-composite normalized difference vegetation indices during 1999 for Korea. The values in the legend show NDVI values. The higher the NDVI values, the better the vegetation vigor.

bility, expressed as SVI, is an estimate of the probability of occurrence of the present vegetation condition (0 < SVI < 1).

$$\text{SVI} = P(Z < z_{ijk}) \qquad [30.3]$$

SVI values were grouped by Peters et al. (2002) into five classes: very poor (0–0.05), poor (0.05–0.25), average (0.25–0.75), good (0.75–0.95), and very good (0.95–1), and were related to drought conditions. The SVI is a good indicator of vegetation responses to short-term weather conditions. High spatial resolution (1 km) and potential for near-real-time evaluation of actual vegetation conditions are the advantages of the SVI. But understanding the strengths and weakness of the SVI is important for determining when and how to use the index because climate conditions other than drought can also cause reduced vigor of vegetation.

In practice, the response of vegetation to precipitation can be examined using NDVI–precipitation relationships. Surface soil wetness derived from satellite images can be related to soil moisture for monitoring local drought conditions.

Drought Mitigation

In the long run, soil and water conservation practices can prepare for and mitigate droughts. NIAST carried out research on multiple functions of paddy farming according to agricultural production conditions to support agricultural issues such as climate change, natural disasters, farm product disaster insurance, and direct payments provided by the Ministry of Agriculture and Forestry of Korea.

The NIAST has quantified the positive impacts of paddy farming, including flood prevention, enhancement of water resources, air and water purification, and soil erosion alleviation (Seo et al., 2001). Negative impacts include the emission of greenhouse gases, nitrogen leaching, and the amount of agricultural water use (Seong et al., 2001). Crop simulation models based on weather data can be used to predict irrigation schedules to aid drought mitigation.

The main activity to mitigate drought in Korea is to compensate for drought damages. First, the severity of drought is assessed by the CACM (Committee of the Agricultural Counter Measure) on the basis of monetary damages. The drought index and precipitation are not legal references. When the funds required for restoring a drought-affected area exceed U.S. $2.5 million for a county, the committee confirms the drought occurrence and recommends that the compensation be provided by the government. Half of the total funds are used to improve water resources by purchasing pumps and other equipments and to pay for electricity and gas, and to dig wells. In addition, the government is obliged to provide aid to those who suffer, to reduce their families' school expenses, reduce the interest rate of official loans, and to provide stored rice for food, under the Counter Measure Law of Agricultural Disaster.

Drought Research Needs

An improved weather information network is a priority for drought monitoring and for developing early warning systems in Korea. Cooperation between neighboring countries for sharing weather information will be required for drought monitoring on a real-time basis.

Process-based simulation models can be used to better understand climate change and its impacts on crop production. Geographic information systems and remote sensing and spatial analysis techniques and models dealing with soil–crop–atmospheric interactions need to be further exploited to improve drought monitoring.

Conclusions

In Korea, drought has always been the most severe natural disaster. Rapid progress has been made to understand this climatically complex phenomenon by developing various drought indices. Cooperative studies between governmental institutions and universities will stimulate efforts to discover the causes of drought. At the national level, the Korea Meteorological Administration has an important role to play by providing quality weather data and drought information. RDA has the important role of studying agricultural droughts and associated soil and crop environments to build a solid research background and provide decision-support information to administrators. Dissemination of drought information to its users will improve agricultural drought studies.

ACKNOWLEDGMENTS Dr. Kennenth A. Sudduth of USDA-ARS, Columbia, Missouri, read an earlier draft of this chapter and offered valuable comments.

References

Byun, H.R. 1996. On the atmospheric circulation that causes the drought in Korea. J. Kor. Meteorol. Soc. 32:455–469.

Byun, H.R., and Y.H. Han. 1994. On the drought on each season in Korea. J. Kor. Meteorol. Soc. 30:457–467.

Byun, H.R., C.H. Jung, and D.K. Lee. 1992a. On the atmospheric circulation in the dry period before Chang-ma. Part I. Existence and characteristics. J. Kor. Meteorol. Soc. 28:71–88.

Byun, H.R., C.H. Jung, and D.K. Lee. 1992b. On the atmospheric circulation in the dry period before Chang-ma. Part II. Comparison with before and after. J. Kor. Meteorol. Soc. 28:89–107.

Byun, H.R., and D.K. Lee. 2002. Defining three rainy seasons and the hydrological summer monsoon in Korea using available water resources index. J. Meteorol. Soc. Jpn. 80:33–44.

Byun, H.R., and S.M. Lee. 2003. The characteristics and predictability of the May drought in Korea. Journal of Climate.

Byun, H.R., and D.A. Wilhite. 1999. Objective quantification of drought severity and duration. J. Climate 12:2747–2756.

Choi, D.H., and S.H. Yun. 1989. Agroclimatic zone and characters of the area subject to climatic disaster in Korea. The Korean Journal of Crop Science 34:13–33.

Ha, K.J., H.M. Oh, and K.Y. Kim. 2001. Inter-annual and intra-annual variabilities of NDVI, LAI, and Ts estimated by AVHRR in Korea. Kor. J. Remote Sens. 17:111–119.

Hayes, M.J., M.D. Svaboda, D.A. Wilhite, and O.V. Vanyarkho. 1999. Monitoring the 1996 drought using the Standardized Precipitation Index. Bull. Am. Meteorol. Soc. 80:429–438.

Jeong, H.K. 2002. Winter heavy rains in Korea. MS. thesis. Pukyong National University, Pakyong, Korea.

Kim, H.Y. 1995. Definitions and Analysis of characteristics of Korean drought. Journal of Korean Society of Civil Engineers 43:23–30.

Koo, J.M., S.Y. Hong, and J.I. Yun. 2001. A sample method for classifying land cover of rice paddy at a 1 km grid spacing using NOAA-AVHRR data. Kor. J. Agric. Forest Meteorol. 3:215–219.

McKee, T B., N.J. Doesken, and J. Kleist. 1993. The relationship of drought frequency and duration to time scale. In: Proceedings of the of Applied Climatology 8th Conference, Anaheim, CA. American Meteorological Society, Boston, MA, pp. 179–184.

MOCT, 2001. WaterVision 2020. Ministry of Construction and Transportation, Seoul, South Korea, p. 32.

Oliver, J.E. 1987. The Encyclopedia of Climatology. Van Nostrand Reinhold, New York.

Palmer, W.C. 1965. Meteorological drought. US. Weather Bureau Technical Paper 45. Washington, DC.

Park, C.-K., and S.D. Schubert, 1997. On the nature of the 1994 East Asian summer drought. J. Climate 10:1056–1070.

Peters, A.J., A.E.L. Walter-Shea, L. Ji. A. Viña, M. Hayes, and M.D. Svoboda. 2002. Drought monitoring with NDVI-based standardized vegetation index. Photogram. Eng. Remote Sens. 68:71–75.

Priestly C.H.B., and R.J. Taylor. 1972. On the assessment of surface heat flux and evaporation using large-scale parameters. Month. Weather Rev. 100:81–92.

Seo, M.C., K.K. Kang, H.B. Yoon, and K.C. Eom. 2001. Assessment of positive function of paddy farming according to productive environmental conditions. NIAST report. National Institute of Science and Technology, Suwon, South Korea.

Seong, K.S., M.C. Seo, K.K. Kang, and H.B. Yun. 2001. Assessment of negative function of paddy farming according to agricultural production conditions. NIAST report. National Institute of Science and Technology, Suwon, South Korea.

Szilagyi, J. 2002. Vegetation indices to aid areal evapotranspiration estimations. J. Hydrol. Eng. 7:368–372.

Tian, S.F., and T. Yasunari. 1998. Climatological aspects and of spring persistent rain over central China. J. Meteorol. Soc. Jpn. 76:57–71.

PART VII

INTERNATIONAL EFFORTS AND CLIMATE CHANGE

CHAPTER THIRTY-ONE

World Meteorological Organization and Agricultural Droughts

MANNAVA V. K. SIVAKUMAR

Adaptation strategies for coping with agricultural droughts must be based on a better understanding of the climatic conditions of the location or region under consideration. As the United Nations' specialized agency with responsibility for meteorology and operational hydrology, the World Meteorological Organization (WMO), since its inception, has been addressing the issue of agricultural droughts. This chapter presents a short overview of the various initiatives undertaken by WMO in this respect.

WMO's Activities in Support of the Combat against Drought

The fight against drought receives a high priority in the long-term plan of WMO, particularly under the Agricultural Meteorology Programme, the Hydrology and Water Resources Programme, and the Technical Cooperation Programme. WMO actively involves the National Meteorological and Hydrological Services (NMHSs), regional and subregional meteorological centers, and other bodies in the improvement of hydrological and meteorological networks for systematic observation, exchange and analysis of data for better monitoring of droughts, and use of medium- and long-range weather forecasts, and assists in the transfer of knowledge and technology. Following is a brief description of various activities undertaken by WMO in the combat against drought.

Research

WMO has been in the forefront of research on interactions of climate, drought, and desertification from its beginnings in the mid-1970s, when it was suggested that human activities in drylands could alter surface features that would lead to an intensification of desertification processes and trends.

The "expanding Sahara" hypothesis paved the way to the description of a biogeophysical feedback mechanism (Charney et al., 1975), proven later by using the numerical climate model of the Goddard Institute for Space Studies (Hansen et al., 1988). Various research programs are being undertaken to study the weather and climate of arid, semiarid, subhumid and other desert-prone areas with a view to predict long-term trends in the general circulation, different rain-producing atmospheric disturbances, and meteorological droughts using statistical and dynamic methods.

Human-induced changes in dryland surface conditions and atmospheric composition can certainly have an impact on local and regional climate conditions because they directly affect the energy budget of the surface and the overlying atmospheric column. These changes to the energy balance have been simulated in many numerical modeling studies covering almost all dryland areas of the world. The outcomes of these studies underscore the need to improve our knowledge of the climate–desertification relationship and, at the same time, call for improvements in the quantity and quality of the data available for further simulations. Accordingly, the WMO World Climate Research Programme (WCRP) launched the Global Energy and Water Cycle Experiment (GEWEX; www.gewex.org) to study the atmospheric and thermodynamic processes that determine the global hydrological cycle and energy budget and their adjustments to global changes, such as an increase in greenhouse gases. Many of the fundamental issues associated with interactions of desertification and climate are receiving attention in the GEWEX objectives.

Empirical studies that had begun with the analysis of historical records received a considerable thrust from improvements in remote-sensing observation and measurement technologies. The International Satellite Land Surface Climatology Project (www.gewex.org/islscp.html) has substantially improved the amount and quality of data available for use in empirical and numerical modeling studies.

WMO is strengthening and intensifying the research on the interactions between climate and desertification. The urgent need to predict interannual climate variations is impelled by the socioeconomic upheavals that have occurred, especially in Africa, over the past few decades. Statistical forecasts of seasonal rainfall up to 3 months in advance are currently being made, but through a coordinated research effort significant improvements in reliability and lead-time could be expected (Cane and Arkin, 2000). Equipped with an improved understanding of the physical mechanisms that govern climate and with more reliable predictive models, countries would be in a far better position to predict the onset of drought and hence mitigate its devastating consequences. Strategic plans could then be developed for capitalizing on the extended predictive capabilities and for converting the forecasts to management decisions that will optimize the use of existing resources.

WMO improves climate prediction capability through the Climate Variability Project (CLIVAR; www.clivar.org) of the WCRP. The prediction of El Niño and associated impacts are becoming possible, with reasonable skill, within time spans ranging from seasons to over one year. The objectives of CLIVAR are first to describe and understand the physical processes responsible for climate variability and predictability on seasonal, interannual, decadal, and centennial time scales through the collection and analysis of observations and the development and application of models of the coupled climate system. The second objective is to extend the record of climate variability over the time scales of interest through the assembly of quality-controlled paleoclimatic and instrumental data sets. A third objective is to understand and predict the response of the climate system to increases of radiatively active gases and aerosols and to compare these predictions to the observed climate record to detect the anthropogenic modification of the natural climate signal. CLIVAR emphasizes the following research themes related to drought: (1) establishing the limits of predictability, taking a regional approach and giving special attention to global ENSO (El Niño/Southern Oscillation; chapter 3) response areas and monsoon circulation systems; (2) assessing the results of GCM (Global Circulation Model) coupled model runs to assess their ability to reproduce the patterns of spatial and temporal variability of rainfall; (3) studying drought- and flood-inducing processes; (4) investigating the predictive capability of SST (sea-surface temperature) and ocean–atmosphere coupling processes in the tropical Atlantic and Indian Oceans using the proposed observational arrays; (5) improving the understanding of tropical mid-latitude interactions and their impacts on predictability for northern and southern Africa; (6) investigating land–sea–air interaction processes using remote and in situ observations and conducting numerical modeling experiments to establish feedback processes; and (7) developing innovative, low-cost solutions to land-based observational and communication needs and to the construction of data bases, identifying centers of existing expertise, and developing assistance programs and projects to ensure the potential for a local base of ongoing research.

Global Climate Change and Dryland Climate

Many scientists and policy-makers believe that climate changes in drylands over the next century will be driven in part by the continued build-up of anthropogenic greenhouse gases. Increases in the global atmospheric concentration of carbon dioxide, methane, nitrous oxides, various halocarbons and other greenhouse gases are well documented, and the upward trends for many of these trace gas concentrations are expected to continue well into the next century. Concern about climate has reached communities in all parts of the globe. As communities have adapted to their local climate, they are sensitive to its variations, and many are threatened by climate change.

The WMO/UNEP Intergovernmental Panel on Climate Change (IPCC) assessments have shown that the observed increases in atmospheric greenhouse gases may lead to global warming, sea-level rise, and space-time changes in the normal patterns of hydrometeorological parameters. Based on evidence from climate models, together with observations from instrumental and other available records, the IPCC (2001) Third Assessment Report concluded that "there is new and stronger evidence that most of the warming observed over the last 50 years is attributable to human activities." Human activities will continue to change atmospheric composition throughout the 21st century. By 2100, the concentration of CO_2 is expected to increase from the present figure of 370 ppm to 540–970 ppm. Over the period from 1990 to 2100, it is projected that the average global air temperature will increase by 1.4 to 5.8°C, and global mean sea-level will rise by 9 to 88 cm. It is reported in the IPCC (2001) assessments that recent regional climate changes, particularly temperature increases, have already affected many physical and biological systems.

As the earth warms, models project that arid and semiarid land areas in southern Africa, the Middle East, southern Europe, and Australia will become even more water stressed than they are today, causing a decrease in agricultural production in many tropical and subtropical areas. Indeed, the year 2000 was the 22nd consecutive year with a global mean surface temperature above the 1961–91 normal temperature and was the 7th warmest year in the past 140 years, despite the persistent cooling influence of the 1997–98 La Niña event.

The predicted increase in temperature would most probably have the effect of increasing potential evapotranspiration rates in the drylands, and in the absence of any large increases in precipitation, many drylands are predicted to become even more arid in the next century. Desertification is more likely to become irreversible if the environment becomes drier and the soil becomes further degraded through erosion and compaction.

Based on experimental research, crop yield responses to climate change vary widely, depending on species and cultivars, soil properties, pests and pathogens, the direct effects of carbon dioxide on plants, and the interactions between carbon dioxide, air temperature, water stress, mineral nutrition, air quality, and adaptive responses. Those countries with the least resources have the least capacity to adapt to climate change and are the most vulnerable. For example, the adaptive capacity of societies in Africa is low due to lack of economic resources and technology, and vulnerability is high as a result of heavy reliance on rainfed agriculture, frequent droughts and floods, and poverty. Grain yields are projected to decrease, diminishing food security, particularly in small food—importing countries. Desertification would be exacerbated by reductions in average annual rainfall, runoff, and soil moisture. Significant extinctions of plants and animal species are projected and would impact rural livelihood.

Early Warning Systems for Drought

At a global level, the World Weather Watch (www.wmo.ch/web/www/www.html) and Hydrology and Water Resources programs, which are coordinated by WMO, provide a solid operational framework on which to build improved warning capacity. The Global Observing System, the analytical capability of the Global Data Processing System, and the ability to disseminate warnings through the Global Telecommunications Systems form the basis of early warning for meteorological- and hydrological-related phenomena. The provision of meteorological and hydrological support to early warning is perhaps the most fundamental aspect of services supplied by NMHS, and it contributes to all four phases of early warning systems: mitigation or prevention, preparedness, response, and recovery. The application of climatological and hydrological knowledge to assess risk and to plan land use contributes to disaster mitigation. The classical role of providing forecast and warnings of severe weather, extreme temperatures, and droughts or floods contributes to preparedness. Updated warnings, forecasts, observations, and consultation with emergency and relief agencies contribute to the response phase. Finally, special forecast and other advice assist recovery operations.

Over the last several decades, major advances have been made in weather forecasting, that is, up to about a week for mid-latitudes, but the understanding of El Niño represents the first major breakthrough in the prediction of the longer term climate.

Climate Information and Prediction Services

Prediction on seasonal and longer time scales, as a basis for warning, is a developing community service. WMO believes that the progress being made in climate research, where appropriate, should be translated into forms that will elevate the socioeconomic well-being of humanity. In 1995 WMO Twelfth Meteorological Congress (WMO, 1995) endorsed the Climate Information and Prediction Services (CLIPS) project to build on research achievements to improve economic and social decisions. WMO has been coordinating or has organized research liaison between the various interested groups through the co-sponsorship of workshops, seminars, and conferences, and through the organization of special training events on El Niño.

In collaboration with partner institutions, WMO has organized several Regional Climate Outlook Fora (RCOF; www.wmo.ch/web/wcp/clips 2001/html/index.html) in many parts of the world. The forums have enabled researchers from various advanced climate prediction centers and regional operational climate centers and experts from NMHSs, to develop consensus or consolidated climate outlook guidance products, together with guidance on interpretation, for dissemination to users. The RCOF proved to be a very effective capacity-building mechanism for the NMHSs in terms of the transfer of knowledge of the current state of development and limitations of seasonal climate prediction science. The forums were therefore used to enhance the regional and climate outlooks and associated impact projections during El Niño events.

Some examples to illustrate the use of El Niño predictions to provide early warnings include: (1) the case of Peru, where the El Niño information has been used for sustainable agricultural production through the alternation of crops (e.g., rice and cotton) during dry years; and (2) in Brazil, agricultural production was enhanced when El Niño information was used in making specific agricultural decisions in 1992 and 1987 when the El Niño event took place.

An important component of the CLIPS project is in the area of training and technology transfer. The objective is to ensure that NMHS have access to the regional and global climate monitoring products (www.wmo.ch/web/wcp/clips2001/html/index.html) that are routinely generated in support of WRCP and that NMHS staff has the necessary training to provide services to aid community decision-making. The CLIPS project is closely linked to the WMO Agricultural Meteorology and Hydrology and Water Resources programs to build on existing linkages with community managers in the land and water resource areas.

Agrometeorological Applications

Major meteorological events such as extended droughts have made high-level government planners in many countries aware of the importance of timely and practical agrometeorological information. Drought monitoring, forecasting, and control have been receiving a high priority in many agencies. The purpose of the Agricultural Meteorology Programme (AGMP) of WMO is to support food and agricultural production and activities. The program assists members by providing meteorological and related services to the agricultural community to help develop sustainable and economically viable agricultural systems, improve production and quality, reduce losses and risks, decrease costs, increase efficiency in the use of water, labor, and energy, and conserve natural resources and decrease pollution by agricultural chemicals or other agents that contribute to the degradation of the environment.

AGMP accords a high priority to implementing activities that combat desertification and to instituting preparedness, management, response, and remedial action against the adverse effects of drought. Since the 1970s, the Commission for Agricultural Meteorology (CAgM; www.wmo.ch/web/wcp/agm/CAgM/CAgMmenu.htm) of WMO has been active in addressing the issue of agricultural drought and made recommendations regarding the role of agrometeorology to help solve drought problems in drought-stricken areas, particularly in Africa. The commission appointed a number of working groups and rapporteurs with expertise relating to drought such as meteorological aspects of drought processes, indices for assessment of droughts, drought probability maps, operational use of agrometeorology for crop and pasture production in drought-prone areas, agrometeorological inputs and measures to alleviate the effects of droughts, assessment of the economic impacts of droughts, and so on. Based on the activities of these working groups and rapporteurs, a number of reports were published and distributed by WMO (www.wmo.ch/web/wcp/agm/CAgM/CAgM menu.htm).

In view of the importance of providing a better understanding of the interactions among climate, drought, and desertification, WMO and the United Nations Environment Programme (UNEP) co-sponsored a study on this subject, and the results were published in *Interactions of Desertification and Climate* (Williams and Balling, 1996). This book helped clarify some of the key issues in the debate on the contributions of climate and anthropogenic activities to desertification. In seeking to understand and quantify the interaction between desertification and climate, this study included a comprehensive analysis of human impact on surface and atmospheric conditions in drylands, the impact of human activities on climate, the impact of climate on soils and vegetation, the impact of climate on the hydrological cycle, human land use in drylands and the influence of climate, linkages between interannual climatic variations in drylands and the global climate system, and future climate changes in drylands and mitigation. Finally, some rehabilitation strategies to combat desertification were recommended as well as an urgent need to establish accurate baseline data relating to dryland degradation using consistent and uniform methods and criteria.

A number of capacity-building activities are also undertaken by AGMP regarding agricultural droughts. Some of the recent activities include:

- WMO/UNSO Training Seminar/Workshop on Drought Preparedness and Management, Banjul, Gambia (September 1995)
- Workshop/Training Seminar on Drought Preparedness and Management for Northern African Countries, Casablanca, Morocco (June 1996)
- Workshop on Drought and Desertification, Bet Dagan, Israel (May 1997)
- Roving Seminar on Agrometeorology related to Extreme Events, Pune, India (April–May 1997)
- Roving Seminars on the Application of Climatic Data for Drought Preparedness and Management of Sustainable Agriculture, Accra, Ghana (November 1999), Beijing, China (May 2001)
- International Workshop co-sponsored by UNDP/UNSO and WMO on Coping with Drought in sub-Saharan Africa: Best Use of Climate Information, Kadoma, Zimbabwe (October 1999)
- Expert Group Meeting on Wind Erosion in Africa, Cairo Egypt (April 1997)
- Expert Group Meeting on Early Warning Systems for Drought Preparedness and Drought Management, Lisbon, Portugal (September 2000).

Proceedings of some of these workshops were published (Sivakumar et al., 1998; UNDP/UNSO and WMO, 2000; Wilhite et al., 2000; WMO, 1997, 1998, 2000; WMO and UNDP/UNSO, 1997).

World Hydrological Cycle Observing System

The World Hydrological Cycle Observing System (WHYCOS) was developed to improve understanding of the status of water resources at a global scale, which is critical to monitoring agricultural droughts. Composed of regional systems implemented by cooperating nations, WHYCOS will complement national efforts to provide information required for wise water resource management. Modeled on WMO's World Weather Watch and using the same information and telecommunications technology, WHYCOS will provide a vehicle not only for disseminating high-quality information, but also for promoting international collaboration. It will build the capacity of National Hydrological Services so that they are ready to face the demands of the 21st century. It will provide a means for the international community to monitor water resources more accurately at the global level and to understand the global hydrological cycle.

Support to Regional Institutions

In 1972, the worsening drought situation in the Sahelian Zone prompted WMO to appoint a consultant to conduct a survey in the area, followed by a group mission of experts, supported by UNDP, to study and recommend further action to strengthen the Meteorological, Agrometeorological and Hydrometeorological Services in the countries concerned. Recommendations of this group led to the formation of the Regional Centre for Agricultural Meteorology and Hydrology (AGRHYMET) in Niamey, Niger.

In the late 1970s and 1980s, droughts caused widespread famine and economic hardships in many countries in eastern and southern Africa. At the request of 24 countries in these regions, in 1989 WMO established two Drought Monitoring Centres (DMCs) in Nairobi, Kenya (chapter 18), and Harare, Zimbabwe, with support from UNDP. WMO, along with the Economic Commission for Africa, sponsored the establishment of the African Centre of Meteorological Applications for Development (ACMAD) in Niamey, Niger, in 1993. ACMAD is the focal point in fostering regional cooperation among the 53 African countries with the rest of the world in climate and environmental concerns with regard to sustainable social and economic development. The center coordinates the activities of the NMHSs of these countries in applying meteorological and hydrological information to important social and economic sectors, such as food production, water resource management, cash crops, pest control, and health. In addition, ACMAD provides operational products such as forecasts and early warnings to extreme climatic occurrences like drought and floods and assists in reducing their adverse impacts in Africa.

Given the importance of the interactions between climate and desertification, WMO has accorded a major priority to this area, and its action plan to combat desertification was first adopted in 1978 at the 30th session of the executive council of WMO and has gone through several revisions.

Priority areas for attention include enhanced observing systems at national, regional, and international levels; promoting the mitigation of the effects of drought and desertification through effective early warning systems; supporting and strengthening the capabilities of members and regional institutions and improved applications of meteorological and hydrological data.

WMO and the International Strategy for Disaster Reduction

The Inter-Agency Task Force for Disaster Reduction (IATF) is one of the most important new institutional arrangements established under the International Strategy for Disaster Reduction (ISDR). According to United Nations General Assembly Resolution 54/219, the IATF is "the main forum within the United Nations for continued and concerted emphasis on natural disaster reduction." WMO chairs the Working Group I on "climate and disasters" of IATF and selected drought as the major topic to improve synergy between the four ISDR working groups. A number of meetings have already been held to develop a common strategy to address the drought issues.

Conclusions

Drought is an insidious natural hazard that is a normal part of the climate of virtually all regions. It should not be viewed as merely a physical phenomenon. Rather, drought is the result of interplay between a natural event and the demand placed on water supply by human societies. It is important to build awareness of drought as a normal part of climate. Improved understanding of the different types of drought and the need for multiple definitions and climatic/water supply indices that are appropriate to various sectors, applications, and regions is a critical part of this awareness-building process. As with other natural hazards, drought has both physical and social component. It is the social factors, in combination with our exposure, that determine risk to society. It is obvious that well-conceived policies, preparedness plans, and mitigation programs can greatly reduce the vulnerability of the farming community and the risks associated with drought. Thus, planners and other decision-makers must be convinced that drought mitigation strategies are more cost effective than post-impact assistance or relief programs. It seems clear that investments in preparedness and mitigation will pay large dividends in reducing the impacts of drought. WMO has a major role to play in drought preparedness and mitigation strategies through its leadership role in early warning systems and preparedness strategies. The strong foundations laid by WMO through its research, applications, and capacity-building initiatives around the world are showing impressive results in dealing with events such as the 1997–98 El Niño. WMO stands ready to assist the agricultural community in developing strategies to coping effectively with agricultural droughts.

References

Cane, M.A., and P.A. Arkin. 2000. Current capabilities in long-term weather forecasting for agricultural purposes. In: M.V.K. Sivakumar (ed.), Climate Prediction and Agriculture. Proceedings of the START/WMO International Workshop, Geneva, September 27–29, 1999. Internacional STARA Secretariat, Washington, DC, pp. 13–37.

Charney, J.G., P.H. Stone, and W.J. Quirk. 1975. Drought in the Sahara: A biogeophysical feedback mechanism. Science 187:434–435.

Hansen, J., I. Fung, A. Lacis, S. Lebedeff, D. Rind, R. Ruedy, G. Russell, and P. Stone. 1988. Global climate change as forecast by the GISS 3-D model. J. Geophys. Res. 93:9341–9364.

IPCC. 2001. Climate Change 2001: The Scientific Basis. Contribution of Working Group I to the Third Assessment Report of the Intergovernmental Panel on Climate Change (J.T. Houghton, Y. Ding, D.J. Griggs, M. Noguer, P.J. van der Linden, X. Dai, K. Maskell, and C.A. Johnson, eds.). Cambridge University Press, Cambridge.

Sivakumar, M.V.K., M.A. Zobisch, S. Koala, and T. Maukonen. (eds.). 1998. Wind Erosion in Africa and West Asia: Problems and Control Strategies. Proceedings of the ICARDA/ICRISAT/UNEP/WMO Expert Group Meeting, 22–25 April 1997, Cairo, Egypt, International Center for Agricultural Research in the Dry Areas (ICARDA), Aleppo, Syria.

UNDP/UNSO and WMO. 2000. Coping with Drought in Sub-Saharan Africa: Better Use of Climate Information. World Meteorological Organization, Geneva.

Wilhite, D.A., M.V.K. Sivakumar, and D.A. Wood. 2000 (eds). Early Warning Systems for Drought Preparedness and Drought Management. World Meteorological Organization, Geneva.

Williams, M.A., and R.C. Balling, Jr. 1996. Interactions of Desertification and Drought. (Published for the World Meteorological Organization and the United Nations Environment Program.) Arnold, London.

WMO. 1995. Twelfth World Meteorological Congress, Geneva, 30 May–21 June 1995. Abridged final report with resolutions. WMO no. 827. World Meteorological Organization, Geneva.

WMO. 1997. Climate, Drought and Desertification. WMO no. 869. World Meteorological Organization, Geneva.

WMO. 1998. Atelier sur la prevention et la gestion des situations de sécheresse dans les pays du Maghreb, 24–28 June 1996, Casablanca, Morocco. World Meteorological Organization, Geneva.

WMO. 2000. Early Warning Systems for Drought and Desertification: Role of National Meteorological and Hydrological Services. WMO no. 906. World Meteorological Organization, Geneva.

WMO and UNDP/UNSO. 1997. Drought Preparedness and Management for Western African Countries. Lectures presented at the Training Seminar, September 4–9, 1995, Banjul, Gambia. World Meteorological Organization, Geneva.

CHAPTER THIRTY-TWO

Food and Agriculture Organization and Agricultural Droughts
ELIJAH MUKHALA

The Food and Agriculture Organization (FAO) of the United Nations was founded in 1945 with a mandate to raise levels of nutrition and standards of living, to improve agricultural productivity, and to improve the condition of rural populations in the world. Today, FAO is the largest specialized agency in the United Nations system and is the lead agency for agriculture and rural development. FAO is composed of eight departments: Agriculture, Economic and Social, Fisheries, Forestry, Sustainable Development, Technical Cooperation, General Affairs, and Information and Administration and Finance. As an intergovernmental organization, FAO has 183 member countries plus one member organization, the European Union.

Since its inception, FAO has worked to alleviate poverty and hunger by promoting agricultural development, improved nutrition, and the pursuit of food security—defined as the access of all people at all times to the food they need for an active and healthy life. Food production in the world has increased at an unprecedented rate since FAO was founded, outpacing the doubling of the world's population over the same period. Since the early 1960s, the proportion of hungry people in the developing world has been reduced from more than 50% to less than 20%. Despite these progressive developments, more than 790 million people in the developing world—more than the total population of North America and Western Europe combined—still go hungry (FAO, 2004).

FAO strives to reduce food insecurity in the world, especially in developing countries. In 1996, the World Food Summit convened by FAO in Rome adopted a plan of action aimed to reduce the number of the world's hungry people in half by 2015. While the proper foundation of this goal lies, among others, in the increase of food production and ensuring access to food, there is also a need to monitor the current food supply and demand situation, so that timely interventions can be planned whenever the possibility of drought, famine, starvation, or malnutrition exists. With an imminent food crisis, actions need to be taken as early as possible because it takes time to mobilize resources, and logistic operations are often hampered by adverse natural or societal conditions, including war and civil strife. The availability of objective and timely information is therefore crucial. Over the years, FAO has successfully carried out timely interventions in cooperation with other United Nations agencies, especially the World Food Programme (WFP).

Several disasters affect agricultural production in many parts of the world. Agricultural drought, defined and discussed in detail in chapter 1, is one of the most devastating in bringing about hunger and food insecurity. Because agricultural droughts recur in many parts of the world, FAO has been involved in agricultural drought monitoring and mitigation for many years, as described in the remainder of this chapter.

Agricultural Drought Monitoring

Different units of FAO monitor drought situations in the world and provide information for appropriate action in case of an impending food crisis. These include the Global Information and Early Warning Service (GIEWS; www.fao.org/giews) in the Economics and Social Department and the Environment and Natural Resources Service (SDRN) in the Sustainable Development Department.

Global Information and Early Warning System

The GIEWS was established in 1975 during the world food crisis of early 1970s and is now one of the leading sources of information on food production and food security for every country in the world, whether they are members of FAO or not. In the past 25 years, the system has become a worldwide network that includes 115 governments, 61 nongovernmental organizations (NGOs), and numerous trade, research, and media organizations. GIEWS has established a unique database on global, regional, national, and subnational food security. GIEWS aims to provide policy-makers and relief organizations with the most up-to-date information available on all aspects of food supply and demand and provides warnings of imminent food crises so that timely interventions can be planned and executed.

The Environment and Natural Resources Service

The Environment and Natural Resources Service comprises Remote Sensing, Agrometeorology, Geographic Information System (GIS), Environment, Energy and Organic Agriculture, as well as the secretariat of the in-house working groups for the international Environmental Conventions and the Global Terrestrial Observing Systems. Its overall mission is to contribute to and promote environmental and natural resources management and conservation in the context of sustainable agriculture, rural development, and food security in the world. The SDRN provides technical support and advisory services in some 50 countries in Africa, Asia, Latin America, the Caribbean, and central and eastern Europe.

Using remote sensing, GIS, and agrometeorological tools, SDRN collects, archives, and processes data on renewable natural resources to provide information on environment and food security. It provides computer-based access to many of the GIEWS information sources and allows, for instance, the rapid display of Advance Real Time Environmental Monitoring Information System (ARTEMIS) imagery and monitors vegetative development over Africa (FAO, 1997).

Through the cooperation with the Global Inventory Modeling and Mapping Studies (GIMMS) group of NASA Goddard Space Flight Center and the Tropical Applications of Meteorology using Satellite group of the University of Reading, an operational processing of METEOSAT thermal infra red (TIR) and the National Oceanic and Atmospheric Administration's Advanced Very High Resolution Radiometer (NOAA-AVHRR) Global Area Coverage (GAC) (http://metart.fao.org/default.htm) data was implemented. METEOSAT data were used to monitor rainfall and normalized difference vegetation index (NDVI) imagery derived from the AVHRR instrument to monitor vegetation development. All products were made available at a common resolution of 7.6 km. The system acquires and routinely processes, in real-time, hourly estimates of rainfall and NDVI (chapters 5 and 6) images, using METEOSAT and NOAA data. The system covers the whole of Africa, and the products are produced on 10-day and monthly bases for use in the field of early warning for food security. Related technology transfer is being implemented through regional remote-sensing projects in the Southern Africa Development Community (SADC) and Inter-Governmental Authority on Development (IGAD) regions with financial assistance from the governments of Japan, The Netherlands, and France, the European Union (2000), and the FAO Regional Office for Africa in Accra, Ghana. ARTEMIS, which assisted SADC in establishing a remote sensing and GIS capacity for its regional food security early warning system, has made significant progress in developing suitable information products, which are now available to various types of users in a timely fashion, using data transmissions through e-mail. In 1988, ARTEMIS generated only imagers (Polaroid prints) for distribution by diplomatic pouch or mail. Since 1998, the quality and geographic coverage of products available with ARTEMIS have greatly improved on account of acquisition of the SPOT-4/VEGETATION data purchased by FAO with the support of the European Union (EU) and developed in close technical cooperation with the Global Vegetation Monitoring Unit of the Joint Research Center of the European Community (http://marsunit.jrc.it/Africa/; FAO, 1997).

Use of Satellite Data

Two main satellite-based variables used in the FAO crop forecasting approach are NDVI and cold cloud duration (CCD). In FAO food security programs, rainfall is often estimated from CCD. Low values of NDVI correspond to sparse or no vegetation (ochre-brown-green), and high values indicate dense vegetation (red-pink-purple). The CCD is an indicator based on high frequency (hourly) thermal infrared observations from geostationary meteorological satellites of the METEOSAT type. It is a measure of the duration, in hours, in which clouds become so cold (below $-40°$ C) that the likelihood of their producing rain is very high. The CCD has been effectively used for agricultural drought monitoring because it serves as a proxy for rainfall. This gives a picture of whether the current season is better or worse than any previous season. However, its greatest potential is to provide and map the estimates of rainfall, which ultimately affects agricultural production in many parts of the world where agriculture is heavily dependent on rainfall, as is the case in the sub-Saharan Africa. The Early Warning System of the SADC has been using El Niño/Southern Oscillation Index values (SOI; chapter 3) to predict the likely outcome of the coming rainy season (FAO, 1996).

Quantitative Application of Satellite Imagery

Satellite imagery has been used for agricultural drought monitoring and other activities for many years. However, less progress has been made in the quantitative use of remote-sensing imagery. For example, the NDVI-based analysis conducted by FAO, at continental and regional scales, has not changed much since 1988. Although data registration and calibration have become much better, the main type of analysis still is a qualitative assessment of the current vegetation situation during the growing season, as compared to previous years or the average, by comparing trends in NDVI.

The use of METEOSAT data for rainfall estimates has also only been partially successful. Although quantitative estimates of rainfall derived from METEOSAT are becoming more refined, the most popular METEOSAT-derived product for early warning and drought monitoring remains the CCD images. These images were originally intended to be an intermediate product only and not suitable for distribution. More recently, METEOSAT data have been merged with data from other sources to improve the quality of the estimates. At the national level, this is often done by interpolating between a large number of observations from meteorological stations using the CCD images as a weighted surface, guiding the interpolation patterns and the ground observations for quantification. At a continental scale, only ground-observed rainfall data through the Global Telecommunication System (GTS) is readily available in real time, which often is of uncertain quality in many food-insecure countries. The sparse density of the meteorological stations reporting to the GTS contributes to the uncertainty. Recent approaches now combine METEOSAT with GTS data and data from microwave images or sounding instruments, the most popular microwave data being Special Sensor Microwave Imager. Although improvements are still needed before remote sensing can provide the quantitative data required for crop yield models, the remote sensing data remain an essential, yet partial, component for monitoring food supply and demand to improve world food security.

Over the years, as the use of satellite-derived information became more and more integrated in the operations of the above programs, there was a growing demand for more and better data, which could only be partially met. The ARTEMIS area coverage was extended first to South and Central America with NOAA GAC-derived NDVI in cooperation with NASA's Goddard Space Flight Center and, later, through cooperation with the Japanese Meteorological Agency. Monsoon monitoring over Asia was also explored, based on Geostationary Meteorological Satellite data. Although this was certainly an improvement, many areas of special interest to GIEWS, such as the Commonwealth of Independent States and North Korea, were still not covered.

In southern Africa, FAO has assisted many SADC countries in rehabilitating the agricultural system after a devastating drought. In such circumstances, FAO makes arrangements for assessing the essential agricultural inputs needed to restore production in the affected countries. FAO also makes an appeal for financial assistance to implement emergency relief, short-term rehabilitation, and preparedness interventions to the international donor community.

FAO Crop Water Requirement Satisfaction Index Model

The FAO crop water requirement satisfaction index (WRSI) has been used extensively, especially in Africa, for crop monitoring for food security. The model can detect the onset of agricultural drought, which is indicated by the crop stress. The model has been used to effectively monitor agricultural drought in many parts of the world.

The WRSI determines a cumulative water balance for each period of 10 days (1 dekad) from planting to maturity. The cycle of each crop is subdivided into successive dekads. For each dekad, using a water-balance approach involving rainfall, evaporation, crop water requirements, and soil water-holding capacity, the cumulative water available (either surplus or deficit) at the beginning of each dekad can be computed (FAO, 1996).

Basically, the water balance is the difference between the effective amounts of rainfall received by the crop and the amounts of water lost by the crop and soil due to evaporation, transpiration, and deep infiltration. The amount of water held by the soil and available to the crop is also taken into account. In practice, the water balance is computed using a bookkeeping approach. The computation is done dekad by dekad and begins before the planting to account for moisture stored in the soil. From the planting dekad, the crop water requirements are calculated as the potential evapotranspiration (PET) times the crop coefficient (KCR). Thus, the available water amount is the difference between the crop water requirements and the working rainfall. These amounts do not consider water stored by the soil. The working rainfall amount reflects the effective water received by the crop and is calculated through a ratio defined by the user on the basis of the type of soil, slope, and so on. Normal rainfall is used in case of missing values. Surplus or deficit result from the water balancing and ranges between the field capacity and the permanent wilting point, depending on the root depth and the soil water-holding capacity. Finally, the WRSI indicates the degree to which cumulative crop water requirements have been met at the growth stage. The WSI represents, at any time of the growing period, the ratio between the actual and the potential evapotranspiration (FAO, 1996).

This WRSI model could be considered a combination of dynamic (water balance) and statistical (calibration of yield function) approaches. In fact, at harvest time, the sum of dekadal water deficit (or stress) suffered by the crop can be used as a variable along with some other relevant variables to forecast crop yield by statistical regression. At present, it is difficult to incorporate, into yield models, the variables derived from soil fertility, technology (mechanization, fertilizer use), varietal differences, and farming practices. It is a characteristic of the FAO approach that these important parameters are considered, along with NDVI, at the next stage of development of yield ("agmet") models.

The yield function is valid for a crop and a group of stations in a homogeneous cropping area. The input data correspond to different geographical units, from weather stations, to pixels (NDVI, CCD: 50 km^2), to administrative units. It is an important step in the forecasting method to convert the data to comparable units (area averaging)—usually administrative areas that are used by planners or decision-makers in the field of food security.

Agricultural Drought Mitigation

FAO attaches particular importance to promoting the production and consumption of drought-tolerant food crops in areas where rainfall is uncertain. This in some cases involves redressing tendencies that were widespread in agricultural development programs during the 1970s and 1980s. During that period, green revolution packages involving high-yield hybrids or composites were promoted in pursuit of higher overall production levels. This strategy often did not identify the risks inherent in adopting varieties that do not perform well under conditions of moisture stress. Shifts in the cropping system may also entail a shift in food consumption habits and raise the need for support to maintain nutritionally balanced diets. Such support may be provided through nutrition education. Support could be furnished, for example, in the context of local-level planning workshops for the formulation of drought mitigation plans.

In many instances, drought-tolerant food plants that were traditionally used have been substituted in recent decades by other crops or varieties, and some are now perceived negatively. Nutrition education programs in this case could include the promotion of underexploited traditional foods (including production/collection, storage, processing, preparation, and consumption) involving a shift in food habits back to formerly familiar foods, rather than introduction of new ones. FAO's Nutrition Programmes Service, in collaboration with Crop and Grassland Service, supports such programs.

FAO's Food Security and Agricultural Projects Analysis Service provides technical assistance to governments and regional organizations to develop drought mitigation plans. Though the objective of this assistance is preventive, it may be a component of broader assistance to develop drought preparedness plans. It typically includes elements such as (1) analysis of the frequency, severity, history, and impacts of drought in the region or country, (2) analysis of responses of different groups of the population to drought, including both longer-term adaptation of livelihood systems and shorter-term coping mechanisms, (3) analysis of measures to undertake drought mitigation measures and identification of the national and regional institutions, including NGOs, private sector, and community organizations that can implement the action plan. Assessment of the strengths and weaknesses of the action plans and identification of the associated risks are also conducted.

FAO supports interinstitutional drought mitigation planning workshops and other activities at national and local levels. These are geared to formulate action strategies and plans. Prominent issues in such strategies are how best to support, strengthen, and supplement existing coping mechanisms. Drought-afflicted populations must minimize the cost of putting such mechanisms into effect. These require a good base of information on local livelihood systems and mechanisms used by different groups to cope with drought, which is part of early warning and food information systems.

FAO has been, and continues to be, actively involved in helping countries prepare for and respond to the adverse impact of El Niño (chapter 3). For several decades, FAO has spearheaded agricultural improvement and rural development in arid, semiarid, and dry subhumid zones ravaged by drought and desertification. These activities involve emergency and rehabilitation actions in the event of agricultural drought.

Long-term Drought Prevention Programs

FAO has assisted countries in implementing long-term preventive measures against agricultural drought. Examples of such measures promoted by FAO include programs to support the construction of wells and small-scale irrigation development programs in southern Africa and Central America, the development of drought-resistant cropping patterns for South Asia, the Sahel, eastern and southern Africa, and the Caribbean, the preparation of a disaster preparedness strategy for the member countries of the IGAD in eastern Africa and the Horn of Africa, and the design and management of strategic food security reserves.

Early Warning and Forecasting Programs

Since March 1997, FAO has intensified the monitoring of weather developments and crop prospects in all parts of the world through GIEWS. The system has issued two reports on the impact of El Niño on crop production in Latin America and Asia. In 1997–98 the focus changed to southern Africa, where the growing season was just commencing at the time. GIEWS discussed with the World Food Program (WFP) the possibility of launching advance emergency operations and sending crop and food supply assessment missions to southern Africa if drought conditions developed. These plans were jointly approved by the director-general (FAO) and the executive director (WFP). The systems assessments provide a lead in initiating food aid and agricultural rehabilitation activities in affected countries.

FAO has helped develop early warning units that work hand in hand with the GIEWS. Among such units are the SADC Regional Early Warning System (REWS), which operates as an integrated system, based in Harare, and autonomous National Early Warning Units (NEWUs) in each of the 14 SADC member countries. Their activities are coordinated by REWS. The main objective of the SADC-REWS is to provide member countries and the international community with advance information on food security prospects in the region through assessments of expected food production, food supplies, and requirements.

The REWU compiles food security data for the SADC region based on the contributions received from the various NEWUs by fax and e-mail and aggregates this information for subsequent publication in a *Quarterly Food Security Bulletin*, supplemented by monthly updates. Similarly, the NEWUs themselves prepare national food security bulletins. Relevant ad-hoc reports are also submitted directly to decision-makers, as required. During the crop-growing season, data are collected on rainfall for every dekad, on crop stages and conditions, and on any adverse effects such as agricultural drought. This information is compiled by NEWU agrometeorologists and is submitted to the REWU for aggregation into 10-day agrometeorological bulletins. Satellite imagery (CCD and NDVI) is used to support and verify ground observations as well as to monitor agricultural drought in the region. FAO has also developed agrometeorological maize-yield forecasting models for each country, based on crop water-satisfaction indices, with a view to forecast preharvest yields (Dorenbos and Kassam, 1976).

Future Directions

FAO recognizes that about 70% of the agricultural areas in the Near East region (chapter 16) are arid or semiarid. Only 20% of the total lands are cultivable. The most serious challenge to agriculture is water scarcity because the average annual rainfall is only 205 mm. Although the region covers 14% of the world's surface, its water resources represent only 2% of the total renewable water resources of the world. FAO has called on the Near East governments to establish a drought watch and early warning system and to support a recently launched Drought Information Network for the Near East and the Mediterranean (http://www.fao.org/WAICENT/OIS/PRESS_NE/english/2002/3084-en.html (FAO, 2002)). The great global challenge for the coming years will be how to produce more food with less water. To highlight this challenge, FAO dedicated the year 2002 as the World Food Day with the theme "Water: Source of Food Security" by bringing together government representatives and civil society organizations in FAO member countries to focus on solutions to the problem of water scarcity and its impact on food security. "The future emphasis must be directed towards increasing the efficiency of water management systems and increasing water productivity, getting more production per drop, as well as to move seriously towards tapping new nonconventional water resources to increase agricultural productivity," FAO Director-General Jacques Diouf told the Near East Agriculture Ministers at the 26th FAO Regional Conference for the Near East (March 9–13, 2002; http://www.fao.org/WAICENT/OIS/PRESS_NE/english/2002/3100-en.html (FAO, 2002)).

Conclusions

Since its establishment in 1945, FAO has contributed a lot to improve the food security situation of rural people, especially in the developing countries. FAO has established a variety of programs for monitoring weather conditions through satellite data and ground-based tools to provide timely information for drought early warning in cases of impending food shortages. It can be said that FAO is fulfilling its purposes. The establishment of the SDRN and GIEWS at FAO has been of great assistance to crop production and agricultural drought monitoring through systematic collection, analysis, and dissemination of food supply and production information in the world.

ACKNOWLEDGMENTS I extend my gratitude to Kennedy Masamvu, the SADC Regional Remote Sensing Unit (RRSU), for granting me the time to prepare this chapter and thanks to Rene Gommes, Michele Bernardi, and Jelle U. Hielkema for recommending me to write the chapter on behalf of FAO. The regional FAO office in Harare provided some material to write the chapter. Special thanks are also due to FAO Sub-Regional Representative Victoria Sekitoleko and to Mark McGuire.

References

Dorenbos, J., and A.H. Kassam. 1976. Agro-meteorological field stations. Irrigation and Drainage Paper no. 27. Food and Agriculture Organization, Rome.

FAO. 1996. Crop forecasting: inputs December 1996. [Available: http://www.fao.org/sd/eidirect/agromet/inputs.htm].

FAO. 1997. Overview of FAO remote sensing, Geographic Information System and agrometeorological activities. Food and Agriculture Organization, Rome.

FAO. 2002. Combating drought, a top-priority for near east countries, FAO says. Press Release March, 2002. Food and Agriculture Organization, Rome.

FAO, 2004. FAO at Work. Available: http://www.fao.org/unfao/about/index_en.html.

CHAPTER THIRTY-THREE

International Activities Related to Dryland Degradation Assessment and Drought Early Warning

ASHBINDU SINGH

Land degradation usually occurs on drylands (arid, semiarid, and dry sub-humid areas). According to the United Nations Convention to Combat Desertification held in Paris in 1994 (UNCCD, 1999), drylands are defined as those lands (other than polar and subpolar regions) where the ratio of annual precipitation to potential evapotranspiration falls within the range of 0.05–0.65.

Land degradation causes reduction in the biological or economic productivity of those lands that may support cropland, rangelands, forest, and woodlands. Land degradation threatens culturally unique agropastoral and silvopastoral farming systems and nomadic and transhumance systems. The consequences of land degradation are widespread poverty, hunger, migration, and creation of a potential cycle of debt for the affected populations.

Historical awareness of the land degradation was cited, mainly at the local and regional scales, by Plato in the 4th century B.C. in the Mediterranean region, and in Mesopotamia and China (WRI, 2001). The occurrence of the "dust bowl" in the United States during the 1930s affected farms and agricultural productivity, and several famines and mass migrations, especially in Africa during the 1970s, were important landmarks of land degradation in the 20th century.

It is estimated that more than 33% of the earth's land surface and 2.6 billion people are affected by land degradation and desertification in more than 100 countries. About 73% of rangelands in dryland areas and 47% of marginal rain-fed croplands, together with a significant percentage of irrigated croplands, are currently degraded (WRI, 2001). In sub-Saharan Africa, land degradation is widespread (20–50% of the land) and affects some 200 million people. This region experiences poverty and frequent droughts on a scale not known anywhere else in the world. Land degradation is also severe and widespread in Asia, Latin America, as well as other regions of the globe. Continuous land degradation is accelerating the loss of agricultural productivity and food production in the world. Over the next 50 years, food production needs to triple in order to provide a nutritionally adequate diet for the world's growing population. This will be difficult to achieve even under favorable circumstances. If land degradation is not checked and reversed, food yields in many affected areas will decline, and, as a result, malnutrition, starvation, and ultimately famine may occur. This chapter provides state-of-the-art information on international activities related to dryland degradation assessment and drought early warning.

Factors Affecting Land Degradation

Many processes, simple or complex, that arise from human activities and habitation patterns are responsible for land degradation. These processes include soil erosion caused by wind and/or water; deterioration of the physical, chemical, biological, or economic properties of soil; and long-term loss of natural vegetation. In addition, overgrazing, excessive irrigation, and intensive tillage and cropping have been among some of the other factors responsible for land degradation. Some of these causes are related to poverty and food insecurity among the people who inhabit the land.

Monitoring Land Degradation

Previous and existing approaches to assessing and monitoring land degradation have not been effective because the system for collecting and disseminating information and adoption of sustainable land use and management practices have been weak in the dryland regions that represent many poor countries. Some countries have databases on the biophysical aspects of land and land resource use and degradation and on various aspects of society and the economy. However, these data are usually not comparable in terms of scale, are often not compatible, and certainly are not integrated and interlinked enough to facilitate decision-making and policy-making. Information about land degradation is rarely standardized or integrated into local planning and management processes. At the international level, this information is not comparable and has insufficient resolution for defining and implementing regional and national action plans (RAPs, NAPs) for international conventions such as the United Nations Conference on Environment and Development (UNCED) held in Rio de Janeiro, Brazil, in 1992.

The earliest assessment of land degradation was biophysical in nature and was derived at the farm level using the universal soil loss equation (Wischmeier, 1976). Early attempts to assess land degradation on larger scales, such as at river basin and bioregional scales, using a combination of remote sensing and ground-based techniques, encountered difficulties mainly due to lack of financial resources and technical limitations. In 1979, the Food and Agriculture Organization (FAO) of the United Nations developed a methodology for assessing soil degradation with detailed criteria for each type of biophysical degradation. Subsequently, FAO conducted a Global Assessment of Progress on Desertification in 1984 and 1992. The global estimate of land degradation or desertification ranged from 2001 million ha in 1984 to 3475 million ha in 1992 (i.e., 13–23% of the earth's surface; FAO, 1979, 1995).

Global Assessment

In 1987 the United Nations Environment Programme (UNEP) requested an expert panel to produce, in the shortest time possible, a scientifically credible method to assess soil degradation at the global scale. The first Global Assessment of Soil Degradation (GLASOD) in the early 1990s provided a systematic qualitative assessment of the extent and severity of land degradation, and its results formed the basis for the *World Atlas of Desertification* at a scale of 1:10 million (UNEP, 1992, 1997). This map is based on input from more than 250 soil scientists in the 21 regions dividing the world. This expert assessment method used a mapping base, a set of semiquantitative definitions for soil degradation, case studies, and a team of national and international experts.

GLASOD is criticized today as inaccurate, subjective, and not appropriate for assessing soil degradation at the country level. Most of the indicators used in the GLASOD were biophysical and not socioeconomic and therefore did not include the institutional and policy driving forces of land degradation. Despite these drawbacks, GLASOD remains the only global database on the status of human-induced soil degradation, and no other data set comes as close to defining the extent of desertification at the global scale (UNEP, 1997). In another approach, Dregne et al. (1991) used results of vegetation degradation estimates for comparison with soil degradation.

Assessment of South and Southeast Asia

In response to requests for more detailed information on soil degradation, in 1993 the Asia Network on Problem Soils (www.isric.nl) recommended preparation of a soil degradation assessment for South and Southeast Asia (ASSOD) at a scale of 1:5 million. The methodology of this assessment reflects comments from the peer review of GLASOD. As a result, ASSOD has a more objective cartographic base and uses the internationally endorsed World Soils and Terrain Digital Database (SOTER) to delineate mapping units (Oldeman, 1988; see also www.isric.nl). Like GLASOD, ASSOD focuses on displacement of soil material by water or wind and in situ deterioration of soil by physical, chemical, and biological processes. ASSOD, however, places more emphasis on trends of degradation and the effects of degradation on productivity.

Although an improvement over GLASOD, ASSOD is not without problems. The assessment of the degree, extent, and past rate of soil degradation is still based on expert opinion, and the scale (1:5 million) is still not adequate to guide national soil improvement policies.

Improving the Land Degradation Assessment

An accurate assessment of land degradation at a flexible scale combining socioeconomic and biophysical aspects and driving forces is needed to plan actions and investments to reverse land degradation to improve socioeconomic livelihoods and to conserve dryland ecosystems and their unique biological diversity. A further advance is the use of the ecosystem approach as a framework for action under the Convention on Biological Diversity (www.biodiv.org) and as a strategy for the integrated management of land, water and biological resources.

Recent advances in participatory planning and management of resources, integrated ecosystem approaches to resources assessment using economic-ecological zoning, remote sensing and Geographic Information System (GIS) techniques, as well as the economic valuation of soil loss and strategic impact assessments of policies and interventions provide an opportunity for developing an improved land degradation assessment system.

In addition to a Global Water Resources Assessment and Global Biodiversity Assessment, the International Institute for Applied Systems Analysis and FAO have developed a system for rational land-use planning on the basis of agroecological zones methodology (Fischer et al., 2000). This methodology can be applied at national, regional, and local levels. The International Food Policy Research Institute and World Resources Institute have undertaken a comprehensive assessment of the earth's ecosystems, such as a Pilot Analysis of Global Ecosystems and Millennium Ecosystem Assessment, sponsored by the Global Environment Facility, the United Nations Foundation, the Packard Foundations, and the World Bank (Reid, 2000). UNEP and FAO have developed guidelines for erosion and desertification control management with particular reference to Mediterranean coastal areas (UNEP/MAP/PAP, 2000). Socioeconomic data sets (Global Farming System Study, Food Insecurity and Vulnerability Mapping systems) have also become increasingly available (www.fao.org).

The vegetation index based on data from the National Oceanic and Atmospheric Administration (NOAA) (chapters 5 and 6), high-resolution satellite data (e.g., Landsat TM, SPOT, RADARSAT, etc.), meteorological satellite (METEOSAT) data (chapters 19 and 32), and GIS techniques that are now available and have been used cost effectively for mapping risk of erosion or land degradation using GIS by overlaying different thematic layers and using a model have been tested in a number of studies, such as the FAO study in Parana, Brazil (FAO, 1997). The method of assessing degradation depends largely on the spatial scale—for example, regional scale versus continental scale, as shown in table 33.1.

Table 33.1 A comparison of various methods for land degradation assessment

Method	Advantages	Problems	Scales applicable	Factors to which applicable	Relative cost
Expert opinion	Rapid, low cost	Subjectivity, reliability	Applied at global, but potential for all scales	Soil, vegetation, water	Low
Remote sensing	Fairly rapid, objective, Large area coverage	Separating change in land use from degradation	Global, regional, national, local	Vegetation, land cover changes	Moderately low
Field monitoring	Directly indicates change in condition of land, quantitative	Slowness and high cost	Local, plus district to national on sampling basis	Soil, vegetation, water, biodiversity	Relatively high
Productivity changes	Degradation is defined in these terms	Variations in management practices	National, district, local	Crop production, animal production	Low at national, high at local levels
Field criteria and talking with farmers	Acquires grass-roots view, links degradation with cause, impact, response	Slowness, subjectivity	Local; district on sampling basis	Soil (including erosion), vegetation, water, biodiversity, socioeconomic indicators	Relatively high, practical only on sampling basis
Modeling	Rapid, low cost, potential for extrapolation	Danger that users confuse modeling with results	All scales	Soil, water	Relatively very low

Source: FAO (2000).

Early Warning Systems

At international levels, there are groups that alert user communities to environmental changes. For example, the World Meteorological Organization's (WMO) early-warning system (www.wmo.org; chapter 31) is composed of weather satellites and thousands of surface monitors. This system has helped predict long-term droughts and provided information for other human-induced disasters.

Another example of early warning at the international scale is FAO's Global Information and Early Warning System (GIEWS; chapter 32). The GIEWS reports the production, stocks, and trade of cereals and other basic food commodities through an analysis of trends and prospects. These reports vary with need and often contain analysis and statistical information on developments in world cereal markets and export prices. These reports also include the impact of El Niño and La Niña on food grain production and reveal trends in any food emergencies around the world.

GIEWS provides an example of how international early warning systems are ultimately truly interconnected to many other international efforts, many benefiting from the efforts and expertise of other groups. GIEWS maintains connections with many other international bodies: (1) the United Nations High Commissioner for Refugees, which supplies data on refugee numbers, (2) the WMO, which provides climate and weather data, (3) the International Labor Organization, which provides information on unemployment and poverty, (4) the United Nations Children's Fund, (5) the International Grains Council, which provides information on the global market, export prices, and freight rates, (6) the Organization for Economic Co-operation and Development, (7) the World Bank and International Monetary Fund, and (8) UNEP's Global Resource Information Database.

The NOAA Climate Prediction Center/ National Centers for Environmental Prediction takes on the task of monitoring El Niño/Southern Oscillation (ENSO) conditions and passing the results and warnings on to user communities (http://www.cpc.ncep.noaa.gov/index.html). As an example, researchers at Clark University have been using normalized difference vegetation index time-series data in conjunction with ENSO data to help predict drought conditions in eastern Africa (www.clarku.edu). One of the early warning systems that provides crucial information to both countries and large regions is the Famine Early Warning System (chapter 19).

Future Directions

A review of international activities relating to land degradation leads to the conclusion that there is a need for international cooperation for (1) developing standardized methods and guidelines for dryland degradation assessment and monitoring; (2) developing a baseline map of dryland land degradation at subregional scale; (3) global assessment of dryland degradation; (4) detailed assessment of land degradation at national level, focusing on areas at greatest risk ("hot spots") and areas where degradation has been successfully reversed ("bright spots"); (5) analysis of the effects of land degradation areas at risk; (6) developing best practices for the control and prevention of land degradation; (7) communicating and exchanging land degradation information and promoting its use in decision-making; and (8) strengthening early warning systems to generate seasonal to interannual climatic predictions to improve effectiveness of programs that aim to mitigate effects of droughts and food shortages on local population.

Some of these issues are being addressed through an international Dryland Degradation Assessment project funded by the Global Environment

Facility as a cooperative activity between FAO, CCD, UNEP, and many other international and national partners (www.fao.org).

Conclusions

There are a number of international, regional, and national initiatives to assess dryland degradation and its impact on agricultural productivity. However, these activities need to be better coordinated to provide optimal benefits to users. Assessing the status and trends of dryland degradation and prediction of droughts is a complex task. It is important not only to monitor biophysical conditions but also to assess human vulnerability to land degradation and droughts. Current policy measures and identification of future priorities should be given equal weight. The dissemination of scientifically credible results to the public and to policy-makers in a simple and understandable form is the key to improving decision-making in regard to land degradation and drought monitoring and prediction.

ACKNOWLEDGMENTS The material for this chapter was heavily drawn from the FAO unpublished background paper for the International Workshop on Dryland Degradation Assessment Initiative, Rome, December 5–7, 2000 and the subsequent project document prepared by the FAO on dryland degradation assessment. The views expressed in this paper are not necessarily those of the United Nations Environment Programme.

References

Dregne, H., M. Kassas, and B. Rozanov. 1991. A new assessment of the world status of desertification. Desertification Control Bulletin 20:6–18.
FAO. 1979. Methodology for Assessing Soil Degradation. A Framework for Land Evaluation. FAO Soils Bulletin 32. Food and Agriculture Organization, Rome.
FAO. 1995. Digital Soil Map of the World (DSMW) and Derived Soil Properties. Version 3.5. CD-ROM. Food and Agriculture Organization, Rome.
FAO. 1997. Land Quality Indicators and Their Use in Sustainable Agriculture and Rural Development. FAO Land and Water Bulletin. Food and Agriculture Organization, Rome.
FAO. 2000. Unpublished Background Paper for the International Workshop on Dryland Degradation Assessment Initiative, December 5–7, 2000, Rome. Food and Agriculture Organization, Rome.
Fischer, G., H. van Velthuizen, and F.O. Nachtergaele. 2000. Global Agro-Ecological Zones Assessment: Methodology and Results. Interim Report IR-00-064. International Institute for Applied Systems Analysis, Laxenbourg.
Oldeman, L.R. 1988. Guidelines for General Assessment of the Status of Human-induced Soil Degradation. Global Assessment of Soil Degradation (GLASOD) Working Paper no. 88/3. International Soil Reference and Reference Centre, Wageningen, The Netherlands.
Reid, W.V. 2000. The millennium ecosystem assessment: strengthening capacity to manage ecosystems sustainably for human well-being. In: S.P. Gawande et al. (eds.), Advances in Land Resource Management for the 21st Century. Soil Conservation Society of India, New Delhi, pp. 23–34.

UNCCD. 1999. United Nations Convention to Combat Desertification in Those Countries Experiencing Serious Drought and/or Desertification, Particularly in Africa. Text with Annexes. United Nations Convention to Combat Desertification, France.

UNEP. 1992. World Atlas of Desertification. (A. Arnold, ed.). United Nations Environment Programme, Nairobi, Kenya.

UNEP. 1997. World Atlas of Desertification, 2nd ed. (N. Middleton and D. Thomas, eds.). United Nations Environment Programme, Nairobi, Kenya.

UNEP/MAP/PAP. 2000. Guidelines for Erosion and Desertification Control Management with Particular Reference to the Mediterranean Coastal Areas. United Nations Environment Programme/Mediterranean Action Plan/Priority Actions Programme, Nairobi, Kenya.

Wischmeier, W.H. 1976. Use and misuse of the universal soil loss equation. J. Soil Water Conserv. 31:371–378.

WRI. 2001. People, Goods and Services in the World's Susceptible Drylands: A Preliminary Set of Indicators. World Resources Institute, Washington, DC.

CHAPTER THIRTY-FOUR

Climate Change, Global Warming, and Agricultural Droughts

GENNADY V. MENZHULIN, SERGEY P. SAVVATEYEV, ARTHUR P. CRACKNELL, AND VIJENDRA K. BOKEN

The climate of a region is a representation of long-term weather conditions that prevail there. Over the millions of years of the existence of the atmosphere on the earth, the climate has changed all the time; ice ages have come and gone, and this has been the result of natural causes. Recently (on geological time scales) the human population has expanded—from half a billion in 1600, to 1 billion in 1800, to almost 3 billion in 1940, and it now stands at about 6 billion. The climate may well now be influenced not only as before by natural events but also by human activities. For example, we are producing vast amounts of carbon dioxide by burning fossil fuels, and this is causing the temperature of the earth to rise significantly. If we argue that we should control our activities to preserve this planet as a habitable environment for future generations, we need to have some scientific knowledge of the effects of our present activities on climate.

In recent years the evidence has been accumulating that on the time scale of decades there is global warming (i.e., the global annual mean surface temperature is increasing). There is also evidence accumulating that part of this increase is a consequence of human activities. The evidence is largely statistical. Within this trend there are bound to be temporal fluctuations and spatial variations. Moreover, in addition to the increase in temperature, it is reasonable to assume that there is, overall, an increase in evaporation of water from the surface of the earth and that there will be a consequent increase in precipitation. But within this overall scenario there are bound to be local variations; some areas may experience more precipitation, but some areas may experience less precipitation. The effect of climate change on the proneness to drought is therefore not uniform but can be expected to vary from place to place. Therefore, whether one is concerned with considering the relation between climate and proneness to drought from the historical evidence or whether one is trying to use models to predict the effect of future climatic conditions, it is necessary to consider the local spatial variations.

To quantify the effects of past human activities on the climate is difficult. The climate is a temporal average taken over a long time scale, whereas reliable detailed observations of the weather (from conventional records or from satellite data) have only been made over a very short and recent period of time. Archaeological or geological evidence that we have for longer periods does exist but is of a much less detailed nature. Consequently it is difficult to determine the extent to which human activities over the last century or two have already affected the climate significantly. For the future, it is extremely important to try to construct models that make use of the environmental parameters (such as the increase in carbon dioxide concentration) that can be quantified and use these models to predict the effects on the climate. Constructing such models is possible, but serious difficulties begin to arise when one tries to use them. This is because the atmosphere/earth system is very complicated and one is trying to extrapolate its behavior over a long period of time, when we cannot even predict the weather more than a few days in advance with any degree of reliability. It is, perhaps, not surprising that different groups of people working with different models often produce quite different predictions of climate change; nevertheless, there is enough agreement that a number of general conclusions have been obtained.

Climate Models

In constructing a climate model one needs to identify the processes that affect the atmosphere, including the interactions between the atmosphere and the surface of the earth (land, oceans, ice masses) (Cracknell, 1994a, 1994b, 2001). The mathematical set of simultaneous integro-differential equations that denote these processes then needs to be constructed. These equations will include various atmospheric parameters, for which values have to be assigned before the equations can be solved. The process of solving these equations has to be programmed for a (large and fast) computer. Such a computer model is described as a General Circulation Model (GCM).

Greenhouse Effect

An important factor is the extent to which solar radiation that reaches the earth is trapped in the atmosphere and the extent to which it is reflected or reradiated to outer space. Various atmospheric constituents, most notably water vapor and carbon dioxide, act in much the same way as the glass in a greenhouse and cause solar energy to be trapped; this is commonly referred to as the greenhouse effect. The greenhouse effect thus causes the atmosphere to be considerably warmer than it would be if these materials were not present in the atmosphere.

It is important to realize that from the point of view of life on earth, the greenhouse effect is good, not bad. The average temperature of the surface of the earth is about 15°C, and in the absence of any greenhouse gases in the atmosphere it would be reduced by about 33°C (i.e., to −18°C). An average temperature of −18°C would make the earth a rather inhospitable place, and any semblance of modern agriculture would only be possible in a few locations near to the equator. A change of only a few degrees in average temperature leads either to ice ages or to melting of the polar ice, and either of these would destroy much of our present civilization. Smaller changes lead to productive agricultural land being turned into a desert. A rise or fall of a few tenths of a degree in the mean temperature may correspond to quite a large change in environmental conditions. The difference in mean temperature between now and one of the recent ice ages is only about 3.5°C. In earlier ice ages the drop in average temperature has been less than 10°C (not 33°C).

What people are concerned with, and what is behind the enormous effort currently being devoted to running climate models on large computers, is the worry about the possibility of substantial climate change (i.e., global warming and changes in precipitation) as a result of changes in the greenhouse effect induced by human activities. It would be most useful if it were possible to identify the human activities that affect the climate and to make reliable predictions of the direct effects of these human activities on the future climate. However, this cannot be done directly because of the natural changes that occur all the time. Unfortunately, the system is so complicated that our present knowledge is somewhat uneven, and our historical knowledge is very sparse indeed. However, in spite of all the difficulties, a great deal of effort has gone into climate modeling in recent years, and some very useful results have been obtained.

Climate Prediction

There are three categories of events that may affect the climate: (1) events that occur outside the earth, (2) natural events on the surface of or within the earth, and (3) human activities.

The source of virtually all the energy that drives the atmosphere is the sun. Thus, any changes in the intensity of the solar radiation arriving at the earth will clearly have a significant effect on the weather and on the climate. The intensity of the radiation arriving at the top of the atmosphere will depend on variations in the intensity of the radiation emitted by the sun, changes in the transmission properties of space between the sun and the earth, and changes in the sun–earth distance. Further details can be found in Cracknell (1994a).

The natural events on the surface of the earth that have an effect on the climate include plate tectonics, variations in the polar ice caps, volcanic eruptions, and ocean circulation. These factors have various time scales associated with them (for further details see, e.g., Cracknell, 1994a).

Human activities that may affect the climate include: (1) increase in the concentration of carbon dioxide in the atmosphere, largely from the consumption of fossil fuels, (2) increase in the concentration of other greenhouse gases in the atmosphere, (3) depletion of the ozone layer in the stratosphere, (4) development of land areas, and (5) other human activities, including rain making, irrigation, and the creation of artificial lakes and reservoirs, change of land use associated with urbanization and aboveground nuclear explosions.

The operational use of numerical models in weather forecasting began in the 1960s. Although their usefulness was originally limited, there have been rapid developments since then, and within 10 years the models were able to provide better forecasts of the basic motion field than could be achieved by an unaided human forecaster. These improvements have come about for two reasons. First, the power of computing facilities available has increased enormously. Second, the amount of data available to describe the present atmospheric conditions has greatly increased in recent years. In addition to conventional surface measurements and radiosonde measurements, there are also satellite observations, including satellite soundings and satellite-derived wind speeds obtained from cloud tracking from geostationary satellite images.

Climate Change

In looking at climate change we are looking at long-term changes compared to averages. Viewed over a century or a millennium, we see that climatic parameters (temperature, rainfall, etc.) are basically stable and vary only slowly. It is the nature of these slow, long-term variations that are of concern in climate studies after local, short-term fluctuations have been smoothed out. It is the stability of the long-term components that makes climate prediction possible.

Sometimes some people (usually politicians) would like to claim, for whatever reason, that there is no evidence for human-induced global warming and that, therefore, there is no need to restrain our behavior in terms of curtailing emissions of greenhouse gases into the atmosphere. Differences among the various models arise at various stages: modeling of the physical processes, choice of grid spacing, boundary conditions and their parameterization, availability of computing power, which increases very rapidly with increasing spatial resolution, and assumptions made about the future.

It is important to realize that many processes occur on a scale that is quite small compared with the grid point spacing. One can therefore never expect to obtain predictions out of model calculations that relate to these processes at these scales. It would therefore be rather risky to rely on absolute predictions made with one particular model.

Intergovernmental Panel on Climate Change

To reach a consensus out of the rapidly expanding mass of conflicting or confusing results obtained from various climate models, the World Meteorological Organization (WMO; chapter 31) and the United Nations Environment Program (UNEP) set up the Intergovernmental Panel on Climate Change (IPCC) in 1988. The IPCC attempted to establish a consensus among the results of calculations from 20–30 different climate models. The IPCC has, over a decade, become more confident about the importance of the role of human activity vis-à-vis the natural variations in the climate. It also warns, however, that regional or local changes may not necessarily all be in the same direction as the general trend. The IPCC's main conclusions can be summarized (Houghton et al., 2001; McCarthy et al., 2001) as follows:

1. Global mean near-surface air temperature increased by about 0.6°C over the 20th century.
2. Temperatures in the lowest 8 km of the atmosphere have risen during 1960–2000 at about 0.1°C per decade.
3. Snow cover and ice extent have decreased.
4. Global mean sea level has risen and ocean heat content has increased.
5. There have been changes in precipitation, with increases in some areas but decreases in some other areas.
6. Most of the warming observed over the last 50 years is attributable to human activities; any contribution from natural factors is small.
7. Concentrations of atmospheric greenhouse gases, including tropospheric ozone, have continued to increase as a result of human activities.
8. Stratospheric ozone depletion and anthropogenic aerosols have a cooling effect, or negative greenhouse effect.
9. Human influences are expected to change the atmospheric composition throughout the 21st century.
10. Projected changes in atmospheric composition arising from human activities, based on various assumptions, will lead to further anticipated temperature and sea level rises throughout the 21st century.
11. There is likely to be an increase in various extreme events.

Climate Change and Agriculture

Temperature and rainfall are the key factors in making decisions about what crops to grow. Thus, agriculture will need to adapt to changes in climate. In nondesert areas droughts arise as fluctuations within the local weather pattern. They cannot be prevented or eliminated, but making use of climate models to obtain reliable predictions of their expected frequency of occurrence can contribute toward wise planning of agriculture to minimize drought losses.

It is useful to consider the global estimates of the area that potentially can be endangered by drought. If from the total global land area we eliminate the territories now unsuitable for agriculture (deserts, tropical woods, polar areas and regions with complex topography and also the territories, where the difference between the annual precipitation and potential evaporation is > 200 mm), the remaining areas may be regarded as agricultural ones. According to the agroclimatological estimates, more than 50% of agricultural land can be endangered by drought.

One of the most important consequences of global warming is degradation of drylands (chapter 33) in different regions of the world. It is important to determine the principal consequences of expected regional climate changes caused by modern global warming in agriculture of different regions and whether they will result in increasing the occurrence and severity of droughts or whether droughts will become less frequent. In the subsequent sections of this chapter we present some preliminary answers to this question.

Wheat Yield Variation and Prediction in Some Countries

Crop prediction in some countries can be studied in relation to recorded climatic change in recent years using prediction models. Wheat yield data were collected from the U.N. Food and Agriculture Organization (FAO) for 55 countries for the last 40 years. Analysis of the long-term changes in wheat yields is interesting for three reasons. First, wheat is the most widely cultivated crop in many countries. Second, the areas of wheat cultivation are often more affected by drought than other main crops. Third, the archive of data on wheat production is the largest and most reliable. We selected relatively small regions or countries because the yields in large territories experience a greater spatial variability. Unfortunately, full information about the crop yields for smaller and climatically homogeneous areas is often not available.

United States

Figure 34.1 shows different yield anomalies, $(y - y_t)/y_t$, for the U.S. states Colorado, North Dakota, and Kansas. These curves show the important regularity that, despite significant distinctions in meteorological conditions in these three U.S. states and their remoteness from each other, the principal features of their annual variability are similar. The five-year smoothed averaged values of relative anomalies were mainly positive during 1900–17, 1940–48, and negative during 1918–39.

In addition to the annual or five-year mean variability, yield variability can also be studied using the hydrothermal coefficient (HTC) and S index that are widely used for drought monitoring in the former Soviet Union (chapter 15). The variations in these indices for the above U.S. states are shown in figure 34.1. The meteorological data used for computing these indices were collected from three meteorological stations: Denver (for Colorado), Bismarck (for North Dakota), and Kansas City (for Kansas) (http://lwf.ncdc.noaa.gov/noaa/climate/).

Wheat, Colorado, USA

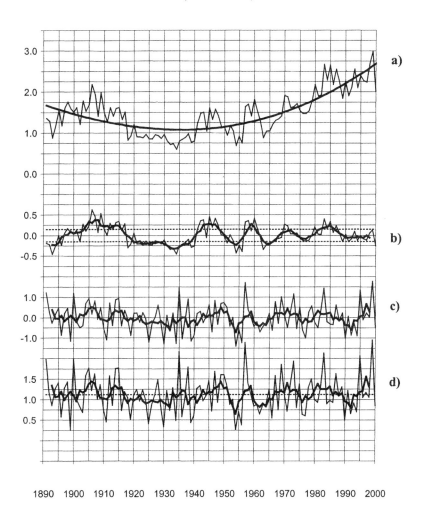

Figure 34.1 The historical variability in wheat yields in Colorado, North Dakota, and Kansas, USA. (a) Annual variability in yield (metric tons per ha) and its trend (solid line), (b) yield anomalies from five-year means (solid line); two horizontal dashed lines show the 15% yield anomalies; (c) S-aridity index and its five-year means (solid line), and (d) hydrothermal coefficient (HTC) and its five-year means (solid line); dashed line shows the HTC normal.

Europe

Figure 34.2 shows wheat yield anomalies for Bulgaria for 1960–2001 after removing the linear (assumed) technological trend (i.e, the yield increase on account of fertilizers, mechanization, and the introduction of high-yield varieties). Such trends were first developed for the United States and the former Soviet Union (Menzhulin and Nikolayev, 1987; Menzhulin et al., 1987). For this purpose fertilizer used (kg/ha), combine harvesters

Wheat, North Dakota, USA

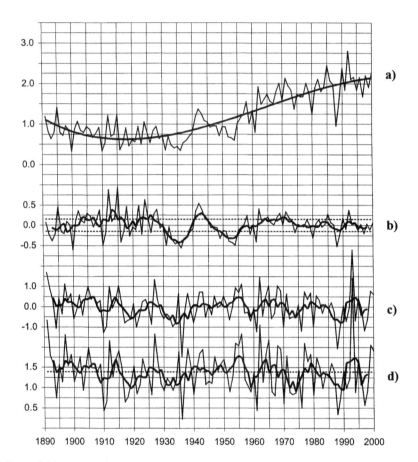

Figure 34.1 *continued*

and tractors (relative numbers/ha) available, electric power, and pesticides and herbicides used have been included in the calculations. In the case of Bulgaria the correlation coefficient between η_5 (5-year smoothed relative yield anomaly) and S_5 (five-year smoothed S index) is 0.51. Wheat yield anomaly was positive from 1970 to 1992. The deviations of annual yields from the technological trend during this period were mainly positive. In one unfavorable year, 1985, the wheat yield in Bulgaria fell to 11% below the trend. After 1983 the yield anomaly trend entered into a declining phase, and after 1992 the wheat yield anomalies were mainly negative. In 1996 a serious failure in wheat production took place, when yield was 47% below the trend.

Figure 34.3 shows wheat yield anomalies during 1960–2001 for several other European countries. By studying the trends in anomalies one can conclude that the 20-year period, from the beginning of the 1970s until

Wheat, Kansas, USA

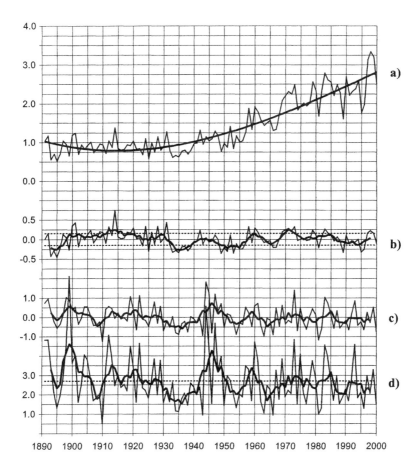

Figure 34.1 *continued*

the beginning of the 1990s, was favorable for many European countries with positive anomalies. Like for Bulgaria during this period, drought (causing 15% or more reduction in yields) was practically absent. The exceptions were only five years: 1979 for Austria, 1991 for Albania, 1993 for Hungary, 1990 for Greece, and 1974 for Malta. Yield anomalies were negative after the mid-1990s.

African Mediterranean and Near East

Figure 34.4 shows wheat yield variation for some African Mediterranean and Near East countries during 1967–95. The yield trend was characterized mainly by negative anomalies. Frequent droughts (causing yield reduction exceeding 20% of the normal) occurred in 1970, 1978, 1983, 1984, and 1985 in Yemen; in 1970, 1971, 1973, 1977, 1979, 1984, and 1989 in Syria;

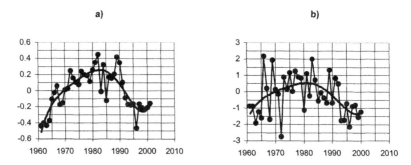

Figure 34.2 Variability in wheat yield in Bulgaria as expressed by (a) S-aridity index and (b) annual variability; bold solid lines represent trends.

in 1968, 1969, 1971, 1978, 1979, 1983, 1984, and 1985 in Lebanon; in 1970, 1973, 1975, 1976, 1977, 1978, 1979, 1981, 1982, 1985, and 1986 in Jordan; in 1970, 1973, 1977, 1982, 1982, 1986, and 1991 in Cyprus; and in 1970, 1976, 1979, 1982, and 1984 in Sudan. However, in Egypt and Iran, a severe drought occurred only once, although the average anomaly of the wheat production in these countries was also negative.

Southern Hemisphere

Figure 34.5 shows the results of the yield anomalies for some Southern Hemisphere countries for last 40 years. It is interesting to note that in these main grain-producing areas of the Southern Hemisphere the changes in the wheat yields followed a common pattern. In South Africa, Namibia, Australia, New Zealand, Argentina, and Uruguay, the beginning and end of the negative phase of the wheat yield anomalies fall approximately in the same years: 1970 and 1995. During this period, drought (causing yield reduction of 20% or more from the normal) occurred in 1978, 1980, 1983, 1985, 1989, 1990, and 1995 in South Africa; in 1977, 1978, 1985, 1986, 1987, and 1988 in Namibia; in 1972, 1977, 1980, 1982, and 1994 in Australia; and in 1970 and 1989 in New Zealand. In Argentina negative wheat yield anomalies occurred during 1970–95, and droughts occurred in 1968 and 1981. In Uruguay drought occurred in 1976, 1977, 1978, 1985, 1986, and 1993.

Having noted that the regularities in the variation of wheat production anomalies in different geographical regions in recent years, we now turn our attention to the question of whether the results obtained could be used for prediction purposes—to extrapolate these trends for the years ahead.

The Possible Impact of Future Climate Changes on the Occurrence of Droughts

For predicting the effect of droughts on agriculture in a given region, one needs to have a completely reliable and detailed climatic forecast for that

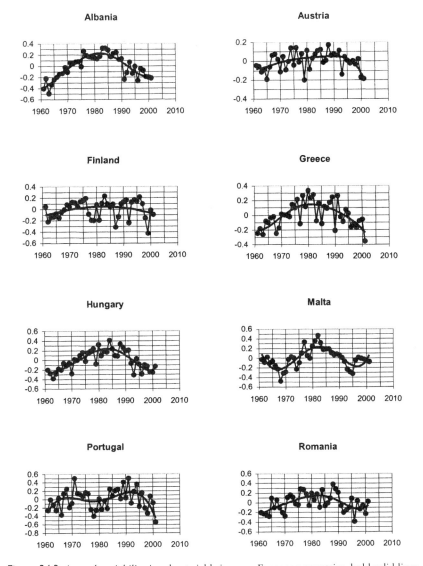

Figure 34.3 Annual variability in wheat yields in some European countries; bold solid lines are the trends.

region and a completely adequate yield model. If these two requirements are satisfied, the technique of agroclimatic analysis can be developed as follows.

At the first stage of the analysis, an available model for the prediction of climate changes is used to carry out the calculations of meteorological and soil dynamics. The second stage of the analysis involves using a crop productivity model with regard to the agrotechnology used. The historical data can be used to give reliable and useful information about the agricultural production losses caused primarily by unfavorable climatic condi-

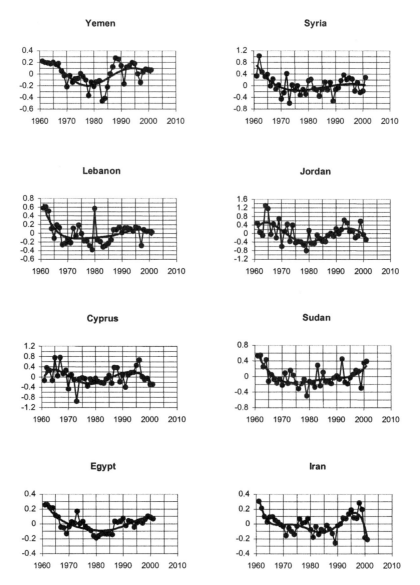

Figure 34.4 Annual variability in wheat yields in some Near East countries; bold solid lines are the trends.

tions. Such an analysis would enable one to see the characteristic tendencies in changes under global warming, which began to be appreciably manifest in regional climate changes. The final result of these two stages should be to provide an estimate of the crop yield and the value of permissible losses due to the unfavorable environmental conditions.

As far as the surface air temperature is concerned, there is a measure of agreement between the results obtained with the use of different GCMs

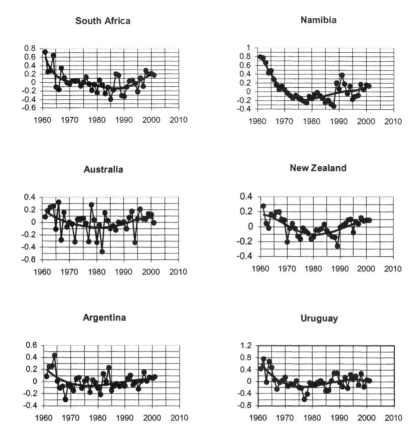

Figure 34.5 Annual variability in wheat yields in some countries of Southern Hemisphere; bold solid lines are the trends.

and the actual data, but when it comes to atmospheric precipitation the agreement is much less satisfactory. This limits the development of complete techniques for analyzing the agricultural consequences of expected climate changes in the manner we have just described. Therefore, any estimates of crop productivity changes that are carried out with the use of different model scenarios will have a rather low accuracy because of the limited reliability of the forecast used.

Trend Analysis

Instead of using climate model predictions, another method of climate change prediction is the empirical extrapolation of trends in modern climatic parameters to the forthcoming decade. In comparison with the methods based on climate models, there is no reliable physical basis for such forecasts. In this case one is assuming that the current global warming masks any natural climatic variability and thus completely controls the main processes of future climate formation. Because of this assumption, such forecasts only have a weak scientific basis, and results obtained with this technique should only be used for the very near future. We would not expect such forecasts to provide reliable results, for example, on the daily dynamics of the meteorological parameters in the vegetation growing season. For this reason the results of calculations with detailed crop productivity models using such a climate change scenario cannot be considered to be reliable. Using parameterized techniques for agrometeorological assessments based on integrated indices would be more justified for this purpose.

Paleoclimatic Reconstruction

The third method of future climate forecasting is paleoclimatic reconstruction, which has been used mainly by Russian researchers. The essence of this method is that the characteristic regularities of the earth's climatic regime in the last warm geological epoch can be assumed to apply to future climatic conditions. Such analogs, reconstructions of temperature and precipitation, are used in the Holocene climatic optimum (about 7000 years ago), in the Riss-Wuerm (Eemian) Interglacial period (about 125,000 years ago), and in the Pliocene Optimum (3–4 million years ago). It is known that the globally averaged surface air temperature in these epochs was about 1.2°C higher (Holocene optimum), about 2°C higher (Riss-Wuerm Interglacial), and 3.5°C higher (Pliocene optimum) than in the middle of the last century (Borzenkova et al., 1987; Budyko et al., 1994). According to modern forecasts based on climate models, these values of the mean surface air temperature can be expected to be reached in 2010, 2030, and 2050, respectively (Houghton et al., 2001). If the paleoclimatic analogs are to be used in agroclimatic assessments, it is necessary to appreciate that they cannot supply the high accuracy of spatial and temporal resolution. At the best, for temperature, paleoreconstructions can provide only semiannual temperature resolution, and for atmospheric precipitation we can only use them to estimate the changes in its mean annual values. Therefore, the agroclimatological techniques using the complex dynamic crop productivity models, which require the day-to-day information for the input meteorological parameters, cannot be directly based on paleoanalog scenarios.

The difficulties involved in all these three methods of climate modeling lead us to the conclusion that, at the present time, the technique of agroclimatological assessments should be to use established submodels based on empirical agrometeorological indices. This is concerned with the assessments of their mean values and in particular the estimation of the abnormal agroclimatic phenomena with which droughts are associated. A major part of such empirical techniques should be the analysis of the actual information about the variation of crop productivity and the relevant meteorological factors, principally the surface air temperature and the precipitation.

Empirical Techniques

With the purpose of developing a forecasting technique for predicting the national wheat production for the leading U.S. states using the 11-year smoothed mean anomaly, η_{11}, the correlation graphs between η_{11} and the S_{11} index and between η_{11} and the HTC_{11} index have been analyzed (figure 34.6).

The calculations have shown that in each specific case both the 11-year smoothed means, S_{11}, and HTC_{11}, can be used as a predictor of this empirical forecasting technique; the choice between them should be determined by the value of the associated correlation coefficient. For wheat production in Colorado State shown in figure 34.6a, the correlation coefficient was .69. These graphs can be used to calculate the changes in the number of agriculturally abnormal years using the forecast of the regional climate changes in any future decade, from which it is possible to obtain the estimate for the change in the S_{11} index. The corresponding estimated change in η_{11} can then be determined from the straight line in figure 34.6a. For example, if in the state of Colorado the S_{11} index becomes equal to -0.1, it corresponds to a change in the decade-mean anomaly of the relative productivity from 0 to -0.12. Using historical data on η values in this area (figure 34.1), it can be concluded that the probability of drought (that is, when the wheat production loss will exceed the standard deviation of the η-indicator, which is 23% for Colorado) can be more than double in the decade. In the case of any other area for which detailed information is not available, we could use the generalized graphs by combining the empirical data of the climatically analogous regions for forecasting purposes. An example of such a correlation graph is shown in figure 34.6b, which generalizes the empirical data of the three selected U.S. states.

A-Indicator

We also find it convenient to introduce an additional parameter, A-indicator, which is the standard deviation of the relative crop yield. We determined A-indicator for wheat producing regions of North America and the former USSR (Menzhulin et al., 1987; Gleik et al., 1990; Menzhulin, 1992). It is commonly assumed that up to the middle of the 1980s global climate changes were insignificant; therefore, the estimates obtained for A-indicator using the data on crop productivity for the previous years can be assumed to correspond to the anthropogenically undisturbed climate. For this analysis the data on annual wheat yields for separate small areas of the two regions for the period 1945–82 have been used.

In analyzing the spatial distribution of the A-parameter for wheat crops over the grain zone of the former USSR, it is important to note that the areas of winter wheat cultivation are mainly located in European Russia. The climate in eastern Russian territories is too severe. For the European part of the former USSR the annual variability of winter wheat production is

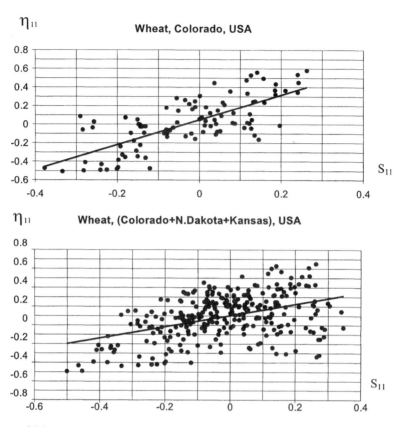

Figure 34.6 Graphs showing relationship between 11-year means of S-aridity index S_{11}, and 11-year means of wheat yield anomalies. $R^2 = 0.32$ (upper graph) and 0.21 (bottom graph). Data period is from 1891 to 2000.

very large (with A-indicator values varying between 0.20 and 0.40). Spring wheat cultivation in the former USSR is much larger than winter wheat. Except for European Russia, spring wheat crops are widely grown in the south of western Siberia and partly in East Siberia and the Far East. The A-indicator ranged from 0.20 to 0.60.

Yield variability in Northern America and Canada was studied during 1934–82. The A-indicator ranged from 0.10 to 0.35, though this range is less than in the case of the former USSR agricultural zone. For winter wheat, annual variability was more pronounced for the southwestern part of the U.S. wheat belt. The unsteady character of the rainfall regime during the plant-growth season results in significant annual variation in crop yields. For the grain areas of New Mexico and Colorado, the A-indicator reached 0.30. The frequent anomalies of wheat yields in these states are usually caused by droughts and dust storms which also cover the adjacent states. Especially large damage to the grain production in these U.S.

areas was caused by strong dust storms in the 1930s. The A-indicator values for winter wheat calculated for a long period for the Oklahoma and Texas grain areas were about 0.23. In central and northern areas of the U.S. wheat belt the stability of winter wheat yield was a little higher, with the A-indicator ranging from 0.15 to 0.20.

The absence of the winter soil freezing and the uniform annual distribution of atmospheric precipitation allows one to obtain rather stable yields of winter wheat in the forest-steppe areas of the U.S. corn belt. For agricultural areas of Ohio, Indiana, Illinois, Missouri, and Iowa, A-indicator ranged from 0.10 to 0.15, though it is also possible to note some tendency toward an increase in the value of this indicator in the western parts of this area where the climatic normal of moistening is lower. For the southern Atlantic U.S. states, the A-indicator was 0.15. In general, the A-indicator for spring wheat in North America remained between 0.10 and 0.30.

Specific Predictions of Future Droughts

Several authors have carried out calculations of possible changes in agroclimatic regime and wheat crop productivity over the former USSR grain belt using different scenarios of global climate changes (Menzhulin et al., 1995; Budyko and Menzhulin, 1996; Menzhulin, 1997). However, it is difficult or even impossible to identify the best model. The climate model recommended by IPCC cannot fully represent the present regime of atmospheric precipitation (Menzhulin, 1998a, 1998b; Houghton et al., 2001).

Figure 34.7 shows the correlation diagram of the statistical dependence between the normal annual precipitation and the A-indicator of the wheat yield abnormality. This is based on the IPCC predictions of the global mean annual temperature and precipitation changes in 2010 and 2050 for regions in the former USSR, European countries, and U.S. states.

In table 34.1 we present the results of our assessments of the changes in the A-indicator for wheat, as applied to some Russian regions, European countries, and U.S. states for climate changes in 2010 and 2050, according to the Holocene and Pliocene optimum analog scenarios, which correspond to global warming by 1°C and 3–4°C, respectively. In table 34.1 the changes in climatic normals of wheat productivity, which were calculated previously with the use of the specially developed crop productivity model, are also presented. This model takes into account the influence of the temperature and precipitation changes as well the atmospheric increase in carbon dioxide on the crop productivity in 2010 and 2050 (Menzhulin et al., 1987). The positive values of the wheat climatic productivity changes indicate a gain of potential productivity (i.e., a favorable change), but the positive values of the A-indicator show an increase of the annual variation in the amplitude of the wheat crop yield (i.e., an unfavorable change).

From table 34.1, it is noticeable that during the period of global warming until 2010, some increase in agricultural drought occurrence is possible in the U.S. states. It also shows a favorable impact of the expected climate

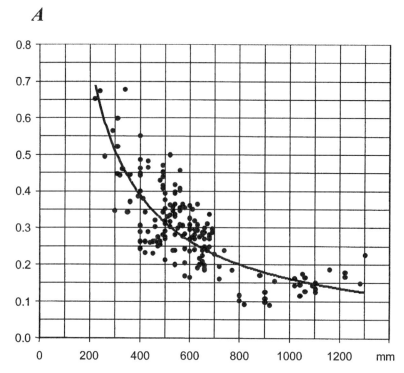

Figure 34.7 The correlation between normal annual precipitation (mm) and A-indicator (A) of wheat yield abnormality.

changes on the climatic normal for wheat productivity in U.S. states, where this impact is usually less than in the Russian regions and in European countries. But later in the phase of higher global warming in the middle of this century, the effect of partial drought that increased in some U.S. states in the first decade of this century will vanish because the atmospheric precipitation in agricultural regions of the United States, Russia, and Europe will increase, according to the Pliocene optimum scenario.

Conclusions

Modern anthropogenic global warming, which has already begun to manifest itself in many of the earth's regions, is the main factor for the occurrence of droughts causing low crop yields in some countries. In the past few decades we observed uniformity in the occurrence of droughts and crop productivity in major wheat-producing regions of the world. In general, in the regions of the moderate latitudes of the Northern Hemisphere, a global warming exceeding 1°C will be favorable for obtaining higher and stable agricultural production. This should promote the planting of more productive thermophilic crops in the northern areas and more hydrophilic cultivars in current arid zones.

Table 34.1 Changes in climatic normals (1961–90) of wheat yield and its annual variability (%) according to paleoclimatic scenarios of temperature and precipitation changes caused by global warming in 2010 and 2050

	2010		2050	
Region	Productivity	Variability	Productivity	Variability
United States				
North Dakota	−1	+20	+35	−20
South Dakota	−4	+22	+31	−19
Nebraska	−5	+18	+29	−24
Colorado	0	+3	+25	−22
Kansas	0	+8	+24	−15
Oklahoma	+4	+1	+21	−11
Minnesota	0	+14	+32	−16
Iowa	−1	+16	+27	−15
Missouri	+1	+8	+19	−12
Illinois	0	+10	+24	−13
Michigan	+2	+11	+23	−9
Ohio	+1	+8	+21	−8
Russia				
St. Petersburg	+19	0	+36	−20
Vorovezh	+7	+5	+26	−21
Toula	+8	+5	+29	−17
Lipetsk	+7	+4	+31	−19
Belgorod	+8	+4	+31	−18
Krasnodar	+17	−17	+31	−31
Volgograd	+12	−18	+52	−35
N. Novgorod	+8	−8	+35	−24
Tatarstan	+8	−7	+38	−25
Orenburg	+17	−18	+58	−37
Chelyabinsk	+13	−13	+42	−35
Altay	+14	−19	+44	−28
Europe				
Poland	+14	+4	+35	−22
Spain	+6	−2	+22	−25
France	+9	−3	+28	−12
Germany	+7	+2	+23	−17
Italy	+4	−5	+16	−20
Romania	+10	−5	+31	−23
Bulgaria	+12	−8	+28	−20
United Kingdom	+17	0	+21	−5
Hungary	+3	−3	+16	−20
Ukraine	+5	−5	+20	−24
Baltic Rep.	+16	+3	+32	−24
Denmark	+16	+3	+26	−11

Considering the perspective of future research in this field, it is necessary to notice that the weakest and the most complex component of such research is the development of the reliable forecasts of regional climate changes. Although recent research work on predicting climate change has made great advances at the global scale, predictions of regional climate change are still unreliable. Until there are considerable improvements in the reliability of predictions of regional climate change, we cannot expect to see much progress in forecasting agroclimatology and long-term drought prediction.

References

Borzenkova, I.I., M.I. Budyko, E.K. Byutner, K.Ya. Vinnikov, Yu.A. Izrael, K.I. Kobak, V.A. Legasov, A.A. Velichko, G.S. Golitsyn, I.L. Karol, I.I. Kuzmin, and G.V. Menzhulin. 1987. Anthropogenic Climatic Changes. Gidrometeoizdat Publishing, St. Petersburg. [English trans. University of Arizona Press, Tucson. 1990.]

Budyko, M.I., I.I. Borzenkova, G.V. Menzhulin, and I.A. Shiklomanov. 1994. Cambios Antropogenicos del Clima en America del Sur (J.J. Burgos, ed.). Serie de la Academia Nacional de Agronomia y Veterinaria, no. 19.

Budyko M.I., and G.V. Menzhulin. 1996. Climate change impacts on agriculture and global food production: Options for adaptive strategies. In: J. Smith et al. (eds.), Adapting to Climate Change: Assessments and Issues. Springer-Verlag, New York, pp. 188–203.

Cracknell, A.P. 1994a. Climate change—the background. In: R.A. Vaughan (ed.), Remote Sensing and Global Climate Change. Springer-Verlag, Berlin, pp. 1–33.

Cracknell, A.P. 1994b. Basis and structure of climate models. In: R.A. Vaughan (ed.), Remote Sensing and Global Climate Change. Springer-Verlag, Berlin, pp. 135–176.

Cracknell, A.P., 2001. Background. In: A.P. Cracknell (ed.), Remote Sensing and Climate Change: The Role of Earth Observation. Springer-Verlag, Berlin, pp. 1–36.

Gleik, P.H., G.V. Menzhulin, C.E. Rosenzweig, and I.A. Shiklomanov. 1990. Climatic change impact on water resources and agriculture. In: M.I. Budyko, M.S. Maccraken, Yu.A. Izrael, and A.D. Heck (eds.), Prospect for Future Climate. Working Group VIII. US/USSR Agreement on the Environment Protection. Lewis Publishers, Boca Raton, FL, pp. 214–259.

Houghton, J.T., Y. Ding, D.J. Griggs, M. Noguer, P.J. Van der Linden, X. Dai, K. Maskell, and C.A. Johnson (eds.). 2001. Climate Change 2001: The Scientific Basis. Contribution of Working Group I to the Third Assessment Report of the Intergovernmental Panel on Climate Change. Cambridge University Press, Cambridge.

McCarthy, J.J., O.F. Canziani, N.A. Leary, D.J. Dokken, and K.S. White (eds.). 2001. Climate Change 2001: Impacts, Adaptation, and Vulnerability. Contribution of Working Group II to the Third Assessment Report of the Intergovernmental Pannel on Climate Change. Cambridge University Press.

Menzhulin, G.V. 1992. The impact of expected climate changes on crop yields: Estimates for Europe, the USSR and North America based on paleoanalog scenarios. In: J.M. Reilly and M. Anderson (eds.), Economic Issues in Global Climate Change: Agriculture, Forestry and Natural Resources. Westview Press, Boulder, CO, pp. 353–381.

Menzhulin, G.V. 1997. Global warming, carbon dioxide increase and the prospects of crop potential: The assessments for Russia using alternative climate change scenarios. J. Agric. Meteorol. Jpn. (special issue) 52, no. 5.

Menzhulin, G.V. 1998a. Prediction of the regional climate changes and the estimation of their principal consequences. In: F.T. Yanshina (ed.), Global Ecological Problems in the Beginning of New Millennium: Materials Presented to the Conference Dedicated to the 85th Anniversary of Academician A.L.Yanshin. Nauka Publishing, Moscow, pp. 197–237.

Menzhulin, G.V. 1998b. Model climate changes scenarios and long-term prediction of global warming consequences: Comparison analysis for some environmental indices, 1999. In: Yu.A. Pykh, D.E. Hyatt, and R.J.M. Lenz (eds.), Environmental Indices Systems Analysis Approach. EOLSS Publishers, Oxford, pp. 116–155.

Menzhulin, G.V., L.A. Koval, and A.L. Badenko. 1995. Potential effect of global warming and carbon dioxide on wheat production in the Commonwealth of Independent States. In: G.A. Peterson (ed.), Climate Change and Agriculture: Analysis of Potential International Impacts. ASA Special Publication no 59. American Society of Agronomy, Madison, WI, pp. 275–292.

Menzhulin, G.V., L.A. Koval, M.V. Nikolayev, and S.P. Savvateyev. 1987. Agroclimatic consequences of modern climate changes: a scenario for north America [in Russian]. Trans. State Hydrol. Inst. 327:132–146.

Menzhulin, G.V., and N.V. Nikolayev. 1987. Calculation of the year-to-year variability and technological trends of crop yields [in Russian]. Trans. State Hydrol. Inst. 327:113–131.

Index

ACMAD (African Center of Meteorological Applications for Development), 408
ADEOS-II (Japanese Advanced Earth Observation Satellite), 96
Advanced microwave scanning radiometer (AMSR)
 passive microwave system, 96–97
 soil moisture retrieval algorithms, 92
Advanced Synthetic Aperture Radar (ASAR), sensor systems, 106
Advanced very high resolution radiometer (AVHRR)
 carrying capacity (CC), 80
 global vegetation index (GVI), 80
 monitoring major droughts, 82–83
 National Oceanic Atmospheric Administration (NOAA), 59, 65, 79, 255, 318–319, 328
 new method and data, 80–82
 normalized difference vegetation index (NDVI), 80–82
 South Korea, 393–394
 vegetation indices, 79–82
Advance Real Time Environmental Monitoring Information System (ARTEMIS), 413
Aerial photographs, remote sensing data, 57

Aerosol-free vegetation index (AFRI), optimized index, 64–65
AET (areal evapotranspiration), South Korea, 393–394
Africa. *See also* Ethiopia; Famine Early Warning System (FEWS) Network; Kenya; Livestock early warning system (LEWS); Near East; Radio and Internet (RANET) system; Southern Africa
 drought monitoring applications, 65, 67–72
 East Africa, 69–71
 El Niño and Southern Oscillation (ENSO) and drought in, 33–34
 livestock early warning system (LEWS) sites, 285
 rangelands, 283–284
 Sahel region, 67–69
 Southern Africa, 71–72
African Center of Meteorological Applications for Development (ACMAD), 408
African Mediterranean countries, wheat yield variability, 437–438
AGMP (Agricultural Meteorology Programme), 406–407
Agricultural drought
 crop yield and planting dates, 6
 crop yield and soil moisture, 6

Agricultural drought (*continued*)
definition, 42–43
factors affecting crop yield, 5
monitoring and predicting, 4–6
problems for Russia, 196
satellite data, 6–7
See also Ethiopia; Latin America; Near East; Russia
Agricultural drought index, Brazil, 161, 162
Agricultural Meteorology Programme (AGMP), 406–407
Agricultural Production Systems Simulator, 7, 381
Agricultural Technology Transfer Station (ATTS), Thailand, 326–328
Agriculture
Australia, 369–370
Bangladesh, 313–314
climate change and, 433–434
impact of drought in Near East, 213–215
India, 297
Indonesia, 330–331
Kenya, 238
Near East, 208, 210
Vietnam, 345–349
West Indies, 147–148
Agrometeorological applications, World Meteorological Organization (WMO), 406–407
Agrometeorological warning system, Latin America, 164, 165
Agronomic models, drought monitoring in Australia, 376
AI. *See* Aridity index (AI)
Algorithms
Satellite Application Facilities (SAFs) in Portugal, 191–192
soil moisture retrieval, 92–94
All-Russian Research Institute of Agricultural Meteorology (ARRIAM), 204
Aman, Bangladesh crop, 314–318
Amazon River basin, 156
See also Brazil; Latin America
AMSR. *See* Advanced microwave scanning radiometer (AMSR)

Annual drought index, Vietnam, 349, 352
Antecedent precipitation index (API), soil moisture, 101–102
Aqua, NASA, passive microwave system, 96–97
Areal evapotranspiration (AET), South Korea, 393–394
Argentina, monitoring droughts, 83, 86, 87
Aridity, Near East, 208–210
Aridity anomaly index, India, 302
Aridity index (AI)
drought in Portugal, 183
Latin America, 160, 161
ARRIAM (All-Russian Research Institute of Agricultural Meteorology), 204
ARTEMIS (Advance Real Time Environmental Monitoring Information System), 413
ASAR (Advanced Synthetic Aperture Radar), sensor systems, 106
Asia. *See also* Near East
soil degradation assessment, 423–424
Atmosphere resistance vegetation index, 64–65
Atmospheric interaction, Ethiopian drought, 230
ATTS (Agricultural Technology Transfer Station), Thailand, 326–328
AussieGRASS modeling framework, 377
Australia
agricultural production, 369–370
Agricultural Production Systems Simulator (APSIM) model, 381
attitudes toward drought, 369
AussieGRASS modeling framework, 377
drought and pasture condition alerts, 380–381
drought monitoring, 379–380
drought monitoring at state level, 375
government's response to drought, 373
impacts of drought, 370–372

incorporation of crop models in modeling framework, 381–382
land-use, 371
livestock numbers, 372
mitigating drought at farm scale, 374
national drought early warning system, 376–382
national wheat yield, 371
operational drought monitoring, 374–376
operational reports, 382
pasture model, 377–378
rainfall, 369–370
rainfall and climate monitoring, 375–376
spatial implementation of models, 378–379
spatial modeling framework, 377
stress index (STIN) model, 381–382
use of agronomic models to monitor droughts, 376
use of satellite imagery, 381
Automation, drought monitoring in Mexico, 138–139
Available water resources index (AWRI), South Korea, 390–392
AVHRR. *See* Advanced very high resolution radiometer (AVHRR)
AWRI (available water resources index), South Korea, 390–392

Bangladesh
  agriculture and economy, 313–314
  drought mitigation, 319–320
  drought monitoring, 318–319
  irrigation, 321
  Kharif drought, 316
  pre-*Kharif* drought, 314–315
  Rabi drought, 316–318
  satellite data monitoring drought, 318–319
Barley production
  Morocco, 214, 215
  North Africa, 214, 216
Bernoulli models
  dependent, 45–47
  independent, 45
Biometeorological time scale model, 7–8

BMG (Bureau of Meteorology and Geophysics), Indonesia, 334–335
BoM (Bureau of Meteorology), drought monitoring in Australia, 374–375
Botswana, monitoring drought, 71
Brazil. *See also* Latin America
  agricultural drought index, 162
  agrometeorological warning system, 164, 165
  aridity index (AI), 160
  better planting techniques, 166
  categorization of drought conditions, 162
  climatic classification and AI, 161
  climatic conditions, 156, 158
  climatic risk zoning, 166
  crop drought index (CDI), 163
  distribution of dry spells, 161
  drought monitoring, 158–164
  Palmer drought severity index (PDSI), 162–163
  precipitation anomaly, 162
  precipitation anomaly from potential evapotranspiration, 159–160
  precipitation distribution, 158
  production losses due to climate anomalies, 159
  soil moisture and potential evapotranspiration, 161
  standardization precipitation index (SPI), 163
  use of satellite data, 163–164
  variation in maize production, 159
  water deficit anomaly, 162
Brightness temperature
  radiation aridity index and, 101
  soil moisture and, 91–92
Bulgaria, wheat yield variability, 435–437, 438
Bureau of Meteorology (BoM), drought monitoring in Australia, 374–375
Bureau of Meteorology and Geophysics (BMG), Indonesia, 334–335

Caribbean Disaster Emergency Response Agency (CDERA), 149
Carrying capacity (CC), 80
Causative factors of drought
  China, 356–358
  Ethiopia, 230–232

Causative factors of drought (*continued*)
  India, 300–301
  Indonesia, 331–334
  Kenya, 238–239
  Latin America, 157–158
  Mexico, 134–135
  Near East, 211–213
  Poland, 171–172
  Russia, 198–200
  South Korea, 387, 389–390
  West Indies, 145–146
CC (carrying capacity), 80
CCD (cold cloud duration), 272, 414–416
CDERA (Caribbean Disaster Emergency Response Agency), 149
CDI (crop drought index), monitoring drought in Latin America, 163
CDMU (central drought management unit), 218–219
CEISS (Centro de Investigaciones Sobre la Sequía), 136–137
Cell expansion, water deficits, 13–14
Central drought management unit (CDMU), 218–219
Centro de Investigaciones Sobre la Sequía (CEISS), 136–137
Cereal crops
  crop water use, 11–12
  drought and water deficit effects on plants, 12–15
CGI (crop growth index), Poland, 177
China
  anomalous normalized difference vegetation index (NDVI) model, 361
  area affected by drought, 367
  causes of droughts, 356–358
  combination model, 361
  crop water stress index, 360
  cultural factors causing drought, 357–358
  distribution of water resources, 357
  drought early warning system, 362–364
  drought monitoring methods, 358–361
  geographic information system (GIS) database, 362–363
  history of droughts in, 354, 355
  meteorological factors causing drought, 356–357
  monitoring droughts, 83, 87
  normalized difference temperature index (NDTI), 360
  normalized difference water index (NDWI), 361
  operational system for drought early warning, 362, 365
  potential drought research needs, 366
  remote sensing, 359–360
  seasonal characteristics of droughts, 355–356
  soil humidity index (SHI), 363–364, 365
  spatial distribution of droughts, 355
  spatial drought pattern of winter drought, 115
  timeslice products, 363–364
  water balance model, 358–359
Chlorophyll, remote sensing of vegetation, 60
Climate agriculture, influence in Portugal, 181–183
Climate change
  agriculture, 433–434
  A-indicator, 443–445, 446
  empirical techniques, 443
  global warming, 445, 447
  human activities, 429–430
  models, 430–431
  paleoclimatic reconstruction, 442
  possible impact of future, and droughts, 438–445
  specific predictions of future droughts, 445–446
  trend analysis, 441–442
Climate index, influence in Portugal, 181, 182
Climate Information and Prediction Services (CLIPS), 405–406
Climate models, greenhouse effect, 430–431
Climate prediction
  change, 432
  human activities, 431–432
  Intergovernmental Panel on Climate Change (IPCC), 432–433

numerical models in weather forecasting, 432
Climate Variability Project (CLIVAR), 402–403
Climatic risk zoning, Brazil, 166
CLIPS (Climate Information and Prediction Services), 405–406
CLIVAR (Climate Variability Project), 402–403
Cold cloud duration (CCD), 272, 414–416
Colorado, wheat yield variation, 434, 435
Combination model, China, 361
Communication, RANET (Radio and Internet) system, 276–278
Community outreach program, livestock early warning system (LEWS), 291–292
Corn, United States and drought, 121, 122
Cotton, droughts in China, 83
Crisis control, Near East, 218–219
Crop calendar, Bangladesh, 315
Crop drought index (CDI), monitoring drought in Latin America, 163
Crop growth index (CGI), Poland, 177
Crop growth models, 7–8
 Portugal, 192
Crop management
 maize, 16–17
 sorghum, 24
 wheat, 21
Crop mixing, India, 306
Crop monitoring
 Indonesia, 336–339
 systems in Ethiopia, 233
Crop planning, India, 304–305
Crop watch group, India, 301–302
Crop water requirement satisfaction index (WRSI) model, 415–416
Crop water stress index China, 360
Crop yield
 dryland bean and maize in Mexico, 140, 141
 dryland corn for Nebraska, 121, 122
 factors affecting, 5
 normalized difference vegetation index (NDVI), 65, 66
 planting dates, 6
 predicting, using standardized precipitation index (SPI), 137–138
 satellite data, 6–7
 soil moisture, 6
Cropping seasons. *See* Bangladesh
Cultural factors, droughts in China, 357–358

DEC (drought exceptional circumstances), Australia, 373
Decision Support System for Agrotechnology Transfer (DSSAT), 7
Deforestation, Ethiopian drought, 231
Degradation. *See* Land degradation
Dekadal index (DI), 259
Dependent Bernoulli model, 45–47
Desertification. *See* Land degradation
Desiccation effects, drought, 14
DI (dekadal index), 259
Difference vegetation index (DVI), 64
Differentiation criteria, drought in Russia, 203
Distribution
 maize, 15
 rice, 17
 sorghum and millet, 21–22
 wheat, 19
DMCN. *See* Drought Monitoring Center in Nairobi (DMCN)
Drought. *See also* Agricultural drought; World Meteorological Organization (WMO)
 definitions, 3–4, 132–134, 145, 196
 early warning systems for, 404–405
 idealized graphs of severity, 14
 impact of, in Australia, 370–372
 impact on Near East agriculture, 213–215
 insurance companies, 134
 maize, 15–16
 Near East, 210–213
 reasons for, in Russia, 198–200
 remote sensing systems for monitoring, 58–60
 rice ecosystems, 17–18
 sorghum and millet, 22–23
 vulnerability in Portugal, 183
 wheat, 19–20

Drought assessment index, Russia, 203
Drought exceptional circumstances (DEC), Australia, 373
Drought frequency, Ethiopia, 228, 229
Drought index
  China, 359
  monitoring in Russia, 201–202
  South Korea, 390–392
  Thailand, 325
  Vietnam, 349, 352
Drought management, Near East, 218–219
Drought mitigation
  Australia, 374
  Bangladesh, 319–320
  Ethiopia, 233–234
  Food and Agriculture Organization (FAO), 416–418
  India, 304–308
  Indonesia, 339, 342
  Kenya, 246–249
  Latin America, 166
  Poland, 177–178
  Portugal, 192
  Russia, 204–205
  southern Africa, 273–274
  South Korea, 395
  United States, 129
  West Indies, 151, 153
Drought Monitor
  drought severity classification system, 128
  maintenance, 126
  map, 126–127
  Mexico, 136–137
  United States, 126–129
Drought monitoring. *See also* Monitoring drought
  Food and Agriculture Organization (FAO), 412
  operational, in Australia, 374–376
  RANET (radio and Internet) system, 276–280
  sugarcane in West Indies, 149–151
  sugar industry, 151
Drought Monitoring Center in Nairobi (DMCN)
  Kenya, 240, 243, 246
  World Meteorological Organization (WMO), 408

Drought preparedness, southern Africa, 273–274
Drought prevention, Food and Agriculture Organization (FAO), 417–418
Drought research needs
  China, 366
  Ethiopia, 234
  Latin America, 166
  Mexico, 138–139
  Poland, 178–179
  South Korea, 396
Drought-resistant crops, Kenya, 246–249
Droughts. *See also* Latin America; Mexico; United States
  Africa, 33–34
  Argentina, 83, 86, 87
  Bangladesh, 314–318
  causative factors in Mexico, 134–135
  China, 83, 87, 354–356
  El Niño and Southern Oscillations (ENSO), 30–35
  Ethiopia, 229–230
  former Soviet Union, 82–83, 85
  impacts in India, 298–300
  impacts in Mexico, 134
  India, 297–300
  Indonesia, 333
  investigations, 97–102
  Latin America, 33
  mainland Portugal, 184–185
  monitoring in Mexico, 136–138
  monitoring major, by AVHRR satellite data, 82–83
  Oceania, 32
  Poland, 171–172
  regional-scale soil moisture dynamics, 98, 99
  soil moisture-related indices, 100–102
  Southeast Asia, 32–33
  South Korea during 1960s—90s, 388
  specific predictions of future, 445–446
  surface and profile relationships, 98, 100
  United States, 34–35, 82, 84
Drought spatialization, Near East, 220–221

Drought stress
  high-resolution synthetic aperture radar (SAR) imaging, 111–113
  large-scale monitoring using scatterometers, 113–114
Drought vulnerability mapping, Near East, 221
Dryland climate, World Meteorological Organization (WMO), 403–404
Dryland degradation. *See* Land degradation
Dry spells, Latin America, 161
DSSAT (Decision Support System for Agrotechnology Transfer), 7
DVI (difference vegetation index), 64

Early warning systems. *See also* Livestock early warning system (LEWS); Warning systems
  China, 362–364, 365
  Ethiopia, 233
  Food and Agriculture Organization (FAO), 418
  Global Information and Early Warning System (GIEWS), 425–426
  Kenya, 246
  national drought, in Australia, 376–382
  Portugal, 192
  Russia, 204
  southern Africa, 273
  World Meteorological Organization (WMO), 404–405
Earth Observing System (EOS), remote sensing systems, 59–60
Earth Resource Technology Satellite (ERTS), drought monitoring, 58
East Africa, monitoring drought, 69–71, 72
Electromagnetic spectrum, 57, 58, 72–73, 90
El Niño
  description, 28–29
  impact on rice and crops in Indonesia, 341
  Indonesia, 333
  regions in world, 36
El Niño and Southern Oscillation (ENSO), 4, 28, 29–30, 426

Australia, 369–370
drought in India, 300–301
drought prediction in Near East, 219–220
droughts, 30–35
droughts in China, 356–357
impact on rainfall variability in Indonesia, 332–334
Latin America, 157–158
Mexico and droughts, 135–136
simple and composite indices, 35–36
southern Africa, 266, 269–270
West Indies, 145–146
Empirical orthogonal function (EOF), South Korea, 392, 393
Employment, India, 308
Enhanced vegetation index (EVI), optimized index, 64–65
ENSO. *See* El Niño and Southern Oscillation (ENSO)
Environmental Policy Integrated Climate, 7
Environment and Natural Resources Service, 412–413
EOF (empirical orthogonal function), South Korea, 392, 393
EOS (Earth Observing System), remote sensing systems, 59–60
Erosion Productivity Impact Calculator, 7
ERTS (Earth Resource Technology Satellite), drought monitoring, 58
Ethiopia
  atmospheric interaction, 230
  crop monitoring, 233
  deforestation, 231
  drought early warning system, 233
  drought frequency, 228, 229
  drought mitigation, 233–234
  drought monitoring systems, 232–233
  establishing drought prediction methods, 234
  food supply systems, 227–228
  history of droughts or famines, 229–230
  land ownership, 232
  major causes of droughts, 230–232
  normalized difference vegetation index (NDVI), 232

Ethiopia (continued)
  overgrazing, 231
  population growth, 231
  potential drought research needs, 234
  rainfall analysis, 232
  recommendations, 234–235
  soil erosion, 230–231
  water balance method, 233
Europe
  wheat yield and global warming, 447
  wheat yield variability, 435–437, 438, 439
European Space Agency Soil Moisture Ocean Salinity Mission (SMOS), 96
Evapotranspiration
  Latin America, 161, 162
  potential (PET), 228
  precipitation anomaly, 159–160
EVI (enhanced vegetation index), optimized index, 64–65

Famine Early Warning System (FEWS), 4–5, 219
  computation of RNCD indices, 259–260
  delineation of base-unit polygons, 259
  evaluation of RNCD, 260–262
  identification of growing season periods, 258–259
  rainfall and NDVI combined-departures method, 258–262
  remote sensing data for, 254–255
  water requirement satisfaction index (WRSI), 255–258
Famines, Ethiopia, 229–230
FAO. See Food and Agriculture Organization (FAO)
Farmers' participation, India, 309
Farming, integrated, India, 306
Federal Emergency Management Agency (FEMA), 121
FEWS. See Famine Early Warning System (FEWS)
FFRI (forest fire index), Portugal, 190–191
Fodder bank, India, 309

Food and Agriculture Organization (FAO), 4, 218
  Advance Real Time Environmental Monitoring Information System (ARTEMIS), 413
  agricultural drought mitigation, 416–418
  agricultural drought monitoring, 412
  crop water requirement satisfaction index (WRSI) model, 415–416
  departments, 411
  early warning and forecasting programs, 418
  Environment and National Resources Service, 412–413
  future directions, 418–419
  global information and early warning system (GIEWS), 412
  long-term drought prevention programs, 417–418
  quantitative application of satellite imagery, 414–415
  use of satellite data, 414–416
  World Food Summit, 411–412
Food for Work program, India, 308
Food Security and Agricultural Projects Analysis Service, 417
Food security and buffer stock, India, 307–308
Forecasting crop production, rice in Indonesia, 336–339
Forecasting programs, Food and Agriculture Organization (FAO), 418
Forest fire index (FFRI), Portugal, 190–191
Former Soviet Union, monitoring droughts, 82–83, 85
Frequency, radar system, 108
Future outlook
  Food and Agriculture Organization (FAO), 418–419
  land degradation, 426–427
  livestock early warning system (LEWS), 292–293

Geographic information system (GIS) database, drought monitoring in China, 362–363

Germination, water deficit effects, 12–13
GEWEX (Global Energy and Water Cycle Experiment), 402
GIEWS (Global Information and Early Warning System), 219, 412, 425–426
GIMMS (Global Inventory Modeling and Mapping Studies), 413
GIS (geographic information system) database, drought monitoring in China, 362–363
Global Assessment of Soil Degradation (GLASOD), 423
Global climate change, World Meteorological Organization (WMO), 403–404
Global Energy and Water Cycle Experiment (GEWEX), 402
Global Information and Early Warning System (GIEWS), 219, 412, 425–426
Global Inventory Modeling and Mapping Studies (GIMMS), 413
Global vegetation index (GVI), NOAA, 80
Global warming
  climate change, 445, 447
  human activities, 429–430
Government, drought in Australia, 373
Grain
  droughts in Argentina, 83
  droughts in China, 83
  droughts in former Soviet Union, 82–83
  droughts in United States, 82
  India, 297–300
  problems for Russia, 196–198
Grazing system model (GRASP), drought monitoring in Australia, 377–380
Greenhouse effect, climate models, 430–431
Green vegetation index (GVI), 64
Groundwater irrigation, Bangladesh, 320, 321
Growing degree days, South Korea, 392–393

History, droughts in China, 354

Human activities, global warming, 429–430
Humidity index, Vietnam, 352
Hydrothermal coefficient, drought monitors in Russia, 200

IATF (Inter-Agency Task Force for Disaster Reduction), 409
Implementation, warning system in Thailand, 328
Incident angle, radar system, 108–109
Independent Bernoulli model, 45
India
  agriculture, 297
  aridity anomaly index, 302
  crop mixing and integrated farming, 306
  crop planning, 304–305
  crop watch group, 301–302
  drought mitigation, 304–308
  drought monitoring methods, 301–304
  drought-prone districts, 297, 298
  droughts, 297
  employment, 308
  food security and buffer stock, 307–308
  impacts of droughts, 298–300
  main causes of drought, 300–301
  moisture adequacy index, 302
  optimizing and improving water resources, 306
  potential drought research or management needs, 308–309
  precipitation deviation, 301
  rainfall, 297
  relief measures, 307–308
  shelterbelts and mulching, 306
  soil and water conservation, 307
  use of satellite data, 303–304
  water use estimation and improving storage, 308–309
  weather forecasting, 303
  weather network, farmers' participation, and fodder bank, 309
  yield modeling, 302–303
India Meteorological Department, 301
Indian Remote Sensing, 59, 304

Indonesia
    agriculture and drought, 330–331
    causative factors of drought, 331–334
    climatology of seasonal rainfall, 332
    crop production forecasting and monitoring system, 336–339
    crop's physical appearance, 338
    drought monitoring and forecasting system, 334–336
    drought vulnerability map, 338–339
    El Niño and drought events, 333
    impact of El Niño on rice and secondary crops, 341
    impact of El Niño/Southern Oscillation rainfall variability, 332–334
    mitigation, 339, 342
    monitoring agricultural drought, 334–339
    national food crops production, 341
    rainfall in rice-producing areas, 335
    rainfall types, 331–332
    rice production loss by district, 340
    satellite data use, 337–338
    using Southern Oscillation Index (SOI), 338
    verification of seasonal prediction, 336
    vulnerability of rice-growing area, 339
Insurance companies, drought definition, 134
Integrated farming, India, 306
Inter-Agency Task Force for Disaster Reduction (IATF), 409
Intergovernmental Panel on Climate Change (IPCC)
    climate prediction, 432–433
    World Meteorological Organization (WMO), 403–404
International Satellite Land Surface Climatology, 402
International Strategy for Disaster Reduction, 409
Internet. *See* Radio and Internet (RANET) system
Intertropical Convergence Zone (ITCZ)
    droughts in Kenya, 238–239

    rainy seasons, 69
    southern Africa, 268
IPCC. *See* Intergovernmental Panel on Climate Change (IPCC)
Irrigation
    Bangladesh, 320, 321
    sorghum, 23–24
    wheat, 20–21
ITCZ. *See* Intertropical Convergence Zone (ITCZ)

Jamaica. *See also* West Indies
    agriculture, 147–148
    drought monitoring, 148
    ecological regions of sugar industry, 147
Japanese Advanced Earth Observation Satellite (ADEOS-II), 96

Kansas, wheat yield variation, 434, 437
Kenya
    agroecological conditions and drought-resistant status of grain crops, 248
    annual rainfall distribution, 242
    arid and semiarid lands, 243
    causes of drought, 238–240
    drought early warning system, 246
    drought mitigation, 246–249
    drought monitoring, 240–246
    growing drought-resistant crops, 246–249
    historical climatic data, 240–243
    human practice contribution to droughts, 239–240
    impact of drought on maize and livestock, 239, 244
    Intertropical Convergence Zone (ITCZ), 238–239
    livestock conditions, 245–246
    monitoring drought, 69–71
    Palmer drought index (PDI), 241–242, 245
    quartile drought index, 242–243, 245
    rainfall variability, 240
    remote sensing, 243, 245
Kharif drought, Bangladesh, 316
Korea. *See* South Korea
Korea Meteorological Administration, drought indices, 390–392

Land degradation
  assessment of South and Southeast Asia, 423–424
  consequences, 421
  early warning systems, 425–426
  factors affecting, 422
  future directions, 426–427
  global assessment, 423
  improving the, assessment, 424, 425
  methods for assessment, 425
  monitoring, 422–424
Land management, Latin America, 157
Land ownership, Ethiopian drought, 232
Landsat series, satellite data, 58
Latin America
  agrometeorological warning system, 164, 165
  causative factors of agricultural drought, 157–158
  distribution of dry spells, 161
  drought mitigation, 166
  drought research needs, 166
  El Niño and Southern Oscillation (ENSO) and drought in, 33
  monitoring agricultural drought, 158–164
  soil moisture and potential evapotranspiration, 161
LEWS. *See* Livestock early warning system (LEWS)
Linear discriminant analysis, 42
Livestock
  conditions in Kenya, 245–246
  drought in Australia, 371–372
  Kenya drought, 239, 244
Livestock early warning system (LEWS)
  acquiring satellite-based weather data, 289–290
  biophysical modeling, 286
  establishment of monitoring sites, 284–286
  geostatistics and spatial extrapolation using vegetation indices, 288, 289
  locations of monitoring sites, 285
  model analysis and verification, 286–287
  Phytomass Growth Simulator model (PHYGROW), 286–288
  rangelands in Africa, 283–284
  spatial characterization, 284
  using meteorological projections, 290–291
Livestock producers, drought mitigation, 129
Long-term drought prevention programs, Food and Agriculture Organization (FAO), 417–418

MAI (moisture adequacy index), India, 302
Maize. *See also* Cereal crops
  Brazil, 156–157, 159
  crop management, 16–17
  distribution and use, 15
  drought and water use efficiency (WUE), 15–16
  global production, 16
  Kenya drought, 239, 244
  mycotoxins, 16
  southern Africa, 268, 273
  Vietnam, 345, 347
  water requirement satisfaction index (WRSI), 259, 260
  yields in north-central Mexico, 138, 140, 141
Management, research and needs in India, 308–309
Mapping system, livestock early warning system (LEWS), 288, 289
Mapping vulnerability, Near East, 221
Markov model, 47–48
Mediterranean climate. *See* Portugal
MEI (multivariate ENSO index), description, 35–36
Meteor, Russian satellite monitoring drought, 204
Meteorological drought, 4
Meteorological factors, drought in China, 356–357
Meteorological projections, livestock early warning system, 290–291
METEOSAT data, Food and Agriculture Organization (FAO), 414–415
Mexican Drought Research Center (CEISS), 136–137
Mexico
  annual temperature variation, 135
  automation for drought monitoring, 138–139

Mexico (*continued*)
  causative factors of drought, 134–135
  density of weather stations, 138
  drought definition, 132–134
  drought monitoring, 136–138
  drought-prone areas, 132
  drought research, 139, 142
  drought research needs, 138–139
  El Niño/Southern Oscillation and drought, 135–136
  impacts of droughts, 134
  monthly mean rainfall values, 135
  predicting crop yields using standardized precipitation index (SPI), 137–138
  SPI correlations with maize and bean yields, 139, 140, 141
  weather variables, 139
Microwave polarization difference index (MPDI), 94
Microwave remote sensing. *See also* Remote sensing
  active systems, 105–107
  advanced microwave scanning radiometer (AMSR), 92, 96–97
  antecedent precipitation index (API), 101–102
  current passive systems, 95–96
  drought-related investigations, 97–102
  future passive systems, 96–97
  microwave measurement and vegetation, 92
  microwave polarization difference index (MPDI) and vegetation parameters, 94–95
  physical basis, 90–92
  radiation aridity index (S), 100–101
  regional-scale soil moisture dynamics using passive microwave sensors, 98, 99
  scanning multifrequency microwave radiometer (SMMR), 94, 95–96
  soil moisture and brightness temperature, 91–92
  soil moisture and reflectivity, 91
  soil moisture-related indices, 100–101, 100–102
  soil moisture retrieval algorithms, 92–94
  Special Satellite Microwave/Imager (SSM/I), 95
  surface and profile relationships, 98, 100
  surface soil moisture, 90–95
  Tropical Rainfall Measurement Mission (TRMM) Microwave Imager (TMI), 95
  Windsat, 97
Millet. *See also* Cereal crops
  distribution, 21–22
  drought and drought resistance, 22–23
  global production, 22
  semiarid region of Africa, 68, 69
Misra's green vegetation index (MGVI), 64
Mitigation. *See* Drought mitigation
Models. *See also* Prediction techniques
  climate change, 430–431
  crop growth, 7–8
  dependent Bernoulli model, 45–47
  drought parameters, 51–52
  independent Bernoulli model, 45
  land degradation assessment, 425
  Markov model, 47–48
  Phytomass Growth Simulator model (PHYGROW), 286–288
  regional drought modeling, 49–51
  spatio-temporal drought, 48–52
  temporal drought, 44–52
  yield, in India, 302–303
Moderate Resolution Imaging Spectroradiometer (MODIS), monitoring drought, 59–60
Modified soil-adjusted vegetation index (MSAVI), 64
MODIS (Moderate Resolution Imaging Spectroradiometer), monitoring drought, 59–60
Moisture adequacy index (MAI), India, 302
Moisture index, drought monitors in Russia, 200
Monitoring drought
  Bangladesh, 318–319
  China, 358–361

Food and Agriculture Organization (FAO), 412
India, 301–304
Indonesia, 334–339
Kenya, 240–246
land degradation, 422–424
Mexico, 136–138
Near East, 215–218
Near East systems, 215–218
Poland, 172–177
Portugal, 185–192
procedures for Russia, 200–204
RANET (radio and Internet) system, 276–278, 276–280
remote sensing systems, 58–60
Russia, 200–204
sites for livestock early warning system (LEWS), 284–286
southern Africa, 271–272
Thailand, 323–325
West Indies, 148–151
Monsoons, India, 304–305
Monthly drought index, Vietnam, 349
Morocco, barley production, 214, 215
MPDI (microwave polarization difference index), 94
MSAVI (modified soil-adjusted vegetation index), 64
Mulching, India, 306
Multivariate ENSO index (MEI), description, 35–36
Mycotoxins, maize, 16

NADAMS (National Agricultural Drought Assessment and Monitoring System), India, 303
NAO. *See* North Atlantic Oscillation (NAO)
NARS (National Agricultural Research System), Bangladesh, 318
NASA Aqua, passive microwave system, 96–97
National Agricultural Drought Assessment and Monitoring System (NADAMS), India, 303
National Agricultural Research System (NARS), Bangladesh, 318
National Climate Center (NCC), Australia, 375–376

National Drought Mitigation Center (NDMC), drought monitoring, 122
National Drought Observatory (NDO), 218
National Early Warning Units (NEWUs), 418
National Environmental Satellite, Data and Information Service (NESDIS), 86
National Meteorological and Hydrological Services (NMHs), 401
National Oceanic and Atmospheric Administration (NOAA) series. *See also* Advanced very high resolution radiometer (AVHRR)
  advanced very high resolution radiometer (AVHRR), 59, 65, 79, 255, 318–319, 328
  AVHRR-based vegetation indices, 79–82
  crop conditions and droughts in Poland, 174–177
  South Korea, 393–394
Naval Research Lab Windsat, passive microwave system, 96–97
NCC (National Climate Center), Australia, 375–376
NDMC (National Drought Mitigation Center), drought monitoring, 122
NDO (National Drought Observatory), 218
NDTI (normalized difference temperature index), China, 360
NDVI. *See* Normalized difference vegetation index (NDVI)
NDWI (normalized difference water index), 73, 361
Near East
  agriculture in countries, 210
  aridity in North Africa and West Asia, 208–209
  barley production fluctuations in North Africa, 216
  barley production fluctuations in West Asia, 217
  barley production in Morocco, 214, 215

Near East (*continued*)
 causes of drought occurrence, 211–213
 central drought management unit (CDMU), 218–219
 climatic moisture regimes in North Africa and West Asia, 209
 deviations of annual precipitation for Syria, 212
 drought, 210–213
 drought monitoring systems, 215–218
 drought research needs, 219–221
 drought vulnerability mapping, 221
 feasibility of drought prediction, 219–220
 Food and Agriculture Organization of United Nations (FAO), 218
 from crisis control to drought management, 218–219
 impact of drought on agriculture, 213–215
 intraseasonal droughts at Syria, 213
 National Drought Observatory (NDO), 218
 North Atlantic Oscillation (NAO), 212–213, 219–220
 precipitation for stations in Syria and Turkey, 211
 region characterization, 208
 spatialization of drought in data-insufficient areas, 220–221
 wheat yield variability, 437–438, 440
Nearest neighbor analysis, 42
Nebraska, corn yield and drought, 121, 122
NESDIS (National Environmental Satellite, Data and Information Service), 86
NEWUs (National Early Warning Units), 418
Niger. *See also* Radio and Internet (RANET) system
 communication by RANET, 279–280
NMHs (National Meteorological and Hydrological Services, 401

NOAA. *See* National Oceanic and Atmospheric Administration (NOAA) series
Normalized difference temperature index (NDTI), China, 360
Normalized difference vegetation index (NDVI)
 crop conditions and droughts in Poland, 174–177
 description, 62, 63, 80
 East Africa, 69–71
 famine early warning system (FEWS), 254–255
 geostatistics and spatial extrapolation using, 288
 India, 303–304
 Kenya, 243
 monitoring drought in Ethiopia, 232
 NDVI model in China, 361
 relationship with crop yield, 65, 66
 Sahel region of Africa, 67–69
 Southern Africa, 71–72
 South Korea, 393–394
 use in India, 303–304
Normalized difference water index (NDWI), 73, 361
North Africa. *See also* Near East
 barley production, 214, 216
North Atlantic Oscillation (NAO)
 drought occurrence in Near East, 212–213
 drought prediction in Near East, 219–220
North-Central Mexico. *See* Mexico
North Dakota, wheat yield variation, 434, 436

Oceania, El Niño and Southern Oscillations (ENSO) and drought in, 32
Office of Disaster Preparedness and Emergency Management (OPDEM), Jamaica, 153
Operational reports, drought in Australia, 382
Orthogonal transformation, vegetation indices, 64
Overgrazing, Ethiopian drought, 231

Paddy, Bangladesh, 314

Palmer drought index (PDI), Kenya, 241–242
Palmer drought severity index (PDSI)
  monitoring drought in Latin America, 162–163
  Portugal, 186–188, 189, 190
  South Korea, 390
  United States, 122–123
Pasture model, drought monitoring in Australia, 377–378
Pattern recognition, drought prediction, 41–42
PD (polarization difference), 93–94
PDI (Palmer drought index), Kenya, 241–242
PDSI. *See* Palmer drought severity index (PDSI)
Period index (PI), 260
Perpendicular vegetation index (PVI), distance-based, 63–64
Phytomass Growth Simulator model (PHYGROW), 286–288
Planting dates
  Bangladesh, 317–318
  crop yield, 6
Planting techniques, Brazil, 166
Poland
  average yield of main cereals, 173
  changes of crop growth conditions, 177
  crop production, 171, 172
  development of drought conditions, 176
  distribution of sensible heat to latent heat index, 178
  drought-mitigating measures, 177–178
  drought monitoring methods, 172–177
  map of distribution of drought conditions, 175
  map of drought-prone areas, 174
  normalized difference vegetation index (NDVI), 175
  potential drought research needs, 178–179
  remote sensing-based crop condition assessment system, 174–177
  spatial distribution of droughts and causes, 171–172
  temperature condition index (TCI), 175–176
  vegetation condition index (VCI), 175
Polarization
  definition, 91
  radar system, 107–108
Polarization difference (PD), 93–94
Population growth
  Bangladesh, 313, 314
  Ethiopian drought, 231
Portugal
  climate index identifying drought-prone areas, 182
  crop growth models, 192
  drought-mitigating measures, 192
  drought monitoring, 185–192
  drought monitoring using satellite data, 189–192
  droughts in mainland, 184–185
  early warning systems, 192
  forest fire index (FFRI), 190–191
  influence of climate agriculture, 181–183
  main causes of drought, 183–184
  normalized difference vegetation indices (NDVI), 189–190
  Palmer drought severity index (PDSI), 186–188
  precipitation, 183–184
  precipitation deviation and deciles, 185–186
  Satellite Application Facilities (SAFs), 191–192
  temperature, 184
  vulnerability to drought, 183
  yield losses of main crops, 185
Precipitation
  drought, 4
  Latin America, 157
  Near East, 210, 211, 212
  Portugal, 183–184, 185–186
  South Korea, 389–390
Precipitation anomaly, 159–160, 162
Precipitation deviation, India, 301
Prediction services, World Meteorological Organization (WMO), 405–406
Prediction techniques. *See also* Models
  climate, 431–433

Prediction techniques (*continued*)
  establishing drought, 234
  feasibility in Near East, 219–220
  future droughts, 445–446
  models, 44–52
  pattern recognition, 41–42
  soil moisture and drought, 89–90
  statistical regression, 40–41
  stochastic or probabilistic analysis, 42–52
  time series analysis, 41
Pre-*Kharif* drought, Bangladesh, 314–315
Preparedness, drought in southern Africa, 273–274
Probabilistic analysis, 42–44
  Vietnam, 349, 350, 351
PVI (perpendicular vegetation index), distance-based, 63–64

Quartile drought index, Kenya, 242–243
Queensland. *See also* Australia
  drought declarations, 379, 380
  drought monitoring, 375

Rabi drought, Bangladesh, 316–318
Radar remote sensing. *See also* Microwave remote sensing; Remote sensing
  agricultural drought, 105–106
  frequency, 108
  high-resolution synthetic aperture radar (SAR) imaging, 111–113
  incident angle, 108–109
  large-scale monitoring using scatterometers, 113–114
  monitoring drought stress, 111–114
  physical fundamentals, 107–111
  polarization, 107–108
  radar frequencies and corresponding wavelengths, 109
  sensor systems, 106–107
  soil, 109–110
  spatial drought pattern of winter drought in China, 115
  system parameters, 107–109
  target parameters, 109–111
  vegetation, 110–111

Radiation aridity index (S), soil moisture-related, 100–101
Radiative index (RI), Thailand, 323–324
Radio and Internet (RANET) system
  configuration, 279
  drought monitoring using, 279–280
  limitations, 280
  origin, 276–278
  technical configuration, 278–279
Rainfall
  drought monitoring in Australia, 375–376
  southern Africa, 266–267
  types in Indonesia, 331–332
  variability in Australia, 369–370
Rainfall analysis, monitoring systems in Ethiopia, 232
Rainfall and NDVI combined departures (RNCD)
  computation of RNCD indices, 259–260
  dekadal index (DI), 259
  delineation of base-unit polygons, 259
  evaluation of RNCD, 260–262
  identification of growing season periods, 258–259
  period index (PI), 260
Rainfall distribution, Kenya, 240, 241, 242
Rainfall index, southern Africa, 272
RANET. *See* Radio and Internet (RANET) system
Rangelands in Africa. *See* Livestock early warning system (LEWS)
Ratio vegetation index (RVI), 61–62
RCOF (Regional Climate Outlook Fora), 405
RDA (Rural Development Administration), growing degree days, 392–393
Reflectivity, soil moisture and, 91
Regional Climate Outlook Fora (RCOF), 405
Regional drought modeling, 49–50
Regional Early Warning System (REWS), 418
Regional institutions, World Meteorological Organization (WMO), 408–409

Relief measures, India, 307–308
Remote sensing. *See also* Microwave remote sensing; Radar remote sensing
  China, 359–360
  crop condition assessment system in Poland, 174–177
  famine early warning system (FEWS), 254–255
  Kenya, 243, 245
  land degradation assessment, 425
  monitoring drought, 58–60
  radar systems, 105–106
  surface soil moisture, 90–95
  vegetation, 60–61
Research needs
  China, 366
  Ethiopia, 234
  India, 308–309
  Latin America, 166
  Mexico, 138–139
  Near East, 219–221
  Poland, 178–179
  South Korea, 396
  World Meteorological Organization (WMO), 401–403
Retrieval algorithms, soil moisture, 92–94
REWS (Regional Early Warning System), 418
Rice. *See also* Cereal crops; Indonesia
  Bangladesh, 313, 314
  distribution, 17
  drought in rice ecosystems, 17–18
  global production, 18
  Indonesia, 330–331, 335
  production loss in Indonesia, 340
  Vietnam, 345, 347, 348
  water production functions, 18–19
RNCD. *See* Rainfall and NDVI combined departures (RNCD)
Rural African communities. *See* Radio and Internet (RANET) system
Rural Development Administration (RDA), growing degree days, 392–393
Russia
  differentiation criteria of drought, 203
  drought causes, 198–200
  drought index, 201–202

drought mitigation, 204–205
early warning system, 204
hydrothermal coefficient, 200
moisture index, 200
procedures for drought monitoring, 200–204
recurrence of severe droughts, 197
water supply index, 200–201
wheat yield and global warming, 447
winter vs. spring wheat yields, 197–198
yield risk estimation, 202–204
RVI (ratio vegetation index), 61–62

SAFs (Satellite Application Facilities), Portugal, 191–192
Sahel region of Africa, monitoring drought, 67–69
SAR. *See* Synthetic aperture radar (SAR)
Satellite Application Facilities (SAFs), Portugal, 191–192
Satellite data. *See also* Africa; Remote sensing; Vegetation indices
  acquisition for livestock early warning system (LEWS), 289–290
  Bangladesh monitoring, 318–319
  drought and crop yield, 6–7
  electromagnetic spectrum, 57, 58
  Food and Agriculture Organization (FAO), 414–416
  future research needs, 72–74
  Indian Remote Sensing Satellite (IRS), 59
  Landsat series, 58
  Moderate Resolution Imaging Spectroradiometer (MODIS), 59–60
  monitoring drought in Latin America, 163–164
  monitoring rice crop growth in Indonesia, 337–338
  National Agricultural Drought Assessment and Monitoring System (NADAMS), 303
  NOAA-advanced very high resolution radiometer (AVHRR), 59
  normalized difference vegetation index (NDVI), 303–304
  radar sensor systems, 106–107

Satellite data (*continued*)
  relationship between NDVI and crop yield, 65, 66
  remote sensing of vegetation, 60–61
  remote sensing systems for monitoring drought, 58–60
  shortwave-infrared (SWIR) wavelengths, 57
  South Korea, 393–394
  Systeme Probatoire pour l'Observation de la Terre (SPOT) series, 58–59
  Thailand, 328
  use in Africa, 65, 67–72
  use in India, 303–304
  use in Portugal, 189–192
  use in Russia, 204
  vegetation indices, 61–65
Satellite imagery, drought assessment in Australia, 381
SAVI (soil-adjusted vegetation index), 64
Scanning multifrequency microwave radiometer (SMMR)
  passive system, 95–96
  polarization difference data, 94
Scatterometers. *See also* Radar remote sensing
  large-scale monitoring using, 113–114
  radar instruments, 106–107
Sea-surface temperatures, southern Africa, 270–271
Shelterbelts, India, 306
SHI (soil humidity index), drought monitoring in China, 363–364, 365
Shortwave-infrared (SWIR) wavelengths, remote sensing, 57, 72–73
SMI (surface moisture index), 73–74
SMMR. *See* Scanning multifrequency microwave radiometer (SMMR)
SMOS (European Space Agency Soil Moisture Ocean Salinity Mission), 96
SOI (Southern Oscillation Index), 29, 338
Soil, radar measurements, 109–110

Soil-adjusted vegetation index (SAVI), 64
Soil conservation, India, 207
Soil degradation assessment for South and Southeast Asia (ASSOD), 423–424
Soil erosion, Ethiopian drought, 230–231
Soil humidity index (SHI), drought monitoring in China, 363–364, 365
Soil moisture. *See also* Microwave remote sensing
  antecedent precipitation index (API), 101–102
  brightness temperature and, 91–92
  crop yield, 6
  indices, 100–102
  Latin America, 161, 162
  microwave emissivity, 93
  microwave measurement and vegetation, 92
  passive microwave remote-sensing, 89–95
  physical basis for remote sensing, 90–92
  polarization, 91
  radiation aridity index (S), 100–101
  reflectivity and, 91
  retrieval algorithms, 92–94
Sorghum. *See also* Cereal crops
  crop management, 24
  distribution, 21–22
  drought and drought resistance, 22–23
  global production, 22
  irrigation, 23–24
  semiarid region of Africa, 69
Southeast Asia, El Niño and Southern Oscillation (ENSO) and drought in, 32–33
Southern Africa
  agriculture, 268
  causes of agricultural drought, 268–271
  cold cloud duration, 272
  drought monitoring techniques, 271–272
  drought preparedness and mitigation strategies, 273–274

early warning system, 273
El Niño/Southern Oscillation
  (ENSO), 266, 269–270
interannual rainfall variability,
  268–269
monitoring drought, 71–72
rainfall, 266–267
rainfall index, 272
sea-surface temperatures, 270–271
vegetation condition index, 272
water requirement satisfaction index,
  272
Zimbabwe's annual rainfall,
  268–269, 270
Southern Hemisphere, wheat yield
  variability, 438, 441
Southern Oscillation Index (SOI)
  crop production in Indonesia, 338
  description, 29
South Korea
  causes of drought occurrence, 387,
    389–390
  climate zones, 386–387
  drought indices, 390–392
  drought mitigation, 395
  drought monitoring methods,
    390–395
  drought research needs, 396
  drought years during 1960s—90s,
    388
  empirical orthogonal function,
    392
  growing degree days, 392–393
  late onset of summer rainy season,
    389
  low precipitation during spring
    season, 389
  low precipitation during winter,
    389–390
  low precipitation in summer rainy
    season, 389
  seasons, 386
  standardized vegetation index,
    394–395
  use of satellite data, 393–394
Soviet Union, former, monitoring
  droughts, 82–83, 85
Space Research and Remote Sensing
  Organization (SPARRSO),
  Bangladesh, 318–319

Spatio-temporal drought models,
  48–52
Special Satellite Microwave/Imager
  (SSM/I), 95
SPI. *See* Standardized precipitation
  index (SPI)
SPOT (Systeme Probatoire pour
  l'Observation de la Terre), satellite
  series, 58–59
SSM/I (Special Satellite Mi-
  crowave/Imager), 95
Standardized precipitation index (SPI)
  advantages over Palmer drought
    severity index (PDSI), 124
  classification of drought categories,
    125
  correlations of dryland bean and
    maize yields in Mexico with, 140,
    141
  drought monitoring in Mexico,
    136–138
  monitoring drought in Latin
    America, 163
  predicting crop yields using, 137–138
  South Korea, 390
  United States, 123–125
Standardized vegetation index (SVI),
  South Korea, 394–395
Statistical regression, drought
  prediction, 40–41
Stochastic analysis, 42–44
Stress. *See* Drought stress
Substrate-mediated phenomena, water
  deficits, 14–15
Sugarcane
  drought monitoring, 149–151
  West Indies, 147–148
Sugar industry
  monitoring agricultural drought,
    151
  worst harvests at Worthy Park, 150,
    152
Surface moisture index (SMI), 73–74
Surface water supply index (SWSI),
  123, 390
SVI (standardized vegetation index),
  South Korea, 394–395
SWIR (shortwave-infrared)
  wavelengths, remote sensing,
  57, 72–73

SWSI (surface water supply index), 123, 390
Synthetic aperture radar (SAR). *See also* Radar remote sensing
   Bangladesh, 319
   high-resolution SAR imaging, 111–113
   radar instruments, 106–107
Systeme Probatoire pour l'Observation de la Terre (SPOT), satellite series, 58–59

Tahiti—Darwin Index (TDI), description, 29
Temperature, drought cause in Portugal, 184
Temperature condition index (TCI), Poland, 174–177
Temporal drought models, 44–45
Thailand
   agricultural drought warning system, 325–328
   Agricultural Technology Transfer Station (ATTS), 326–328
   agriculture and economy, 323
   drought index, 325
   drought monitoring, 323–325
   implementation, 328
   input variables computing drought severity, 326
   radiative index (RI), 323–324
   regions, 324
Time series analysis, drought prediction, 41
Timeslice products, drought monitoring in China, 363–364
Transformed soil-adjusted vegetation index (TSAVI), 64
Transformed vegetation index (TVI), 62–63
Tropical Rainfall Measurement Mission (TRMM) Microwave Imager (TMI), 95
Turgor-mediated phenomena, water deficits, 13–14
Turkey. *See also* Near East
   climatic moisture regions, 208, 209

Uganda. *See also* Radio and Internet (RANET) system
   communication by RANET, 279–280
United Nations, Food and Agriculture Organization (FAO), 411
United Nations Environment Program (UNEP), 4, 423
   Intergovernmental Panel on Climate Change (IPCC), 432–433
   World Meteorological Organization (WMO), 403, 407
United States
   Colorado and wheat yield variability, 435
   drought mitigation, 129
   Drought Monitor map, 127
   Drought Monitor's drought severity classification system, 128
   dryland corn yields for Nebraska, 121, 122
   El Niño and Southern Oscillations (ENSO) and drought in, 34–35
   Kansas and wheat yield variability, 437
   monitoring droughts, 82, 84
   North Dakota and wheat yield variability, 436
   Palmer Drought Severity Index (PDSI), 122–123
   Standardized Precipitation Index (SPI), 123–125
   U.S. Drought Monitor, 126–129
   wheat yield and global warming, 447
   wheat yield variation, 434, 435, 436, 437

Vegetation
   microwave measurement and, 92
   radar measurements, 110–111
   remote sensing, 60–61
VEGETATION (VEG) sensor, 59
Vegetation condition index (VCI)
   China, 363–364, 365
   crop conditions and droughts in Poland, 174–177
   southern Africa, 272
Vegetation health index, 81–82, 87
Vegetation indices
   aerosol-free vegetation index (AFRI), 64–65

atmosphere resistant vegetation
    index (ARVI), 64–65
  distance-based, 63–65
  enhanced vegetation index (EVI), 64
  geostatistics and spatial
    extrapolation using, 288
  land degradation, 424
  normalized difference vegetation
    index (NDVI), 62, 63
  optimized indices, 64–65
  orthogonal transformation, 64
  perpendicular vegetation index (PVI),
    63–64
  ratio vegetation index (RVI), 61–62
  relationship between NDVI and crop
    yield, 65, 66
  slope-based, 61–63
  soil-adjusted vegetation index
    (SAVI), 64
  transformed vegetation index (TVI),
    62–63
Vegetation stress, indices characterizing, 81–82, 87
Vietnam
  agricultural regions, 345, 346, 352
  annual drought index, 349, 352
  drought indices, 349, 352, 353
  general characteristics of regions,
    346–349
  humidity index, 352
  monthly drought index, 349
  probability of agricultural drought,
    349, 350, 351
  rice area lost due to droughts, 348
Vulnerability, drought in Portugal, 183
Vulnerability mapping
  Indonesia, 338–339
  Near East, 221

Walker circulation, 29
Warning systems. *See also* Early
    warning systems; Livestock early
    warning system (LEWS); Thailand
  agricultural drought, in Thailand,
    325–328
  agrometeorological, 164, 165
Water balance equation, 12
Water balance method, monitoring
    systems, 233
Water balance model, China, 358–359

Water conservation, India, 207
Water deficit anomaly, 162
Water requirement satisfaction index
    (WRSI)
  famine early warning system (FEWS),
    255–258
  model by Food and Agriculture
    Organization (FAO), 415–416
  southern Africa, 272
Water resources
  droughts in China, 357
  improving in India, 306
Water supply index, drought monitors
    in Russia, 200–201
Water use
  estimation and storage in India,
    308–309
  South Korea, 387, 389–390
Water use efficiency (WUE)
  definition, 12
  maize, 15–16
  rice, 17
  sorghum and millet, 22–23
  wheat, 19
WCRP (World Climate Research
    Programme), 402
Weather. *See also* Climate prediction
  data for livestock early warning
    system (LEWS), 289–291
  islands in West Indies, 146
Weather forecasting, India, 303
Weather network, India, 309
Weather stations, density in Mexico,
    138
Weather variables, drought monitoring
    in Mexico, 139
West Asia. *See also* Near East
  barley production, 214, 217
West Indies
  agriculture, 147–148
  causative factors of drought,
    145–146
  drought mitigation, 151, 153
  drought monitoring, 148–151
  drought monitoring for sugarcane,
    149–151
  ecological regions of Jamaican sugar
    industry, 147
  geography, 144
  map, 145

West Indies (*continued*)
  monitoring agricultural drought in sugar industry, 151
  Office of Disaster Preparedness and Emergency Management (OPDEM), 153
  rainfall, 144–145, 146
  regional drought monitoring, 149
Wheat. *See also* Cereal crops
  African Mediterranean, 437–438
  biometeorological time scale model, 7–8
  Colorado, 434, 435
  crop management, 21
  distribution, 19
  drought and water use efficiency (WUE), 19–20
  drought in Australia, 370, 371–372
  droughts in United States, 82
  Europe, 435–437
  global production, 20
  irrigation, 20–21
  Kansas, 434, 437
  Near East, 437–438
  North Dakota, 434, 436
  Portugal, 185, 186
  productivity and variability with global warming, 447
  Southern Hemisphere, 438
  United States, 434–435
  yield problems for Russia, 196–198
  yield variation and prediction, 434–438
WHYCOS (World Hydrological Cycle Observing Station), 408
Windsat, Naval Research Lab, passive microwave system, 96–97
WMO. *See* World Meteorological Organization (WMO)
World Climate Research Programme (WCRP), 402
World Food Program, Food and Agriculture Organization (FAO), 418
World Food Summit, Food and Agriculture Organization (FAO), 411–412

World Hydrological Cycle Observing Station (WHYCOS), 408
World Meteorological Organization (WMO), 4
  Agricultural Meteorology Programme (AGMP), 406–407
  agrometeorological applications, 406–407
  Climate Information and Prediction Services (CLIPS), 405–406
  Climate Variability Project (CLIVAR), 402–403
  early warning systems for drought, 404–405, 425–426
  expanding Sahara hypothesis, 402
  global climate change and dryland climate, 403–404
  Global Energy and Water Cycle Experiment (GEWEX), 402
  Intergovernmental Panel on Climate Change (IPCC), 403–404
  International Satellite Land Surface Climatology Project, 402
  Regional Climate Outlook Fora (RCOF), 405
  research, 401–403
  United Nations Environment Programme (UNEP), 407
  World Climate Research Programme (WCRP), 402
WRSI. *See* Water requirement satisfaction index (WRSI)
WUE. *See* Water use efficiency (WUE)

Yield modeling, India, 302–303
Yield risk estimation, drought monitoring in Russia, 202–204

Zimbabwe. *See also* Southern Africa
  agriculture, 268
  drought, 266–267
  El Niño and Southern Oscillation (ENSO) and drought in, 33–34
  monitoring drought, 71–72
  rainfall, 269–270